MEMBRANES, DISSIPATIVE STRUCTURES, AND EVOLUTION

ADVANCES IN CHEMICAL PHYSICS

VOLUME XXIX

ADVANCES IN CHEMICAL PHYSICS

MEMBRANES, DISSIPATIVE STRUCTURES, AND EVOLUTION

Edited by
G. NICOLIS AND R. LEFEVER

Faculté des Sciences
Université Libre de Bruxelles
Bruxelles, Belgium

Volume XXIX

AN INTERSCIENCE ® PUBLICATION
JOHN WILEY AND SONS
New York · London · Sydney · Toronto

Library of Congress Cataloging in Publication Data:

Membranes, dissipative structures, and evolution.

(Advances in chemical physics; v. 29)
"An Interscience R publications."
"Many of the papers published in this volume were
presented at an EMBO conference held in Brussels, November 22–24, 1972."
Includes bibliographical references.
1. Biological control systems—Congresses. 2. Biological physics—Congresses. 3. Biological chemistry—
Congresses. I. Nicolis, G., 1939–   ed. II. Lefever, R., 1943–   ed. III. European Molecular Biology Organization   IV. Series.

QD453.A27   vol. 29   [QH508]   541'.08s [574.8'75]
ISBN 0-471-63792-0   74-23611

Printed in the United States of America

10 9 8 7 6 5 4 3 2 1

PREFACE

I am very pleased to present this special volume devoted to "Membranes, Dissipative Structures and Evolution." During the last few years these subjects have made spectacular advances: On the one hand, our experimental knowledge of membranes and biochemical aspects of cooperative processes has been rapidly increasing; on the other hand, new theoretical tools have become available.

From the theoretical point of view the great surprise is certainly the essential difference that appears between equilibrium and near-equilibrium situations on one side and far-from-equilibrium situations on the other. This is the basis of the distinction between equilibrium structures and dissipative structures. We see now that this distinction applies as well to fluctuation theory. New elements enter into fluctuation theory whenever we are interested in fluctuations around far-from-equilibrium situations described by nonlinear equations. In general, we can really speak of a new physical chemistry corresponding to far-from-equilibrium situations.

This development leads to a series of fascinating mathematical problems related to stability theory and bifurcation analysis. Some of these problems are now amenable to analytical study. Others have led to fascinating computer simulations.

Many of the papers published in this volume were presented at an EMBO conference, "Membranes, Dissipative Structures and Evolution," held in Brussels, November 22–24, 1972. Other papers have been added to give a more complete view of the subject.

I believe that this volume presents the first attempt to give a unified view of such diverse problems of basic interest in physical chemistry and biology.

I. PRIGOGINE

CONTRIBUTORS TO VOLUME XXIX

H. Besserdich, Akademie der Wissenschaften der DDR, Zentralinstitut für Molekularbiologie, 1115 Berlin-Buch, D.D.R.

A. Boiteux, Max-Planck-Institut für Erhährungsphysiologie, 46 Dortmund, Germany

Liana Bolis, Institute of General Physiology, University of Rome, Italy

Jean-Pierre Boon, Faculté des Sciences, Université Libre de Bruxelles, Bruxelles, Belgium

H. Busse, Max-Planck-Institut für Erhährungsphysiologie, 46 Dortmund, Germany

J. Chanu, Laboratoire de Thermodynamique des Milieux Ioniques et Biologiques, Université Paris VII, Paris, France

J. Charlemagne, Laboratoire de Thermodynamique des Milieux Ioniques et Biologiques, Université Paris VII, Paris, France

M. Delmotte, Laboratoire de Thermodynamique des Milieux Ioniques et Biologiques, Université Paris VII, Paris, France

J. L. Deneubourg, Faculté des Sciences, Université Libre de Bruxelles, Bruxelles, Belgium

G. Gerisch, Friedrich-Miescher-Laboratorium der Max-Planck Gesellschaft, 74 Tübingen, Germany

B. C. Goodwin, School of Biological Sciences, University of Sussex, Brighton, Sussex, England

B. Hess, Max-Planck-Institut für Erhährungsphysiologie, 46 Dortmund, Germany

J. Julien, Laboratoire de Thermodynamique des Milieux Ioniques et Biologiques, Université Paris VII, Paris, France

E. Kahrig, Akademie der Wissenschaften der DDR, Zentralinstitut für Molekularbiologie, 1115 Berlin-Buch, D.D.R.

Kazuo Kitahara, Chimie Physique II, Faculté des Sciences, Université Libre de Bruxelles, Bruxelles, Belgium

Y. Kobatake, Faculty of Pharmaceutical Sciences, Hokkaido University, Sapporo, Japan

P. Läuger, Department of Biology, University of Konstanz, Germany

R. Lefever, Faculté des Sciences, Université Libre de Bruxelles, Bruxelles, Belgium

E. Margoliash, Department of Biological Sciences, Northwestern University, Evanston, Illinois

A. M. Monnier, Université Paris VI-9, Quai St. Bernard, Paris, France

G. Nicolis, Faculté des Sciences, Université Libre de Bruxelles, Bruxelles, Belgium

B. Novak, Department of Pharmacology, Biozentrum der Universität Basel, Klingelbergstrasse 70, Switzerland

P. Ortoleva, Department of Chemistry, Massachusetts Institute of Technology, Cambridge, Massachusetts

I. Prigogine, Faculté des Sciences, Université Libre de Bruxelles, Bruxelles, Belgium

J. Ross, Department of Chemistry, Massachusetts Institute of Technology, Cambridge, Massachusetts

Daniel Thomas, E.R.A. $n = 338$ du C.N.R.S., Université de Rouen, France

Jack S. Turner, Center for Statistical Mechanics and Thermodynamics, The University of Texas, Austin, Texas

Charles Walter, Department of Biomathematics, University of Texas, M.D. Anderson Hospital, Houston, Texas

L. Wolpert, Department of Biology as Applied to Medicine, The Middlesex Hospital Medical School, London, England

CONTENTS

STABILITY AND SELF-ORGANIZATION IN OPEN SYSTEMS

I. PRIGOGINE* AND R. LEFEVER

Université Libre de Bruxelles, Faculté des Sciences,
Bruxelles, Belgium

CONTENTS

I. INTRODUCTION

In several papers included in this volume use is made of stability theory and of nonequilibrium thermodynamics. It seems appropriate, therefore, to present a discussion of some general aspects. Very recently a review dealing with the historical development of nonequilibrium thermodynamics has been written by Prigogine and Glansdorff.[1] Since the earliest publications on this subject, the idea of nonequilibrium as a possible source of order has been repeatedly emphasized.[2-4]

*Also Center for Statistical Mechanics and Thermodynamics, The University of Texas at Austin, Austin, Texas, U.S.A.

However, the existence of a nonequilibrium environment, although necessary, is far from being a sufficient condition of self-organization in open systems.

In fact, near equilibrium the state of matter may be obtained through extrapolation from equilibrium. In some cases constraints that prevent the system from going to equilibrium may increase the entropy; in others they may decrease the entropy. Examples may be found in previous publications.[3] In such situations, we can really not speak about self-organization.

On the contrary it has now been shown that certain open systems when switched from equilibrium to *far* from equilibrium conditions become unstable and undergo a complete change of their macroscopic properties.[6,7] In particular it was shown that the symmetry-breaking instabilities in respect to diffusion earlier described by Turing,[8] and whose conditions of occurrence had also been investigated in a general, formal way,[9,10] entered within this class of phenomena. The kinetic properties responsible for it and the various behaviors appearing beyond such instabilities were studied on model systems,[11–15] and it was also shown that the properties of several well-known biochemical processes were indicative of systems functioning beyond such a point of instability.[16,17] In fact the number of such manifestly unstable biological systems is steadily enlarged, and in the chapter by Nicolis (p. 29) some examples are given, chosen at levels as different as enzymatic regulation and biological evolution.

The main point to remember is that in all these cases, there appears to exist a thermodynamic threshold for self-organization that corresponds to a clear distinction between the class of *equilibrium structures** and those structures that have been called *dissipative structures* because they only appear as a spontaneous response to *large* deviation from thermodynamic equilibrium. This response corresponds to the amplification of fluctuations, which are then stabilized by the flows of energy and matter. Such fluctuations trigger the appearance of organizations whose probability of occurrence at equilibrium would be negligible. Thus in order to achieve a satisfactory understanding of systems that function in the nonlinear domain of thermodynamics and exhibit a self-organizing character, we require a detailed knowledge of two sets of parameters: (1) molecular parameters, which determine the dynamic processes going on inside the system and also the nature of the fluctuations which might

---

*This class comprises the structures that are obtained by a continuous modification of the equilibrium regime when the boundary conditions steadily deviate from their equilibrium value. These structures form what has also been called "the thermodynamic branch."

occur, and (2) the boundary conditions, which are imposed on the system and control its interactions with the outside world; these are parameters that refer to the intensity of the fluxes through the system and to its geometry. The influence of this latter set of parameters should never be regarded a priori as spurious or simply contingent. Far from equilibrium, even with a *given* set of molecular kinetic properties, a large variety of dissipative structures may be generated simply by changing the properties of the boundaries. For example, while at equilibrium a chemical reaction is insensitive to the dimensions of the reaction vessel, out of equilibrium, on the contrary, by a simple change in dimension of the reaction volume, one may switch from a time-independent steady state to a regime periodic in time.

All these remarks clearly illustrate the essential role played by stability considerations. This is why in Section II we present some general definitions of stability, introduce the concepts of Lyapounov function, and enumerate some basic problems of structural stability.

Before we switch to the thermodynamic stability theory, we first discuss, in Section III, the dynamic meaning of entropy. Recent developments in nonequilibrium statistical mechanics have made clear the microscopic meaning of entropy and entropy production. Consequently, we can make more precise the nonequilibrium region to which a thermodynamic description may be consistently applied.

In Section IV we discuss thermodynamic stability theory and show that in many cases we can construct a Lyapounov function that has a simple physical signification. We then briefly discuss, as an example of structural stability, the coupling between pattern formation and cellular division as it has been described recently by Martinez.

The last sections have a largely explorative character. We present some preliminary considerations on the relation between the structure (or topology) of chemical networks and entropy production. We show that modifications of chemical networks due to catalytic effects may have a simple physical meaning in terms of "adaptation" of the open system to external conditions in the sense that they permit the propagation of the values of chemical potentials given by external conditions inside the reacting system. For reasons indicated briefly (see also the chapter by Nicolis), this effect should be of importance in the thermodynamic interpretation of prebiological evolution.

II. DEFINITIONS OF STABILITY, LYAPOUNOV FUNCTIONS, AND STRUCTURAL INSTABILITIES

We shall limit ourselves here to a few definitions and results in stability theory, which are relevant to the next sections. A very complete presentation of stability problems in ordinary differential equations and methods of investigation can be found in the classical textbooks by Andronov et al.,[18] Minorsky,[19] or Cesari.[20] For a general review on stability problems in biological systems see the papers by Nicolis.[21,22]

A. Stability in the Sense of Lyapounov and Orbital Stability

Let us consider a systems described by the set of differential equations:

$$\dot{x}_i = F_i(x_1, x_2, \ldots, x_n), \qquad i = 1, \ldots, n \tag{2.1}$$

The resting state x_i^0, such that all $F_i = 0$, is stable in the sense of Lyapounov if any solution $U(t)$ of (2.1) (i.e., any series of functions $x_i = U_i(t)$ that satisfies (2.1) identically, taking into account the initial and boundary conditions) that at some time instant t_0 is close to x_i^0 remains in its neighborhood for all times $t \geqslant t_0$. More precisely, if given $\epsilon > 0$ and t_0, there exists a neighborhood $\eta(\epsilon, t_0)$, such that any solution $U_i(t)$ for which $|U_i(t_0) - x_i^0| < \eta$ satisfies $|U_i(t) - x_i^0| < \epsilon$ for $t \geqslant t_0$, then the resting state x^0 is said to be stable. It is *asymptotically stable* if, moreover, $\lim_{t \to \infty} |U_i(t) - x_i^0| = 0$. These definitions can straightforwardly be extended to motions.

When the system (2.1) is autonomous (i.e., when time does not appear as an explicit variable on the right-hand side of the equation), those motions that follow closed trajectories or orbits in the phase space are of special interest; indeed, they correspond to a periodic behavior of the x_i variables in time. It should be noticed that because of the translational invariance of autonomous systems, an infinity of solutions differing from each other by their phase correspond to a given closed trajectory C. One then has the following definitions of stability: A given orbit C of (2.1) is *orbitally stable* if given $\epsilon > 0$, a representative point P of a trajectory that at time t_0 is within a distance $\eta(\epsilon)$ of C remains within a distance ϵ of C for all $t \geqslant t_0$. Otherwise C is said to be orbitally unstable. Moreover, if the distance between P and C tends to zero when t tends to infinity, then C is said to be *asymptotically orbitally stable*.

It is important to distinguish stability in the sense of Lyapounov from orbital stability. The conditions for orbital stability only require that two orbits C and C' remain in the neighborhood of each other; stability in the sense of Lyapounov or of a solution also, requires that the representative points P and P' associated with different solutions remain close to each other if they initially were. A supplementary distinction must also be made concerning asymptotic stability. Clearly a periodic solution cannot be asymptotically stable if there exist other periodic solutions in its neighborhood; indeed, a periodic function cannot be asymptotic to another periodic function. However, if $U(t)$ is a periodic solution of (2.1), then $U(t+\tau)$ is another solution of the same system for all values of τ. There thus exist periodic solutions infinitesimally close to $U(t)$, and as a result a periodic solution of an autonomous system cannot be asymptotically stable. An orbit that is asymptotically orbitally stable is called a *limit cycle* according to the terminology of Poincaré.

From these definitions it should be clear why stability theory is so important in macroscopic physics. We deal always with systems that are formed by many interacting entities (such as electrons, atoms, or molecules). Therefore fluctuations are unavoidable. The description of physical situations in terms of differential equations such as (2.1) neglects these fluctuations. Therefore the condition of validity of this description is precisely that these equations satisfy stability conditions.

B. Lyapounov Functions

From these stability definitions, a criterion of stability can straightforwardly be deduced. If we consider the positive definite function x^2 (the square of the distance in the space of states between two neighboring solutions) and are able to show that its time derivative obeys the inequality

$$(\dot{x}^2) \leqslant 0 \qquad (2.2)$$

for all values of t, then the evolution of (2.1) is stable ($\leqslant 0$) or asymptotically stable (< 0). It must be noticed that (2.2) constitutes only a *sufficient* condition of stability; indeed, for example, an oscillatory behavior of x^2 would be within the general definition of stability without obeying (2.2) for all times.

A quadratic function like x^2 that leads to a stability condition of the form (2.2) is called a *Lyapounov function*. Although the existence of a Lyapounov function leads only to a sufficient condition for stability, it is easy for autonomous differential equations to indicate necessary and sufficient conditions of stability.

C. Linear Stability

Suppose that we have a system described by a set of autonomous ordinary differential equations (i.e., such that time does not appear as an explicit variable on the right-hand side of a system like (2.1)) that admits a steady-state solution $\{x_i^0\}$. Linear stability of this solution will be insured if any small deviation δx_i of a variable i ($|\delta x_i|/x_i^0 \ll 1$) from its steady-state value regresses when $t \to \infty$. In the neighborhood of $\{x_i^0\}$

the response to arbitrary small fluctuations $\{\delta x_i^0\}$ is given by the linear set of equations

$$\frac{d}{dt}\delta x_i = \sum_{j=1}^{n} \alpha_{ij}\, \delta x_j \qquad (2.3)$$

where α_{ij}, defined as $\alpha_{ij} = (\partial F_i/\partial x_j)_0$, is evaluated at the steady state. Equation (2.3) admits solutions of the form

$$\delta x_j = \delta x_j^0 e^{\omega t} \qquad (2.4)$$

which are called normal-mode solutions. Stability implies that the eigenvalue $\omega = \omega_r + i\omega_i$ corresponding to each normal mode possesses a negative real part ω_r. This condition is fulfilled whenever all the roots of the secular equation

$$|\omega \delta_{ij} - \alpha_{ij}| = 0, \qquad \delta_{ij} = \begin{cases} 0 & (i \neq j) \\ 1 & (i = j) \end{cases} \qquad (2.5)$$

obtained by substituting (2.4) in (2.3), satisfy the condition

$$\omega_r < 0 \qquad (2.6)$$

An explicit form of this stability condition (2.6) corresponds to the well-known Hurwitz-Routh condition (see, for example, Cesari).[20] It insures a *necessary* and *sufficient* stability condition, in the sense of Lyapounov, of the steady-state solution $\{x_i^0\}$.

It is, however, important to notice that this type of linear analysis furnishes no indication concerning the *amplitudes* of the fluctuations. We must not expect, therefore, that it always satisfactorily predicts the behavior of open systems, particularly if fluctuations of *finite* amplitude have a nonvanishing probability of occurring. This has been illustrated on the following chemical example[23]:

$$A + X \rightleftarrows 2X$$

$$X + E \rightleftarrows C \qquad (2.7)$$

$$C \rightleftarrows E + B$$

A and B are initial and final products whose concentration is maintained constant, and furthermore, one has the conservation relation $E + C = \text{const}$. Plotting the steady-state value of X as a function of A with B constant, we can obtain a curve of the form represented in Fig. 1. Clearly there exist values of A for which three steady-state solutions are possible; on the basis of the linear stability analysis, it is easily found that the lower branch of states (from $A = 0$ to P) and the upper branch (from P' to $A \rightarrow \infty$) are stable, while on the contrary the intermediate states between P and P' are unstable. A hysteresis loop can thus, in principle, be described by the system when A is varied in such a way that the system is successively switched from P to P'; it is expected that the jumps between the lower and upper branch occur at P and P'.

In numerical calculations,[23] which take into account a description of fluctuations, it is, however, found that no such hysteresis loop seems to exist. In fact the system always jumps at some intermediate point ξ between P and P'. This is an excellent example of the effect of fluctuations on the macroscopic behavior. This result should also be related to the properties of nonequilibrium systems, which have been described by Kobatake,[28] Schlögl,[24] and Kitahara,[25] and for which the existence of a construction analogous to the Maxwellian construction for the vapor pressure of a Van der Waals gas is possible.

D. Stability in Respect to "Small Parameters"

As we have already noticed, stability problems are of crucial importance in all systems that involve many degrees of freedom and that as a result are continuously submitted to the effect of *fluctuations*—that is, to random and spontaneous deviations from the average deterministic laws of evolution. Roughly speaking, the effect of fluctuations may manifest itself in two ways:

1. Fluctuations may simply correspond to random perturbations whose regression or amplification is described by the laws of evolution

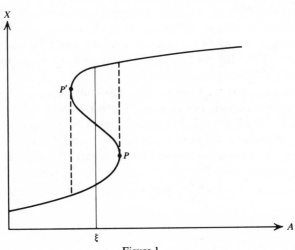

Figure 1.

which govern the average motion of the system in the absence of fluctuations.

2. Fluctuations may also correspond to an alteration of the average deterministic laws of evolution of the system. We have already considered an example of the amplification of fluctuations on the macroscopic level. Another aspect corresponds to the possibility that new pathways are opened to the time evolution of the system as the result of fluctuations. *Small parameters* that would normally play no role in the macroscopic evolution become essential as the result of fluctuations. This situation may be considered as a special case of structural stability, which will be studied in the next section. We do not discuss here particular models, which may be found elsewhere (see Ref. 26), but remain as general as possible.

Suppose that we have a set of chemically interacting species X_i $(i=1,\ldots,n)$, which all exist in relatively abundant quantities. The evolution in time of the $\{X_i\}$ is given by a set of differential equations:

$$\dot{X}_i = F_i^e(X_1,\ldots,X_n) + \langle F_i \rangle(X_1,\ldots,X_n) \qquad (2.8)$$

among the concentrations of the constitutents X_i. Here F_i^e is a supply term related to the exchanges of substrates and products with the external environment; $\langle F_i \rangle$ describes the chemical processes going on inside the system. It is assumed that the distribution of all concentrations in space inside the system is kept uniform and that there exists at least one asymptotically stable steady-state solution of (2.8), in the neighborhood of which the system is functioning. In agreement with our stability definitions, this means that any small deviation $\delta X_i = X_i(t) - X_i^0$ of the concentration of a chemical species i from its steady-state value X_i^0 will regress in time in such a way that $\delta X_i \to 0$ as $t \to \infty$. Furthermore, in agreement with the linear stability analysis, the relation between δX_i and time is of the form $\delta X_i(t) \propto e^{\omega t}$, where ω is a typical normal mode whose real part here is negative.

Suppose now that we have to take into consideration the formation, by random fluctuation (mutations, or error copies), of new products from the X_i, which we call Y_j $(j=1,\ldots,m)$. The number of differential equations describing the joint evolution of X and Y will increase by m. Let us take $m=1$ and write the kinetic equations for Y and X in the following form:

$$\epsilon \frac{dY}{dt} = G(\{X\}, Y, \epsilon) \qquad (2.9a)$$

$$\frac{dX_i}{dt} = F_i^e + F_i(\{X\}, Y, \epsilon) \qquad (2.9b)$$

where F_i and G are functions of the old and new variables. ϵ is a parameter characteristic of the properties of the enlarged system, and such that in the limit $\epsilon \to 0$ the original set of equations (2.8) is recovered. In other words, for $\epsilon = 0$, there exists a value $\langle Y \rangle(\{X\})$ furnished by the condition $G(\{X\}, \langle Y \rangle, 0) = 0$ and such that it satisfies the identity

$$F_i(\{X\}, \langle Y \rangle(\{X\}), 0) \equiv \langle F_i \rangle(\{X\}) \qquad (2.10)$$

Mathematically, the systems (2.9) can be constructed in such a way that the steady state $\{X^0\}$ of (2.8) becomes unstable upon the addition of a new substance Y which follows kinetics of the form (2.9a) and satisfies (2.10).* In the limit $\epsilon \to 0$ the effect of the new substance on the evolution of $\{X\}$ disappears. For ϵ not equal to zero but small, the stability of the enlarged system will depend on the regression of $n+1$ types of fluctuations in the concentrations characterized by $n+1$ normal modes, n of which have values differing only by correction terms of order ϵ from the original normal modes of the systems (2.8); the supplementary mode introduced by the new substance Y is the only one, therefore, that can destabilize the old system. When this happens, it means that the solutions of the equations (2.9) do not remain for all times in an ϵ neighborhood of the solutions of (2.8). We have evolution through unstable transitions of the original system upon addition of new substances. This evolution may then lead, for example, to a new state of high Y concentration, dominated by the new substances.

Situations of this type have been analyzed within the context of prebiotic evolution for systems in which the new substances introduce a drastic change in functional properties.[26] ϵ is then the inverse of some typical kinetic constant k characterizing the new function and the new time scale of evolution imposed to the system. The essential property of the instability is that it occurs for $\epsilon \to 0$ (i.e., $k \gg 1$); it corresponds to an increase in the interactions with the environment and in the specific dissipation (i.e., dissipation per unit mass). In other words, this instability, which is triggered by nonequilibrium environmental conditions ($F_i^e \neq 0$ and large) maintaining a continuous energy dissipation, increases the level of dissipation further and thereby creates conditions that are favorable to the appearance of other instabilities. This behavior has therefore been called an "*evolutionary feedback.*" We consider this notion in more detail in Section 6.

E. Structural Stability

We have seen that fluctuations may lead to a different macroscopic evolution described by altered differential equations (see (2.9) and

---

*Notice also that in (2.9a), there is no flow of mass term included from the surrounding: Y is produced solely from $\{X\}$ and is therefore essentially a closed system.

(2.10)). This leads to a new and more general formulation of stability theory. We may indeed ask if a motion is stable when new *small* terms are added to the differential equations independently if such terms may arise as the effect of fluctuations. This is the basic question studied in the theory of structural stability.[18]

Let us consider a simple example. Suppose that we have a system whose motion in time is a solution of the following set of differential equations:

$$\frac{dx}{dt} = P(x,y), \qquad \frac{dy}{dt} = Q(x,y) \qquad (2.11)$$

The question then is whether this motion is stable, *when* $\epsilon \to 0$, with respect to the motion described by the new equations

$$\frac{dx}{dt} = P(x,y) + \epsilon p(x,y), \qquad \frac{dy}{dt} = Q(x,y) + \epsilon q(x,y) \qquad (2.12)$$

A simple example[18] will show that qualitatively important stability changes might result from the passage (2.11) to (2.12) even in the limit $\epsilon \to 0$. Take the following equations:

$$\frac{dx}{dt} = by, \qquad \frac{dy}{dt} = -bx \qquad (2.13)$$

which are a linearized form of the Lotka-Voltera system. In the (x,y) phase space (2.13) yields an infinite set of closed *orbitally* stable trajectories surrounding the point $(x=0, y=0)$. Compare now the system

$$\frac{dx}{dt} = ax + by, \qquad \frac{dy}{dt} = -bx + ay \qquad (2.14)$$

it is found that as soon as the parameter a is different from zero, however small, the point $x=y=0$ becomes *asymptotically* stable: it is then a focus to which all trajectories in phase space converge. In other words Equations (2.11) are structurally unstable with respect to fluctua-

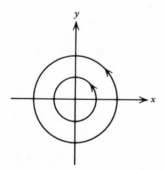

Figure 2.

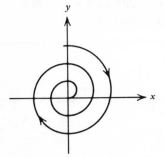

Figure 3.

tions that introduce additional terms as in (2.12). This exemplifies the fact that one has to be very careful in using differential equations that might become structurally unstable, since very slight alterations will produce enormous effects. Indeed, with equations like (2.11), it is possible for the system to remain for all times on a closed trajectory— perhaps even one situated at an infinite distance from the point $x=y=0$. On the contrary, when $a \neq 0$ all motions in time will converge toward the origin, which is the only stable state of the system.

As a second example, suppose that in a chemical system described by equations of the form (2.11) we now allow for fluctuations that break spatial homogeneity. In order to take into account the response of the system to such disturbances of its average state, we have to add diffusion terms to (2.11) which in their simplest form are given by Fick's law. We thus have

$$\frac{\partial x}{\partial t} = P(x,y) + D_x \frac{\partial^2 x}{\partial r^2} ; \qquad \frac{\partial y}{\partial t} = Q(x,y) + D_y \frac{\partial^2 y}{\partial r^2} \qquad (2.15)$$

It is then essential to know if the solution of Equations (2.11) is structurally stable or not in respect to the addition of diffusion terms. If this is not so even in the limit $D_x, D_y \rightarrow 0$, spatial inhomogeneities due to fluctuations may be amplified and stabilized. A completely different distribution of matter and functional organization appears inside the system, which is no longer in first approximation predictible on the basis of (2.11). In this regard, a broad class of partial differential equations of the second order has been studied by Fife,[36] who has been able to describe their behavior by analytic methods. On the other hand, several model systems have also been investigated in great detail. For example, in the case of the system [Fig. 2(a)] reported in the chapter by

Nicolis, the stability conditions with respect to diffusion of the *homogeneous* steady-state regimes can be expressed simply as

$$B < B_c = \left[1 + A \left(\frac{D_X}{D_Y} \right)^{1/2} \right]^2$$

When $B = B_c$, the system will develop a spatial pattern whose characteristic length λ is given by the relation

$$\lambda^2 = \frac{(D_X D_Y)^{1/2}}{A}$$

Clearly, even for very small values of D_X and D_Y, B_c may remain finite and in an accessible range of concentration values. Similar results can be predicted in biological systems; particularly, the enzymatic reaction of phosphofructokinase with ATP and ADP has been studied in great detail and shown to be structurally unstable in respect to diffusion for a range of values of the parameters.[17,40]

III. NONEQUILIBRIUM THERMODYNAMICS—IRREVERSIBILITY AS A SYMMETRY-BREAKING PROCESS

In the first section of this chapter we have emphasized the importance of "distance from equilibrium." This is obviously a new thermodynamic parameter characterizing the state of the system. In equilibrium, the state of the system is determined by quantities such as pressure, temperature, and so on. In nonequilibrium, supplementary variables have to be added.

The second law of thermodynamics postulates the existence of a state function S, the entropy. The change of entropy dS during a time dt can be split into two parts: the entropy production $d_i S$ caused by changes inside the system and the flow of entropy $d_e S$ caused by the interaction with the outside world:

$$dS = d_e S + d_i S \tag{3.1}$$

The entropy production $d_i S$ caused by changes inside the systems is never negative:

$$d_i S \geqslant 0 \tag{3.2}$$

There are nonequilibrium situations where the macroscopic evaluation of entropy production and the entropy flow is possible. This is so when there exists within each small mass element of the medium a state of *local equilibrium* for which the local entropy s is the same function of the local macroscopic variables as at equilibrium. As discussed in more detail elsewhere, this implies that dissipative processes are sufficiently dominant to exclude large deviations from statistical equilibrium. We may therefore apply the local-equilibrium assumption to transport processes described by linear laws as well as to not-too-fast chemical reactions (such that the activation energy is large compared to the thermal energy). When these conditions are satisfied, we obtain for the entropy production per unit time

$$P = \frac{d_i S}{dt} = \sum_\rho J_\rho X_\rho \qquad (3.3)$$

where J_ρ and X_ρ are the conjugate thermodynamic fluxes and forces. In the case of chemical reactions, J_ρ and X_ρ are simply equal to v_ρ and $\mathcal{C}_\rho T^{-1}$, respectively the rate and affinity of the chemical reaction. It is only very recently that a general formulation of nonequilibrium statistical mechanics has been achieved, which leads in turn to a microscopic definition of entropy of generality comparable to that of the phenomenological definition.

In order to do so the dynamic nature of irreversibility had to be elucidated. We have now shown that irreversibility may be viewed as a symmetry-breaking process arising in well-defined classes of dynamic systems formed by many interacting units.[37]

We have to limit ourselves here to a short summary of some of the ideas involved.

Let us start from classical or quantum dynamics as expressed by Hamilton's equations of motion

$$\frac{dq}{dt} = \frac{\partial H}{\partial p}, \qquad \frac{dp}{dt} = -\frac{\partial H}{\partial q} \qquad (3.4)$$

or Schrödinger's equation

$$i\frac{\partial \psi}{\partial t} = H\psi \qquad (3.5)$$

Both descriptions may be unified through the Liouville–von Neumann equation for the distribution function (or density matrix[38])

$$i\frac{\partial \rho}{\partial t} = L\rho \qquad (3.6)$$

with

$$L = \begin{cases} -i\{H,\rho\} & \text{(Poisson bracket)} \\ [H,\rho] & \text{(commutator)} \end{cases} \qquad (3.7)$$

The solution of equation (3.6) requires a more precise specification of the problem in which we are interested. We suppose known the value of $\rho(t)$ at $t=0$ and calculate $\rho(t)$ for $t \geqslant 0$. This corresponds to an *initial-*

value problem. If, on the contrary, we are interested in $\rho(t)$ for $t \leq 0$, we have a *final-value problem*. Thermodynamic behavior is obtained when the initial-value problem leads to different solutions from the final-value problem. The entropy may be expressed in terms of the density matrix ρ, and it may be shown that it increases for *all initial-value problems*. In other words, the second law of thermodynamics holds for a class of (and not for all) solutions of the Liouville equation (3.6). We see that we can associate irreversibility with "symmetry breaking." If the solution for the initial-value problem coincides with those for the final-value problem, no entropy that satisfies the second law can be defined. The important point is that we have now established an exact general condition that a system has to satisfy to present thermodynamic behavior. We have called this condition the dynamic condition of dissipativity (it is related to the analytic continuation of the resolvent $1/(L-z)$ in the complex plane[37]). It can be shown to be satisfied for systems, such as interacting gases or anharmonic lattices, to which we may expect to apply the second law of thermodynamics. We may say that the thermodynamic description is the expression of the breaking of the original dynamic groups as described by Equation (3.6) into two semigroups: one for the "retarded" solutions corresponding to the initial-value problem, and the other to the "advanced" solutions corresponds to the final-value problem.

For *isolated* systems, the entropy is a Lyapounov function in the sense defined in Section II. Thermodynamic equilibrium is then stable against arbitrary perturbations. However, this Lyapounov function has in general no simple macroscopic meaning (see Ref. 39). It cannot be expressed in terms of observables such as temperature. Using a somewhat paradoxical formulation, we could say that the second law is in general a theorem in the dynamics of large systems and not a theorem in thermodynamics considered as macroscopic physics. It is only *near* equilibrium that we can express this Lyapounov function in terms of macroscopic variables, and then we are back to our conception of local equilibrium. But even in this restricted range we can also introduce the distance from equilibrium as a parameter and study what happens when we impose stronger and stronger constraints on the system.

IV. THERMODYNAMIC STABILITY THEORY

Let us start with the classical Gibbs-Duhem stability theory applicable to closed systems (which do not exchange matter with the outside world). The entropy flow [see (3.1)] is then given by

$$d_e S = \frac{dQ}{T} \qquad (4.1)$$

where dQ is the heat received by the system and T the absolute temperature.

Using (3.1), (3.2), and (4.1), we obtain immediately from the first principle of thermodynamics

$$T d_i S = T dS - dE - p dV \geqslant 0 \qquad (4.2)$$

where dS is the total entropy variation of the system. From (4.2) it is clear that an equilibrium state is stable provided any fluctuation of the variables satisfies the inequality

$$\delta E + p \delta V - T \delta S \geqslant 0 \qquad (4.3)$$

We thus have, for isolated systems (E, V constant), the stability criterion $\delta S \leqslant 0$. If any fluctuation around the equilibrium state of maximum entropy leads to a negative entropy variation, then by virtue of the second principle, the equilibrium state is asymptotically stable in the sense of Lyapounov: all fluctuations will be damped. Alternatively, if V, T or p, T is maintained constant, (4.3) leads to an analogous stability condition for the thermodynamic potential F or G. Most stability problems of equilibrium states can appropriately be described in terms of these classical thermodynamic potentials E, F, G. Confronted with nonequilibrium situations, this method encounters a major difficulty: the extremal properties of F or G require a control of fluctuations in variables like p or T not only on the boundaries of the system, but also at each point inside it. This is hardly realizable, particularly in open systems, where most of the time only the boundary conditions can be controlled. The results and methods of classical thermodynamics cannot, therefore, in general be transposed to the nonequilibrium domain. Such an extension needs a different type of approach, such as was carried out in recent years by Glansdorff and Prigogine.[5] It provides a new formulation of the equilibrium stability theory for problems with given *boundary values* and is also applicable to a very broad class of nonequilibrium situations. In fact, the range of validity of this theory extends to all situations for which the local-equilibrium assumption holds. Let us simply sketch the main results.

A. Stability Condition of the Equilibrium State

For a system in a stable equilibrium state and subject to *fixed boundary conditions*, it can be deduced from the entropy balance that the excess entropy $\delta^2 S$, obtained by expanding the entropy around the equilibrium state

$$S = S_e + (\delta S)_e + \frac{(\delta^2 S)_e}{2} \qquad (4.4)$$

is a quadratic negative definite expression*

$$T\delta^2 S = -\left[\frac{C_V}{T}(\delta T)^2 + \frac{\rho}{\chi}(\delta v)^2_{N_\gamma} + \sum_{\gamma\gamma'}\mu_{\gamma\gamma'}\delta N_\gamma \delta N_{\gamma'}\right] < 0 \qquad (4.5)$$

and moreover that its time derivative is equal to the entropy production P of the system:

$$\frac{1}{2}\frac{\partial \delta^2 S}{\partial t} = P > 0 \qquad (4.6)$$

In other words, $\delta^2 S$ is a Lyapounov function, and its behavior around equilibrium states demonstrates the damping of all fluctuations inside the system.

This result is much more general than that of the chemical Gibbs-Duhem theory, as it is independent of the existence of any thermodynamic potential. It should, however, be noticed that it applies only to *small* fluctuations, as higher-order terms have been neglected in (4.5).

Let us now go over to systems submitted to *nonequilibrium* boundary conditions and ask under which conditions such systems still exhibit stability properties analogous to the equilibrium systems. Alternatively, the question we would like to investigate is whether or not, out of equilibrium, there exists a Lyapounov function such that the behavior of systems still can be understood in terms of an equilibrium type of physical chemistry.

B. The Linear Range

The central quantity of nonequilibrium thermodynamics is the entropy production P (cf. Section III), which is a positive definite quantity that can be zero only at equilibrium, that is, when $J_\rho = 0$ and $X_\rho = 0$ simultaneously. Therefore, we can define the *linear range* of nonequilibrium thermodynamics as the domain close to equilibrium in which linear relations,

$$J_\rho = \sum_{\rho'} L_{\rho\rho'} X_{\rho'} \qquad (4.7)$$

hold between the fluxes and forces. The $L_{\rho\rho'}$ are phenomenological

---

*$C_V, \chi, \rho, N_\gamma$ are, respectively, the specific heat at constant volume, the isothermal compressibility, the density, and the molar fraction of γ. ψ_γ is the chemical potential of γ and $\psi_{\gamma\gamma'} = \partial\psi_\gamma/\partial N_{\gamma'}$.

coefficients that satisfy the well-known Onsager reciprocity relations

$$L_{\rho'\rho} = L_{\rho\rho'} \tag{4.8}$$

Substituting (4.7) into (3.3) and taking (4.8) into account, we can demonstrate that:

$$\frac{dP}{dt} \leqslant 0 \tag{4.9}$$

where the equality corresponds to the steady state. We thus find that the steady states that belong to the linear range are characterized by an extremum principle, according to which entropy production is at a minimum for the steady state compatible with the constraints imposed on the boundaries. Furthermore, P is a Lyapounov function, and we consequently see that not only around the equilibrium state, but also around the steady states immediately close to it, the fluctuations will regress in time.

C. The Nonlinear Range

The range of validity of linear thermodynamics for which (4.8) holds is more restrictive than the local-equilibrium assumption. Even for situations corresponding to very large deviations from the equilibrium state, most of the time we still may suppose that locally the equilibrium state is preserved at each point of the system, so that locally $\delta^2 s < 0$. The inequality is satisfied when the local-equilibrium state is stable, as we always assume in what follows. This suggests an approach to stability problems based on the excess entropy $\delta^2 s$ as a Lyapounov function. In this way, whenever the following condition on the excess entropy production:

$$\frac{\partial \delta^2 s}{\partial t} = \sum_\rho \delta J_\rho \delta X_\rho > 0 \tag{4.10}$$

is fulfilled, the steady state considered is stable. In (4.10), δJ_ρ and δX_ρ are the fluctuations of the fluxes and forces around the nonequilibrium steady state, compatible with the boundary conditions.

It is interesting to notice that far from equilibrium, the stability properties are characterized by using the concept of a Lyapounov function in another way than in the equilibrium case. Indeed, at equilibrium the conditions of stability are deduced by imposing the sign of $\delta^2 s$, while far from equilibrium, on the contrary, it is the time derivative of $\delta^2 s$ that furnishes a criterion of stability. We thus see that there exists a possibility even for purely dissipative systems (i.e., systems without hydrodynamic motions) to escape from the type of organization and functional properties that characterize the equilibrium regimes or the states close to them. *Nonequilibrium* may be the *source order* producing new extremely coherent behaviors in space and in time, provided that the condition in (4.10) cannot be fulfilled for the steady states that extrapolate the properties of the equilibrium regime. In the case of chemical reactions, it can easily be verified that the kinetic properties underlying a breakdown of the inequality (4.10) are of an autocatalytic or cross-catalytic nature (see the examples mentioned by Nicolis, p. 00). One of the most important characteristics of such instabilities is that the relation between order and fluctuations is much more complex than at equilibrium, where everything is determined by the strict properties of the thermodynamic potentials. In the nonlinear range, not only are fluctuations the trigger for the appearance of new structures, but also different fluctuations may correspond to different structures, and several stable nonequilibrium regimes may coexist. Therefore, in contrast with the ordering principles of equilibrium based on the classical potential, we may say that far from equilibrium there prevails a type of order that we have called *order through fluctuations* and whose description would require a new extension of chemical physics.

V. SUCCESSION OF INSTABILITIES, THE MARTINEZ MODEL

An interesting example where a succession of instabilities is involved has recently been discussed by Martinez.[27] This author has shown how, in the course of development, a system of cells can evolve by a succession of instabilities through a set of distinct states. The important point to realize is that in order to regulate such a passage, not only must the system be unstable, but also the instabilities must in some way regulate the boundary conditions of the system. Indeed, if given *invariant* boundary conditions are imposed on a system, one expects that beyond a point of instability, the system chooses, according to the fluctuations initially present, one stable regime among the set of such states compatible with the boundary conditions. It then remains in this state as long as this set of boundary conditions is maintained. The interesting idea in the paper of Martinez is that the instabilities control their own boundaries, or, more exactly, the relation between boundary values and system dimensions, via *cellular division*. In its simplest version, this model, which combines growth and compartmentalization, can be summarized in the following way: It consists of a linear array of cells, each of which divides only if the concentration of some substance Z within it reaches a threshold value Z_t. The synthesis of Z depends on the concentration of two morphogens X and Y, which react according to scheme [Fig. 2(a)] mentioned by Nicolis (see p. 00). Y is allowed to diffuse between the cells. Now characteristically scheme 2(a) may present different spatial patterns according to the value of the ratio D/l^2, where D is the diffusion coefficient of the substance and l the length of the system. As a result, when such a pattern appears, neighboring cells will have different X and Y concentrations and accordingly different rates of synthesis of Z; the threshold Z_t will be reached in only some of the cells, which then will divide and consequently change the ratio D/l^2. This latter effect destabilizes the original pattern, and a new distribution of matter among the cells appears, which initiates further division of other cells. This process then keeps going on; very complicated and organized situations may result if growth is considered in two or three spatial dimensions.

Although this is a highly idealized model, involving chemical steps that are not likely to occcur in cells, it nevertheless indicates very strikingly that "self-organizing automata" may exist on a purely chemical basis. There is no doubt, on the other hand, that similar results could be obtained with certain more realistic, but also more complicated, enzymatic reactions, with which it has already been established that the occurrence of spatial patterns of *supracellular dimension* is possible (see, for example, Ref. 17).

An important feature of self-organization and development, viewed as a succession of instabilities, which must be stressed, is the fundamentally irreversible character of the whole process: Each time the system reaches a point of instability, it spontaneously and *irreversibly* evolves toward a new structural and functional organization; furthermore, these jumps can occur only at given time instants after the beginning of the whole process; in other words, the system has a natural time scale of irreversible aging associated with its own internal properties. In some sense, one could be tempted to think that we have here a first approximation of what biologists (see F. Jacob[29]) describe, in the evolution toward highly organized states, as the successive integration of levels of increasing structural and functional complexity.

VI. ENTROPY PRODUCTION AND EVOLUTIONARY FEEDBACK

The model of Martinez has already permitted us to note that a self-organizing system cannot be described simply by the succession of states of increasing complexity that it presents; moreover, a typical time order is always associated with every process of self-organization and constitutes an intrinsic manifestation of the dynamic processes by which the system jumps from one state to the other.

It is therefore interesting, when one considers a phenomenon like biological evolution, to look at the time intervals elapsing between the appearance of the different major stages of organization and to inquire whether this would not give some hints on the mechanism of evolution itself. In this respect several authors (for example, recently, de Duve[30]) have pointed out that there is no proportionality between the increase in organization corresponding to an evolutionary step and the time it requires to occur. de Duve attributes to biological evolution a time course of at least 3.2×10^9 years, half of which was required to see the appearance of the first protistes; thereafter half of the remaining time was necessary for the evolution of the first invertebrates, half of that for the first vertebrates, and finally, some 200 millions years ago, the first mammals. Man himself is probably "only" 2 million years old.

In other words, even though the organization and complexity of living systems can hardly be evaluated quantitatively, these figures indicate an acceleration of evolution in the course of time. In its most *macroscopic* manifestations evolution is an *autocatalytic* phenomenon: Any progress is followed by another one that has a greater chance to occur. If one further considers (*1*) that, as is likely, the molecular mechanism of evolution has not appreciably changed in going from the primitive to the evolved organisms—that is, the frequency of random mutations in proteins and nucleic acids has remained the same and independent of the functional activity, and (*2*) that the time interval between generations increases for higher organisms, then one realizes that this acceleration really is a paradoxical phenomenon; it cannot be explained by simple molecular considerations on reproduction and self-replication of macromolecules. Clearly we need something more than self-replication; we need something that plays the role of an *increasing selection pressure* in favor of the more organized states.

We do not know, of course, the exact nature of this driving force; but it certainly involves the interactions of living systems with the external world and suggests a relation between structure and energy dissipation which can be summarized by the following *"evolutionary feedback"* (for more details see Refs. 26,31). Each time that an instability is followed by a higher level of energy dissipation, one may consider that the driving force for the appearance of further instabilities has been increased. Some irreversible processes taking place inside the system are functioning more intensely and have accordingly increased their depar-

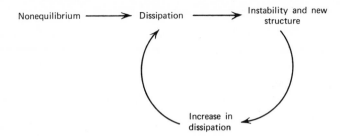

ture from the equilibrium state; the probability that there exists a class of fluctuations with respect to which these processes are unstable has become higher. On the contrary, if the result of the instabilities is to decrease the level of dissipation, one gets closer to the properties of an isolated system at equilibrium, that is, closer to a regime where all fluctuations that bring about a deviation from the state of maximum disorder are damped.

These considerations clearly indicate that in order to describe the behavior of such self-organizing systems, we need some index related to the level of dissipation whose variation would be characteristic of the organizing tendency manifested by the system. If we admit that the source of order basically lies in the intensity of the purely dissipative irreversible processes going on, we are lead to the suggestion that this role could be assumed by the specific entropy production (s.e.p.) of the system (the entropy production per mole). Any change in mass, energy content, or rate or affinity of reaction inside the system would be reflected by this quantity. Particularly, within the context of evolution, the behavior of this quantity has already been investigated on model systems that present instabilities due to the appearance of template effects in a polymerization reaction (see Refs. 31, 32). The characteristic result is that when the polymerization reaction is switched from a regime in which it occurs mainly by simple condensation of the monomers to a regime in which they are assembled on a template, a discrete and important jump of the s.e.p. is observed. In other words, it is found that an instability by which the system manifestly reaches a more complex stage of functional organization, at the same time creates conditions which favor a subsequent evolution of this state through nonequilibrium instabilities.

Within the remaining part of this section, we will briefly consider the relation between functional organization and s.e.p. in some very simple models.

A. Simple Linear Chemical Networks*

Let us first consider the transformation of some initial substrate A into K intermediate products X_k and a final product B, and investigate the behavior of the steady-state entropy production per mole, σ, as a function of the topology of the reaction scheme, that is, as a function of the reaction connections between the chemical species. It is assumed that the reactions follow linear kinetics, and that all forward and also all backward kinetic constants have been set equal. We first take a simple linear chain of reactions:

$$A \underset{l'}{\overset{l}{\rightleftarrows}} X_1 \rightleftarrows X_2 \rightleftarrows \cdots \rightleftarrows X_K \rightleftarrows B \qquad (6.1)$$

where l and l' are, respectively, the forward and backward kinetic constants. When $l/l' = 1$ it is easily verified that σ as a function of the ratio $x = A/B$ and of the number K of intermediates X_k is given by the expression

$$\sigma = \frac{2(x-1)}{(K+1)(K+2)(x+1)} \ln x \qquad (6.2)$$

We thus find that, independently of the value of the overall affinity of reaction, given by $\sim\ln(A/B)$, such a chain of reactions sees its s.e.p. decrease when its length increases—that is, when more intermediate steps have to be taken into account between the initial and final products A, B.

From a practical point of view, one could regard scheme (6.1) as a reaction of polymerization. Each step k of the chain corresponds to the addition of a monomer to the intermediate X_{k-1} in such a way that the concentration A_k of monomers remains constant in the medium and that l represents the product of A_k with the kinetic constant of condensation ($l = k_a A_k$). It then can be verified that Equation (6.2) states that when the length of the final polymer B increases, its rate of production, which is proportional to the steady-state value

$$X_K^0 = \frac{A + KB}{K+1} \qquad (6.3)$$

tends to diminish even if the overall affinity is triggered to infinity (cf., for example, Equation 6.3 with $A = \text{const}$ and $B = 0$). Alternatively one could also say that when the chain synthesizes longer polymers, its

*More details will be found in Ref. 31.

consumption of monomers per unit time tends to diminish. One has

$$\frac{dA}{dt} \sim -k\left(A - X_1^0\right) \sim -k\left(A - \frac{KA}{K+1}\right) \to 0 \qquad \text{as } K \to \infty \qquad (6.4)$$

Let us now compare these results with those obtained in the case of a linear scheme having a topology as different as possible from that of scheme (6.1); this is the case when there are no reactions between the K intermediate products, so that one has

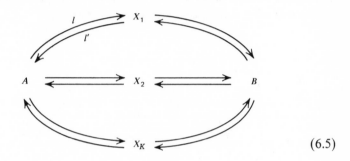

$$(6.5)$$

It is readily shown that when $l = l'$, σ is given by

$$\sigma = \frac{K(x-1)}{(K+2)(x+1)} \ln x \qquad (6.6)$$

One sees that in contrast with what happens with Equation (6.2), the value of σ calculated with (6.6) increases with K and tends toward a finite constant value. (For $x \gg 1$ one has simply $\sigma \sim \ln x$.)

In summary, these results could be interpreted in the following way: Starting with a linear chain of reaction of the type (6.1), involving a number K of intermediate compounds, any process of evolution by which it was progressively transformed into (6.5) would be associated with an overall increase of σ. It must be pointed out that this variation of σ does not simply correspond to an increase in the number of loops that exist in parallel between A and B, or in the number of chemical reactions taking place, but really reflects the nature of the functional changes introduced by the loops in the system. In this respect, one may consider that two competing antagonistic effects occur when a new loop appears: (*1*) the variation of the total mass of the system, and (*2*) the change in the distribution of the chemical-potential difference $\mu_A - \mu_B$ along the reaction paths. If the ratio A/B is sufficiently big, any new loop will lead to an increase of mass, which tends to decrease σ. This effect, however, is compensated if the distribution of the chemical-potential difference among the intermediates, which corresponds to an overall *constant* affinity of reaction, can be associated with a higher rate of the overall transformation $A \to B$. For a given value of the boundary conditions one may then say that the system has managed to interact more strongly with the external world in such a way that it is traversed by a higher flux of matter. It can easily be verified that one can distinguish on this basis between transformations occurring in reaction chains that may at first seem very close—for example:

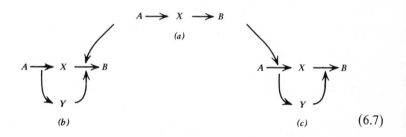

$$(6.7)$$

In the case $(a) \to (b)$ one observes a decrease of σ, while with $(a) \to (c)$ one has the reverse effect.

B. Steady-State Entropy Production S.E.P. of a Simple Catalytic Model

It is interesting to interpret these results with respect to the effect of catalysis on a reaction chain such as (6.1). Some years ago, one of us (I. P., Ref. 33) considered the following catalytic scheme.

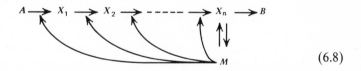

$$(6.8)$$

It was verified that provided $A/B \to \infty$, (6.8) necessarily has a higher level of dissipation than (6.1). It is also interesting to observe the way in which this result is achieved. In the case of system (6.1), we find that between A and B there is a continuous degradation of the chemical potential,

$$\mu_A > \mu_{X_1} > \mu_{X_2} > \cdots > \mu_{X_n} > \mu_B$$

If $n \gg 1$, one has $\mu_{X_i} - \mu_{X_{i+1}}$ tending to zero, and as a result the overall rate of transformation of A into B is very slow. In case (6.8), on the contrary, one finds that $\mu_A \cong \mu_{X_1} \cong \mu_{X_2} \cong \cdots \cong \mu_{X_n}$ but $\mu_{X_n}/\mu_B \gg 1$; in other words, the effect of catalysis is to propagate the chemical potential without degradation along the chain and to sustain between X_n and B a very large affinity of reaction. It is thus clearly by keeping one of the reactions very far from equilibrium that the system becomes more efficient. In fact, because of the catalysis, it behaves as if the chain were much shorter and involved only one intermediate compound.

VII. PHYSICAL BASIS OF SELF-ORGANIZATION

In any discussion of self-organization, one is confronted with the problem of defining and quantifying the notion of order. The description of a self-organizing process, indeed, begins with the recognition of some gradient of order or hierarchy among the successive stages that a system presents in the course of its time evolution. In the next step, one is confronted with the problem of finding a mechanism that accounts for the transition from one degree of organization in the hierarchy to the next one.

If one thinks of biological evolution on a very broad scale, there is no great difficulty in recognizing some hierarchy among the variety of living forms. When one goes from the molecular level to the level of cellular organites, cells, organs, and systems, there is an obvious gradation, which corresponds to a continuous increase in structural and functional complexity. This duality between structural and functional aspects really is the essential property of biological order. It also makes the notion of order in biology fundamentally distinct from the notion in physics, which is usually related to entropy. To characterize biological order even in its molecular manifestations, entropy is clearly not an entirely adequate function. An order criterion based on entropy differences between various systems would make no great distinction between most arbitrary amino acid sequences coming out of some type of Monte Carlo game; it would be helpless to determine those sequences that have some functional interest. Similarly the evaluation of the enormous information content of biological molecules is rather useless as long as it does not give us any indication of the mechanism by which it has emerged. What must be taken into account is the fact that the purpose of biological structures is to accomplish certain functions with the best efficiency possible. In other words, the degree of organization of a biological system is always evaluated with respect to what biologists would call its "project," that is, to see for an eye, to hear for an ear, and so on. The better this project is realized, the more organized the system is. On the other hand, the transition toward the most elaborate forms of the project is then explained by the existence of a "selection pressure," which merely corresponds to the advantages gained by the organism over its environment when the project is realized with more efficiency.

Recently Eigen[34] has put forward a general theory of self-organization describing the evolution of macromolecules, which formulates these considerations in a very elegant fashion: In that case the "project" considered is self-replication of the molecules; the organization of a system is evaluated by a value function which expresses the fidelity with which the set of replication processes is performed; the gradient of organization defined in this way thus corresponds to the increasing chances of perpetuation of molecules that better minimize random mutations. The selection pressure is created by coupling this process with a constant influx of monomers and precursors. In a somewhat different approach, Kuhn[35] looked for a possible pathway of emergence of the genetic apparatus, consisting of many little steps, any one of these steps being obtained from the previous one by looking at the possible modes of evolution of the system in its environment, estimating the time required for each mode, and then choosing the fastest possibility as the next step. The project here is to evolve faster, and it tends to favor at a molecular level an autocatalytic behavior of the type stressed for biological evolution in general by de Duve.

Results such as those reported in the preceding section have thus suggested to us a point of view that may be regarded as complementary to the one of Eigen and Kuhn. Within the perspective of evolution, one would have to look for instabilities or dissipative structures that increase the departure of an open system from equilibrium. Consequently a higher level of interaction with the outside world could be used as the source of order. Instabilities and kinetic properties would thus be divided into two classes according to whether or not they correspond to an increase in dissipation. The index or value function for such a *gradient of evolution* would be the entropy production measured intensively.

The results, of course, are very fragmentary; however, they give some hope that it is not unrealistic to try to associate with "*la logique du vivant*" a general physical quantity whose variation continuously follows the process of self-organization. Almost thirty years ago, one of us wrote,[3]

Ce n'est donc ni l'entropie, ni d'ailleurs aucun potentiel thermodynamique, qui permet de caractériser l'évolution irréversible d'un système, mais seulement la production d'entropie qui s'approche de sa valeur minimum.

Dans certains cas cette valeur minimum de la production d'entropie ne peut être atteinte qu'en augmentant l'hétérogénéité, la complexité du système. Peut-être que cette évolution spontanée qui se manifeste alors vers des états à hétérogénéité plus grande pourra-t-elle donner une impulsion nouvelle à l' interprétation physico-chimique de l'évolution des êtres vivants.

In this passage already we have the duality between creation of structure and maintenance of structure. The situation at present seems to indicate that these two aspects can be characterized by different behavior of the entropy production: While the creation step leads to an increase in σ, the steps of maintenance, on the contrary, seem to follow the theorem of minimum entropy production.

References

1. I. Prigogine and P. Glansdorff, *Bull. Acad. Roy. Ch. Sci.* **59**, 672 (1973).
2. E. Schrödinger, *What is Life*, Cambridge University Press, London, 1945.
3. I. Prigogine, *Etude Thermodynamique des Phénomènes Irréversibles*, Dunod, Paris and Desoer, Liège, 1947.
4. I. Prigogine and J. M. Wiame, *Experientia*, **2**, 415 (1946).
5. P. Glansdorff and I. Prigogine, *Thermodynamic Theory of Structure, Stability and Fluctuations*, Wiley Interscience, New York, 1971.
6. I. Prigogine and G. Nicolis, *J. Chem. Phys.*, **46**, 3542 (1967).
7. I. Prigogine, "Structure, Dissipation and Life," in *Theoretical Physics and Biology*, M. Marois, Ed., North Holland, Amsterdam, 1969.
8. A. M. Turing, *Phil. Trans. Roy. Soc. (London)*, **B237**, 37 (1952).
9. J. I. Gmitro and L. E. Scriven, in *Intracellular Transport*, K. K. Warren, Ed., Academic, New York, 1966.
10. H. G. Othmer and L. E. Scriven, *Ind. Eng. Chem.*, **8**, 302 (1969).
11. I. Prigogine and R. Lefever, *Cj. Chem. Phys.*, **48**, 1695 (1968).
12. R. Lefever, *J. Chem. Phys.*, **49**, 4977 (1968).
13. R. Lefever and G. Nicolis, *J. Theor. Biol.*, **30**, 267 (1971).
14. B. Lavenda, G. Nicolis, and M. Herschkowitz-Kaufman, *J. Theor. Biol.*, **32**, 283 (1971).
15. M. Herschkowitz-Kaufman and G. Nicolis, *J. Chem. Phys.*, **56**, 1890 (1972).
16. I. Prigogine, "Dissipative Structure in Biological Systems," in *Theoretical Physics and Biology*, M. Marois, Ed., North-Holland, Amsterdam, 1971.
17. I. Prigogine, R. Lefever, A. Goldbeter, and M. Herschkowitz-Kaufman, *Nature*, **223**, 913 (1969).
18. A. A. Andronov, A. A. Vitt, and S. E. Khaikin, *Theory of Oscillators*, Pergamon, Oxford, 1966.

19. N. Minorsky, *Nonlinear Oscillations*, Van Nostrand, Princeton, N.J., 1962.
20. L. Cesari, "Asymptotic Behavior and Stability Problems," in *Ordinary Differential Equations, Arg. Mathem. New Series*, **16**, Springer, Berlin, 1962.
21. G. Nicolis, *Adv. Chem. Phys.*, **19**, 209 (1971).
22. G. Nicolis and J. Portnow, *Chem. Rev.*, **73**, 365 (1973).
23. J. Turner, *Adv. Chem. Phys.*, this volume.
24. F. Schlögl, *Z. Phyzik*, **253**, 147 (1972).
25. K. Kitahara, Ph. D. thesis, Univ. of Brussels (1974).
26. I. Prigogine, G. Nicolis, and A. Babloyantz, *Physics Today*, **25**, 12 (1972).
27. H. Martinez, *J. Theor. Biol.*, **36**, 479 (1972).
28. Y. Kobatake, *Physica*, **48**, 301 (1970).
29. F. Jacob, *La logique du Vivant*, Gallimard, Paris, 1970.
30. C. de Duve, "La Biologie au XXème siècle," Communication au Colloque *Connaissance scientifique et Philosophie* organisé à l' ocaasion du bicentenaire de l'Académie Royale de Belgique.
31. G. Nicolis and R. Lefever, in preparation.
32. A. Goldbeter and G. Nicolis, *Biophysik*, **8**, 212 (1972).
33. I. Prigogine, *Physica*, **31**, 719 (1965).
34. M. Eigen, *Naturwissenschaften* **58**, 465 (1971).
35. H. Kuhn, "Selforganisation of Nucleic Acids and the Evolution of the Genetic Apparatus," in *Synergetics*, H. Haken and B. G. Teubner, Eds., Stuttgart, 1973.
36. P. C. Fife, "Singular perturbation by a quasilinear operator," in *Lecture Notes in Mathematics, No. 322*, Springer, Berlin, 1972.
37. I. Prigogine, *Nature*, **246**, 67 (1973).
38. I. Prigogine, *Nonequilibrium Statistical Mechanics*, Wiley-Interscience, New York, 1962.
39. I. Prigogine, C. George, F. Henin, and L. Rosenfeld, *Chemica Scripta*, **4**, 5 (1973).
40. A. Goldbeter and R. Lefever, *Biophys. J.*, **12**, 1302 (1972).

DISSIPATIVE INSTABILITIES, STRUCTURE, AND EVOLUTION

G. NICOLIS

Faculté des Sciences, Université Libre de Bruxelles,
Bruxelles, Belgium

CONTENTS

I. INTRODUCTION

Our principal goal in this chapter will be to comment on the physicochemical basis of certain aspects of biological order related to both the origin and the evolution of prebiological matter as well as to the problem of the maintenance of this order in living organisms.

The emergence and maintenance of biological order at successively higher levels of complexity is also the central theme on which this book is centered. Our program leads us from the most "elementary" level of the spatiotemporal organization of simple biochemical networks (see the chapter by B. Hess et al., in p. 137) to the most "macroscopic" level of the functioning of specific biological structures such as membranes (see the chapters by B. C. Goodwin, p. 269; by P. Läuger, p. 309; by Y. Kobatake, p. 319: and by R. Lefever and J. L. Deneubourg, p. 349). Between these two extremes we have chemical evolution, the spontaneous formation of spatial patterns, and the problem of development (see the chapters by L. Wolpert, p. 253: and by E. Margoliash, p. 191).

What is the relation between these phenomena and first principles of physics, in particular the laws of thermodynamics? Is it possible to describe, in precise physicochemical terms, phenomena like regulation, information sotrage, and the transmission of genetic information during development? These are some of the main questions we shall be concerned with in this paper.

We start, in Section II, with a brief account of some recent developments in the thermodynamics of irreversible processes, which will help us greatly to better formulate the problem of the emergence and maintenance of biological order. This point is developed in more detail in the preceding paper by I. Prigogine and R. Lefever. Next, we compile in Section III a number of results obtained from the analysis of model systems, which will tell us, in a sense, *what type of behavior one can expect in principle* from nonlinear chemical networks operating far from equilibrium, and whether this behavior bears some analogies with biological order. In Section IV we make a few remarks about the status of the deterministic and stochastic descriptions in biology. And finally, we present briefly, in Section V, a number of biological problems where the previous considerations can be applied. This point is also discussed further in other chapters in this volume, especially those by P. Ortoleva and J. Ross (p. 49), by D. Thomas (p. 113) and by B. Hess et al. (p. 137).

II. THERMODYNAMICS OF IRREVERSIBLE
PROCESSES. DISSIPATIVE STRUCTURES

The failure of classical equilibrium concepts to interpret biological order is nowadays well recognized. It suggests that one should try to extend the concept of order to *nonequilibrium situations*, for systems in which the appearance of ordered structures would be very unlikely at thermodynamic equilibrium. In recent years, this question has attracted a great deal of attention. The principal results are as follows[1] (see also the preceding chapter by I. Prigogine and R. Lefever):

1. Linear systems, in particular systems close to equilibrium, evolve always to a disordered regime corresponding to a steady state that is stable with respect to all disturbances, provided equilibrium itself is stable.* In other words, if a system deviates from the steady state owing to the action of a random perturbation, a mechanism will be developed tending to damp this perturbation. Stability can be expressed in terms of

---

* This implies that the system is far from a region of ordinary phase transitions.

a variational principle of *minimum entropy production*, which asserts that for time-independent boundary conditions the dissipation inside the system (measured by the produced entropy per unit time) attains the least possible value compatible with the boundary conditions.

2. Consider now an open system obeying *nonlinear kinetic laws* and that is driven, in addition, far from thermodynamic equilibrium.* As the distance from equilibrium increases, and provided that the boundary conditions remain fixed and independent of time, the steady states close to equilibrium are shifted continuously to a branch of states showing an equilibrium-like behavior (see Section III for examples). We shall call this branch the *thermodynamic branch*.

The difference with equilibrium, however, is that the laws of thermodynamics can no longer guarantee the stability of this branch. The system may therefore have to leave this branch and evolve to a new stable regime, which *may*—but does not have to—correspond to a spatially or temporally organized state.

We have called these regimes *dissipative structures* to indicate that, contrary to what common intuition would suggest, they are created and maintained by the dissipative, entropy-producing processes inside the system. Let us emphasize that the arguments advanced here do not constitute an "existence theorem" establishing the occurrence of order through dissipation. Rather, they provide a sufficient condition that must be realized in order to have spontaneous emergence or ordered structures.

3. Finally, in spite of the diversity of the situations that may arise in nonlinear systems, one can demonstrate a few general thermodynamic theorems underlying these nonequilibrium order phenomena.[1] As one would expect, there seems to be no variational principle in this case. Nevertheless one can show that when a nonequilibrium state becomes unstable, the average amount of dissipation (measured by the produced entropy) introduced by the disturbance (which is assumed to be small) becomes negative. In this way, the stability properties are connected with thermodynamic functions of direct experimental interest.

The results we have just mentioned have been obtained by means of a very general thermodynamic analysis.[1] The main conclusion has been that in certain cases, nonequilibrium can act as a source of order. On

---

* One may recall that nonlinearity is a general rule in biology. It appears in various forms such as: cooperativity at the molecular level, direct autocatalysis or activation, inhibition, and so on. Intuitively, the occurrence and the role of nonequilibrium constraints in biology is also quite obvious. It will be elucidated further in the examples of Sections V and VI.

the other hand, ordered behavior—especially in the form of time periodicities—has long been known in certain fields of mathematics, especially in the theory of nonlinear differential equations which had been worked out by Poincaré and later by the Andronov school.[2] Since the pioneering work of Lotka and Volterra on the oscillatory behavior of competing biological species,[3] the methods of nonlinear mathematics have been applied extensively to the study of models simulating certain aspects of chemical kinetic and biological processes. Among the most representative pieces of work in this area one should mention the models of the functioning of open chemical reactors,[4] Turing's work on morphogenetic patterns,[5] and Thom's general formulation of morphogenesis in terms of the topological theory of ordinary differential equations.[6]

For a long time these two types of approach, although convergent, remained practically independent. One of the aims of the next section, which is devoted to the analysis of models, is to show the complementarity of the two points of view. The intimate relation between the two approaches is already clear when one observes that the notion of *dissipative structure*, which is one of the notions that recur frequently in this volume, implies both a thermodynamic condition (a critical distance from equilibrium) *and* a mathematical condition, namely the bifurcation of a new solution arising beyond the instability of the thermodynamic branch.

III. MODEL DISSIPATIVE SYSTEMS

In a first discussion of model systems, we shall adopt a macroscopic description. Consider a reacting mixture of n species. The system is *open*, that is, it is subject to the flow of chemicals from the outside world, which are converted within the reaction volume $\mathcal{V}$ into the intermediates $X_1, \ldots, X_n$ (Fig. 1). For simplicity, we assume that the system is *isothermal*, at *mechanical equilibrium*, and subject to *time-independent boundary conditions*. Moreover, diffusion within $\mathcal{V}$ is approximated by Fick's law, and the diffusion coefficient matrix is taken to be diagonal. All these conditions can, of course, be relaxed. The main reason we adopt them is first, that the mathematics becomes somewhat more transparent, and second, that most of the effects we neglect are indeed spurious for the study of the biological phenomena we want to understand here.

Under the condition just listed, the instantaneous state of the system will be described by the composition variables $\{X_i\}$ $(i = 1, \ldots, n)$, denot-

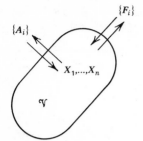

Figure 1. An open system subject to the flow of initial and final chemicals $\{A_i\}$ and $\{F_i\}$, respectively, and containing n intermediates $X_1,\ldots,X_n$, within the reaction volume $\mathcal{V}$.

ing, for example, the average mole fractions or the average partial densities of the intermediate chemicals. The time evolution of these variables is given by the well-known conservation-of-mass equations[1]:

$$\frac{\partial X_i}{\partial t} = v_i(\{X_j\}) + D_i \nabla^2 X_i \tag{1}$$

The source terms v_i describe the effect of chemical reactions, and the D_i's are assumed constant. According to the well-known laws of chemical kinetics, in the most general case the v_i's are nonlinear (usually polynomial) functions of X_j's. Thus we are confronted, from the very beginning, with a set of nonlinear partial differential equations (which should be supplemented, or course, with appropriate boundary conditions). At present there exists no mathematical theory of these equations comparable in generality to the corresponding theory for ordinary differential equations. One sees clearly the necessity of studying models, in order to illustrate the type of behavior one can expect from the solutions of Equations (1). We now discuss briefly some of these phenomena. We remain as general as possible, but from time to time we refer to two models that have been developed by our group, namely, a simple autocatalytic trimolecular model[7] and an allosteric dimer model with positive feedback[8] (see Fig. 2). The individual kinetic steps for the trimolecular model are as follows:

$$A \rightarrow X$$

$$B + X \rightarrow Y + D$$

$$2X + Y \rightarrow 3X$$

$$X \rightarrow E$$

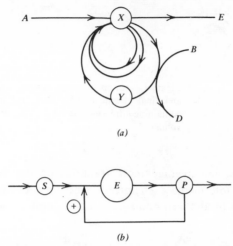

(a)

(b)

Figure 2. (a) A simple autocatalytic model. Circled symbols represent intermediates. It is assumed that the concentrations of all other chemicals are maintained constant in time. (b) An allosteric enzyme model with positive feedback from the product. The enzyme E can exist in several configurations, some of which bind P preferentially.

The principal results are as follows:

1. At or near equilibrium one finds a unique steady-state solution, which is always stable with respect to arbitrary perturbations. This agrees with the general results discussed in the previous section.

2. Suppose now the system is driven far from equilibrium. For the model described in Fig. 2(a) this can be achieved by removing products D and E as soon as they are produced. For time-independent boundary conditions, one can define the branch of steady states referred to in the previous section as the *thermodynamic branch*. A typical profile of the intermediates in this case is shown in Fig. 3 [corresponding to the model of Fig. 2(a)], where we assume that diffusion takes place along a single spatial dimension. The difference with equilibrium is, however, that if one carries out a linearized stability analysis of the equations describing the model [which are, of course, of the type shown in Equation (1)] one finds that, owing to the nonlinearity and to the nonequilibrium constraints, states on this branch are not necessarily stable. The new stable regime that the system will subsequently attain in this case depends on:

●The values of the boundary conditions.
●The values of the parameters, such as the rate constants, the diffusion coefficients D_i, and the length L. The last two types of parameter can

be combined into a single one: D_i/L^2. Notice that the rate constants measure the *coupling between intermediates*, whereas D_i/L^2 measure the intensity of the *coupling between neighboring spatial regions*.

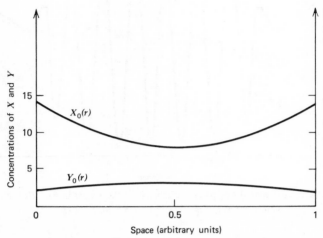

Figure 3. Steady-state solutions on the thermodynamic branch for the model in Fig. 2(a). Numerical values: $D_A = 1.97 \times 10^{-1}$, $D_x = 1.05 \times 10^{-3}$, $D_Y = 5.25 \times 10^{-3}$, $A(0) = A(1) = \bar{A} = 14$, $X(0) = X(1) = \bar{X} = 14$, $Y(0) = Y(1) = \bar{Y} = 1.86$, $B = 26$. All forward rate constants are set equal to unity.

●In certain cases, also on the type of perturbations acting on the system.

We now describe some of the most characteristic types of regime found from various models.

3. Beyond the instability of the thermodynamic branch, the system can attain a new *steady state*, which differs from the previous one (see Fig. 3) by the spontaneous breaking of spatial symmetry. An example referring to the model in Fig. 2(a) is shown in Fig. 4. We obtain a low-entropy, regular spatial distribution of the chemical constituents. This solution has some quite unexpected properties. In the first place, the possibility of creating high concentrations of the chemicals in limited regions of space gives to these states interesting *regulatory properties*. More surprising is the fact that the final configuration depends, to some extent, on the type of initial perturbation. This primitive *memory* makes these structures capable of storing information accumulated in the past. Finally, for certain values of the parameters these

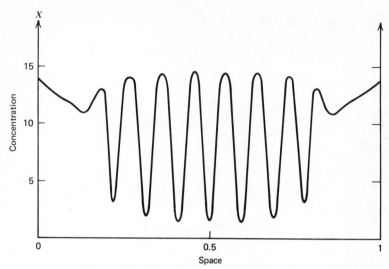

Figure 4. Localized spatial dissipative structures for the model in Fig. 2(*a*). Numerical
values are as in Fig. 3. except for $B = 30$, $\overline{Y} = 2.14$.

spatial dissipative structures may take the form shown in Fig. 5. We
have here an interesting *duplication* of a structure within the reaction
volume. This type of behavior is likely to be of some interest in the
problem of prebiotic evolution as well as in that of development.

4. For different ranges of values of the parameters, corresponding to
a diffusive coupling between spatial regions which is weaker than in the
previous case, systems showing the behavior described in the last
paragraph can evolve to a regime that not only depends on space but is
also periodic in time. This wavelike behavior can correspond to a
propagating or to a standing wave. An example referring to model in
Fig. 2(*a*) is shown in Figs. 6 and 7. In the course of one period there
appear wavefronts of composition, which propagate within the reaction
volume—first outwards, and then, after reflecting on the boundaries,
into the interior. At each point in space the chemical concentrations
undergo sustained oscillations. Thus, the overall phenomenon can be
considered as the result of a *coupling between nonlinear oscillators*. This
space-time dissipative structure provides, therefore, a primitive
mechanism for *propagating information* over macroscopic distances in
the form of chemical signals arising from cellular metabolism. For
further comments on composition waves we refer the reader to the
chapter by B. Hess et al. (p. 137).

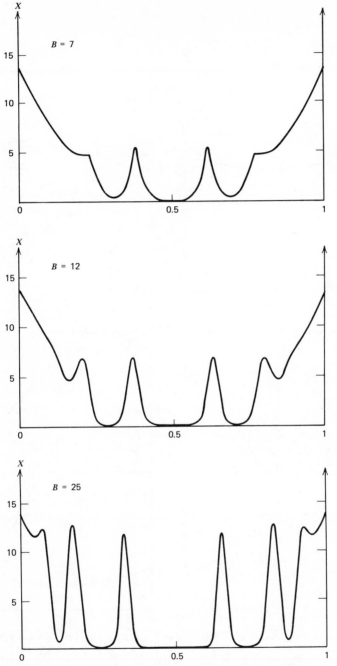

Figure 5. Duplication of a localized dissipative structure for $D_A = 0.26 \times 10^{-1}$ and for the same values of the other parameters as in the previous figures.

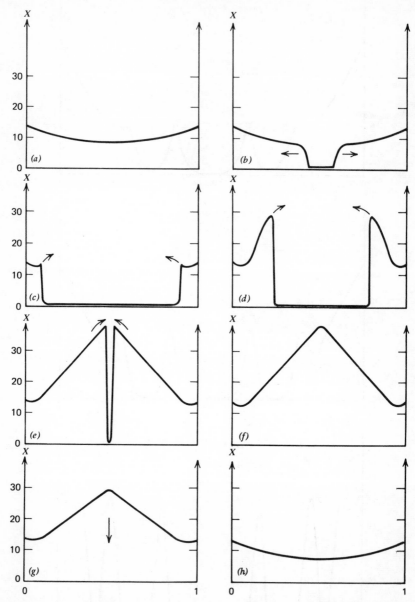

Figure 6. Characteristic steps of evolution of X when the model in Fig. 2(a) exhibits a propagating-wave solution. Numerical values: $D_A = 1.95 \times 10^{-1}$, $D_X = 1.05 \times 10^{-3}$, $D_Y = 0.66 \times 10^{-3}$, $B = 77$, $\overline{X} = \overline{A} = 14$, $\overline{Y} = 5.5$.

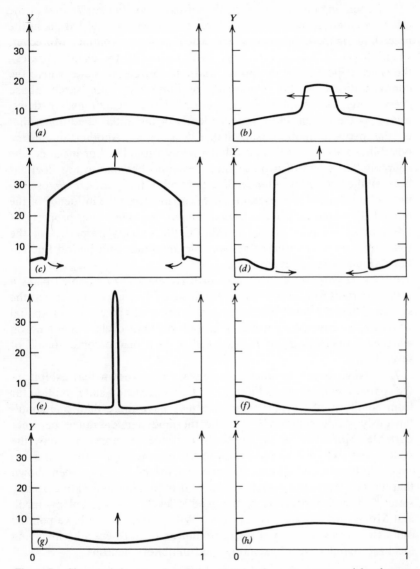

Figure 7. Characteristic steps of evolution of Y for the same system and for the same numerical values as in Fig. 6.

5. Depending on the values of the diffusion coefficients D_i and of the other parameters, the above structures can either be localized (as in the preceding figures), or occupy the whole reaction volume. Moreover, when all (D_i/L^2)'s are very large (with respect to the chemical rates), the space dependences disappear, and one can obtain states where the concentrations of the chemicals oscillate with the same phase everywhere. The amplitude and period of the oscillations are determined by the system itself, independently of the initial conditions; that is, the periodic motion is a stable *limit cycle*. Another interesting possibility arises in systems involving many variables. For instance, the coupling "in series" of two nonlinear reactions like that in Fig. 2(a) has been studied recently.[9] One finds a possibility for *quasiperiodic* oscillations of one half of the reaction scheme arising from the influence of the other half, which acts like a periodic input. An interesting property of this type of behavior is that it exhibits the stability properties of the usual limit cycles as well as appreciable amplitude and period fluctuations around some average value.

6. It is remarkable that the diffusive coupling D_i/L^2 (which plays a very important role, as we have seen) depends strongly on the *size* of the system. Thus, the same biochemical network could give rise to a spatial structure, a propagating wave, or a uniform limit cycle, depending on whether it operates in an isolated cell or in a macroscopic "field" of several cells.

7. Finally, several nonlinear models are now known that exhibit, in the spatially homogeneous limit, more than one steady-state solution far from equilibrium.[1] In particular, one may have more than one simultaneously stable solution *before* the thermodynamic branch becomes unstable. Moreover, in the presence of three or more solutions, the system may exhibit a phenomenon of the *hysteresis* type. In addition to simple autocatalytic systems showing this behavior, it has been shown that certain regulatory systems may also present alternative steady states[10] through repression at the genetic level, as suggested by Jacob and Monod. In a quite different area, Wilson and Cowan have pointed out[11] that hysteresis may arise in nonlinear neural networks involving localized populations of excitatory and inhibitory neurons.

The coupling between multiple steady-state or sigmoidal kinetics and diffusion may also lead to several unexpected properties. Thus, one may obtain symmetry-breaking transitions of a new type, leading to situations where, for example, two stable states of the homogeneous limit prevail in different regions of space. One has the feeling that phenomena of this type should be of interest in the problem of development, and we say a few words on this in the final part of the paper.

Most of the results we have presented so far have been obtained by simulation of the differential equations on a CDC 6400 computer. However, one begins now to see analytically the mathematical mechanisms responsible for certain phenomena such as lack of uniqueness, spontaneous breaking of symmetry, localizaion, and so on.[12]

In conclusion, we have seen that in certain types of systems there is an inherent possibility for evolution beyond a threshold corresponding to appropriate values of the boundary conditions, the size, and certain local parameters (rate constants, etc.). Obviously, this evolution and the subsequent stabilization to a dissipative structure is *not* a universal phenomenon in chemical kinetics, but requires some very stringent conditions. Our next task is therefore to see if these prerequisites for self-organization are also shared by living systems. But before we go to this, we make a few remarks concerning some general aspects of the evolution of self-organizing systems through unstable transitions.

IV. STOCHASTIC AND DETERMINISTIC ASPECTS OF EVOLUTION

In the preceding discussion we have ignored the mechanisms by which a system is driven to the new regime beyond the instability. By analogy with well-studied problems in physics such as phase transitions, one would be tempted to attribute the onset of evolution to *internal fluctuations*. Indeed, fluctuations—the spontaneous deviations from some average regime—are a universal phenomenon of molecular origin and are always present in a system comprising many degrees of freedom. Thus, a system that on the average is in a steady state slightly below, or at the threshold for instability will always have a nonvanishing probability of reaching the unstable region through fluctuations. As a result, certain types of fluctuations will be amplified, and they will drive the system subsequently to the new regime. In addition to the fluctuations, a physical system is also subject to systematic or random *external* disturbances, which may also trigger evolution by a similar mechanism.

Now, the formation of a fluctuation of a given type is fundamentally a stochastic process. The response of the system to this fluctuation is a deterministic process obeying the macroscopic laws [such as Equations (1)], as long as the system can damp the fluctuation. But in the domain of formation of a new structure, fluctuations are amplified and drive the average values to the new regime. Thus, in this region one should expect that the macroscopic description in terms of averages will break down and that the evolution will acquire an essentially statistical character. This conjecture has been verified on several models, which we analyzed recently using the stochastic theory of chemical kinetics.[13]

We have then a picture of a system evolving through instabilities: In the neighborhood of a stable regime, evolution is essentially deterministic in the sense that the small fluctuations arising continuously are damped. But near the transition threshold the evolution becomes a stochastic process in the sense that the final state depends on the probability of creating a fluctuation of a given type. Of course, once this probability is appreciable, the system will eventually reach a unique stable state, provided the boundary conditions are well specified. This state will then be the starting point for further evolution. For further comments on the role of fluctuations we refer the reader to the chapter by K. Kitahara (p. 85) and by J. Turner (p. 63).

V. BIOLOGICAL ILLUSTRATIONS

We do not discuss here the nonbiological experimental examples of dissipative structures, such as the Belousov-Zhabotinski reaction,[1] for which the reader is referred to the paper by B. Hess et al. (p. 137). We focus instead on some representative biological problems where the ordered behavior we discussed previously in models plays a prominent role, and in particular:

●Regulatory processes.
●Excitable systems.
●Cell communication and development.
●Evolution.

A. Regulatory Processes

We mention briefly two examples that have been worked out recently in our group.

The first problem is an application of the allosteric dimer model that we mentioned earlier. It has been shown that sustained oscillations in the product and substrate concentrations can occur for experimentally acceptable values of the allosteric and rate constants.[8] This model has been applied to phosphofructokinase, the enzyme that is believed to be responsible for glycolytic oscillations, which presents similar regulatory properties to the model in Fig. 2(b) (see the chapter by B. Hess et al., p. 137). Qualitative agreement has been obtained with the experimental observations concerning glycolytic self-oscillations. For instance, the amplitude of the product oscillation as a function of the substrate injection rate goes through a maximum, in agreement with experimental observations by Hess et al. A further prediction of the model is that glycolysis can generate chemical signals in the form of concentration waves that propagate over *supercellular* distances.[14] Professor Hess has undertaken some preliminary steps aiming at an experimental verification of this conjecture.

The second example is the problem of β-galactosidase induction in $E.$ $coli.$[15] Bacterial populations grown at low concentration of inducer in a fixed medium contain cells that are in either of two possible states: maximally induced or noninduced. The bacteria can be shifted from one state to the other by varying the environmental conditions. This "all-or-none" character is attributed, in part, to the existence of galactoside permease, which is induced at the same time as β-galactosidase and is, moreover, a part of the inducer (e.g. the lactose) transport system. We are therefore in the presence of an autocatalytic feedback phenomenon. A theoretical model of the galactosidase-permease system has been worked out and has given rise to qualitatively similar behavior.[15] Moreover, it turns out that in the induced bacteria, the inducer entry is assured by active transport. The dissipative instabilities that are pertinent here belong to the type of multiple steady-state transitions.

B. Excitable Systems

All-or-none phenomena appear also in the problem of membrane excitation of nervous cells, which switch abruptly from a rest state of low ionic permeability to an excited state of high permeability, which is closer to thermodynamic equilibrium. The influence of nonequilibrium contraints in this type of behavior has been analyzed recently by Lefever et al.[16] and is discussed more fully in the chapter by Lefever and Deneubourg in this volume (p. 349).

C. The Problem of Development

As we pointed out earlier, certain features of the model systems analyzed in Section III bear strong analogies with what one expects to happen during the early stages of the development of multi-cellular organisms. The problem is of course so large and so diverse that it can be looked upon from different points of view. On the purely macroscopic level, a tempting mechanism for obtaining spatial and temporal patterns in morphogenesis has been proposed by Martinez.[17] He considers a chemical reaction chain similar to that in Fig. 2(a), which describes the inhomogeneous prepattern of spatial configuration of, say, two morphogens x and y. When x and y reach some prescribed threshold concentrations, growth and cell division occur locally. As a result, the morphogen prepattern is no longer a stable steady state. The system then evolves to a new inhomogeneous stable configuration, which again is perturbed by division, until the final "adult" state is reached.

A more fundamental approach to development from the standpoint of dissipative structures could be based on models of genetic regulation giving rise to a combination of multiple steady-state or sigmoidal kinetics and of spatial structures. A preliminary step in this direction has been undertaken by Edelstein.[18] Further work has been reported by Babloyantz and Hiernaux.[19]

D. Evolution

Finally, one expects that the emergence of order through a succession of unstable transitions arising from continuous perturbations should be a central theme in the problem of the prebiological evolution of matter. An interesting formulation of this problem has been proposed by Eigen,[20] who pointed out the analogy between such perturbations and the random mutations giving rise to "error copies" in the process of biopolymer replication. Recently we have discussed the thermodynamic meaning of prebiotic evolution.[21] Our guiding idea has been that in a multistage evolution, each step leading to more ordered macromolecular configurations should already contain in itself the "germs" of the next transition leading to further evolution. One way to achieve this is to increase, under suitable nonequilibrium constraints, the interactions between the system and the outside world (i.e., ultimately, the nonlinearity of the process). This fact should also be translated by an increase of the amount of dissipation per unit mass inside the system. We may visualize this *evolutionary feedback* as follows:

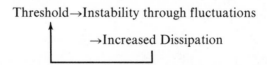

We have shown that there exist large classes of systems describing the synthesis of biopolymers that exhibit this evolutionary feedback, as a result of the fact that transitions increasing the populations of polymer copies have certain autocatalytic growth properties. This behavior contrasts strikingly the commonly observed tendency (see Section II) of physico-chemical systems near equilibrium toward a state of minimum dissipation. For further remarks on thermodynamics of evolution we refer to the chapter by I. Prigogine and R. Lefever (p. 1).

VI. CONCLUDING REMARKS

The main theme of this chapter has been that the coherent behavior induced by a deviation from thermodynamic equilibrium is directly related to biological order. One is now tempted to ask the following question. What is the interest of the "converse" of this statement—in other words, having agreed once and for all that biological order is a thermodynamic order, what is the insight one gains by applying the concepts and methods of irreversible thermodynamics to the study of living systems? We do not try to give an exhaustive answer to this question. Rather, we compile a few ideas that may partly clarify the problem.

1. In the first place, it is instructive to point out the decisive role of nonequilibrium constraints in the examples outlined in the previous section. In the phosphofructokinase reaction, nonequilibrium arises primarily from the irreversible entry of ATP into the system, from the irreversible decomposition of various enzymatic complexes, and from the irreversible departure of ADP. An even more obvious constraint in the form of a chemical-potential gradient across the membrane is present in the β-galatosidase–permease system, as well as in the problem of membrane excitation. Finally, as Eigen has pointed out already,[20] if the synthesis of prebiotic biopolymers via template action proceeded close to equilibrium, autocatalysis would not be sufficient to ensure the survival and the stability of a given polymer copy against the errors arising in the replication process.

2. Next, one should recall that most of the theoretical work done nowadays in biology is based on mathematical models, which are assumed to describe correctly the phenomenon to be studied. In many cases, the details of the kinetics of all biochemical reactions is not fully elucidated. Thus, most of the models comprise necessarily a number of drastic assumptions and simplifications. Thermodynamics is helpful in this case, because it enables one to eliminate all models that are incompatible with the laws of macroscopic physics. For instance, any reasonable biophysical model should be such that coherent behavior (e.g., in the form of sustained oscillations) is ruled out close to equilibrium.

3. In certain problems such as prebiotic evolution, the kinetics is so complex and so poorly known that one cannot be sure about the precise form of the rate equations. Thus, it is very helpful to be able to deduce in this case general thermodynamic theorems such as the theorem about the increase of the rate of dissipation per unit mass.

4. As we have stressed repeatedly in this chapter, in all problems involving evolution beyond an unstable transition, fluctuations play a crucial role. This is particularly true in problems like those mentioned in Section III, where the final dissipative structure depends on the way the

system was perturbed initially. Now fluctuations are a purely statistical-mechanical and thermodynamic concept. The probability of a certain type of fluctuation depends explicitly on thermodynamic quantities such as the excess entropy.[13] Thus the theory of nonlinear differential equations cannot by itself yield information about fluctuations unless it is supplemented with thermodynamic concepts.

5. From a more general viewpoint, irreversible thermodynamics, in particular the concept of dissipative structures, provides a general framework and an appropriate language for analyzing, in a unified fashion, complex and extremely diverse phenomena related to the multiple aspects of biological order.

Naturally, as we saw throughout this paper, for every particular problem the thermodynamic analysis must be supplemented with mathematical models based on precise biological data. One can hope that this strong interaction between biological observations, mathematical methods, and thermodynamic concepts will prove very fruitful in modeling complex phenomena that have escaped quantitative study so far.

References

1. P. Glansdorff and I. Prigogine, *Thermodynamics of Structure, Stability and Fluctuations*, Wiley-Interscience, New York, 1971.
2. A. A. Andronov, A. A. Vitt, and S. E. Khaikin, *Theory of Oscillators*, Pergamon, Oxford, 1966.
3. V. Volterra, *Leçons sur la théorie mathématique de la lutte pour la vie*, Gauthier-Villars, Paris, 1931.
4. See, for example, D. Cohen, in *Nonlinear Problems in the Physical Sciences and Biology*, I. Stakgold, D. Joseph and D. Sattinger, ed. Springer, Berlin, 1973, p. 15.
5. A. M. Turing, *Phil. Trans. Roy. Soc. London*, **B 237**, 37 (1952).
6. R. Thom, *Stabilité Structurelle et la Morphogenèse*, Benjamin, New York, 1972.
7. See, for example, I. Prigogine and G. Nicolis, *Quart. Rev. Biophys.*, **4**, 107 (1971); M. Herschkowitz-Kaufman, Ph. D. Dissertation, University of Brussels, 1973.
8. A. Goldbeter and R. Lefever, *Biophys. J.*, **12**, 1302 (1972).
9. J. Tyson, *J. Chem. Phys.*, **58**, 3919 (1973).
10. A. Babloyantz and G. Nicolis, *J. Theor. Biology*, **34**, 185 (1972).
11. H. R. Wilson and J. D. Cowan, *Biophys. J.*, **12**, 1 (1972).
12. G. Nicolis and G. Auchmuty, *Proc. Nat. Acad. Sci. (U.S.A.)* **71**, 2748 (1974).
13. G. Nicolis and I. Prigogine, *Proc. Nat. Acad. Sci. (U.S.A.)*, **68**, 2102 (1971); G. Nicolis, *J. Stat. Phys.*, **6**, 195 (1972).
14. A. Goldbeter, *Proc. Nat. Acad. Sci. (U.S.A.)*, **70**, 3255 (1973).
15. A. Babloyantz and M. Sanglier, *FEBS Letters*, **23**, 364 (1972).
16. R. Lefever and J. P. Changeux, *C. R. Acad. Sci. (Paris)*, **275D**, 591 (1972); R. Blumenthal, J. P. Changeux, and R. Lefever, *J. Membr. Biol.*, **2**, 351 (1970).
17. H. Martinez, *J. Theor. Biology*, **36**, 479 (1972).
18. B. Edelstein, *J. Theor. Biology*, **37**, 221 (1972).

19. A. Babloyantz and J. Hiernaux, *Proc. Nat. Acad. Sci. U.S.A.* **71**, 1530 (1974); submitted in *Bull. Math. Biology*.
20. M. Eigen, *Naturwissenschaften*, **58**, 965 (1971).
21. I. Prigogine, G. Nicolis, and A. Babloyantz, *Physics Today* **25**, No. 11 and 12 (1972).

STUDIES IN DISSIPATIVE
PHENOMENA
WITH BIOLOGICAL APPLICATIONS

P. ORTOLEVA AND J. ROSS

*Department of Chemistry, Massachusetts Institute of
Technology,
Cambridge, Massachusetts*

CONTENTS

I. INTRODUCTION

We have considered a variety of nonequilibrium phenomena in systems undergoing various combinations of reactions, transport and irradiation. Many of these we believe to have biological applications. We emphasize here the qualitative aspects of the analyses, referring the reader for details to articles and to the fundamental text of Glansdorff and Prigogine.[1]

II. REACTING SYSTEMS BOUNDED BY
MEMBRANES OF VARIABLE PERMEABILITY

For some purposes a crude model of a cell or cell organelle is simply a homogeneously reacting system contained in a membrane of variable permeability and placed in contact with a bath. The dependence of the permeability of the membrane on concentration admits the possibility of feedback loops, which lead to a variety of distinctively nonequilibrium phenomena, including multiple steady states and sustained

49

oscillations consistent with a given bath of time-independent composition. The biological role of the multiple steady states arising from such feedback mechanisms was first considered by Rashevsky[2] and more recently by Babloyantz and Sanglier.[3] We discuss here some general aspects of both the multiple steady states and oscillatory phenomena.[4]

For each chemical species i we introduce the concentration n_i and n_i° in and outside the volume of the reacting system, the permeative flux J_i^p across the membrane, and the rate of production J_i^r due to reaction. With this we have a general equation for the system dynamics,

$$\frac{dn_i}{dt} = J_i^p(\mathbf{n}, \mathbf{n}^\circ) + J_i^r(\mathbf{n}) \qquad (2.1)$$

The flux J_i^p in general may depend on the concentrations of all species on both sides of the membrane. J_i^r depends only on the concentrations $\mathbf{n}$ within the reaction volume. In the simplest case the flux J_i^p may be written $J_i^p = h_i(\mathbf{n}, \mathbf{n}^\circ)(n_i^\circ - n_i)$, where $h_i(\mathbf{n}, \mathbf{n}^\circ)$ does not diverge as $n_i \to n_i^\circ$ (i.e., cross effects such as the flux of i driven by the concentration difference of $j(\neq i)$ are neglected). The function $h_i(\mathbf{n}, \mathbf{n}^\circ)$, and the more general set of coefficients that includes cross effects, will be collectively referred to as the permeability $\mathbf{H}$.

The system may have feedback either in the homogeneous reaction mechanism (i.e., product activation or repression of an enzyme, autocatalysis, or complex cross-netted enzymatic schemes) or due to the composition dependence of the membrane permeability. The presence of the latter may be shown in general to lead to instability of steady-state solutions ($dn_i/dt = 0$) in the limit of rapidly varying permeability ($|\partial \mathbf{H}/\partial n_j| \to \infty$ for some j). The instability is, in this limit, essentially independent of the details of the reaction mechanism; also there is greater potentiality for instability with an increasing number of intermediate species in the reaction mechanism. The analysis proceeds by making an asymptotic expansion of the characteristic equation arising in the stability analysis of the steady states in terms of those contributions to $|\partial \mathbf{H}/\partial \mathbf{n}|$ that are large.

Beyond the linear analysis, simple models reveal the existence of multiple steady states and limit-cycle behavior. Furthermore, transitions between these various states *may or may not* be reversible with variations in the concentration of one external chemical component, depending on the values of the others. In the case of reversible transitions hysteresis may be observed. For example, in a two-species system S, P with the reaction $S \to P$ occurring in the reaction volume, $J_S^p = h_S(P)$ $[S^\circ - S]$, $J_P^p = h_P[P^\circ - P]$, and $dh_S/dP > 0$, multiple steady states may be obtained for certain nonequilibrium ranges of $S°$ and $P°$ when dh_S/dP is sufficiently large. Transitions between the stable steady states induced by changes in $P°$ are irreversible for a certain range of values of $S°$. In a similar system with the Michaelis-Menten scheme $S + E \rightleftharpoons A \rightarrow E + P$ and $dh_S/dP < 0$, one obtains limit cycles when $S°/P°$ and $|dh_S/dP|$ are sufficiently large.

There are many examples of biological membranes whose permeability depends on concentration. In yeast the decreasing hexose permeability with increasing glucose-6-phosphate (G6P) and the enzymatic production of G6P from hexose bear strong resemblance to the simple Michaelis-Menten model discussed above (see Ref. 4 and references cited therein). Thus, the oscillatory tendency in the hexose-transport-G6P-production feedback loop may account for the difference in the phase relationship between oscillatory glycolytic intermediates in experiments on yeast cell extract and intact yeast cells.[5-7]

The existence of multiple steady states in systems composed of a reaction volume bounded by a membrane of variable permeability may have important implication for cellular differentiation. Applications have been given for yeast cells[2] and E. coli.[3] In crown gall tumor cells[8] there appears to be a positive feedback loop involving the production of inositol from mineral salts and sucrose (supplied from the culture medium) and the enhanced permeation of ions by internal inositol.[9] Such a positive feedback loop could yield multiple stable states of inositol production and thus may be an underlying mechanism distinguishing between the most autonomous tumor-cell type (capable of growing on mineral salts and sucrose) and normal and less autonomous tumor cells, which have an exogenous inositol requirement.

It is important in developing a theory of cellular differentiation to distinguish between potentially reversible transformations, which conserve genetic information, and those that are not information conserving. The concept of multiple stable states of reaction and transport consistent with a given environment can provide a basis only for the reversible, information-conserving transformations. As we have pointed out, however, a certain degree of irreversibility may exist with regard to transitions between steady states upon variations of certain culture-medium concentrations at otherwise constant culture conditions.

III. LOCALIZED INSTABILITIES

There are many examples in biology where nonequilibrium reactions are localized in a particular region with a medium that itself is in a state of net reaction and transport. A partial list of important localized sites of reaction in the cellular architecture includes the cell membranes, mitochondria, nucleus, and ribosomes. Oscillations have been observed in suspensions of mitochondria prepared from pigeon heart[10] and rat liver.[11] In a plasmodia such as certain slime-mold species and striated muscles, the nucleus is an important well-localized reaction site. Also, there are the amoeboid cells themselves in a dispersion of aggregating, oscillating slime-mold cells.[12] If the medium between the localized sites contains no essential nonlinearity (i.e., if, in the absence of localized sites, it may be linearized about some stable steady-state solution to the equations of reaction and transport), then the set of nonlinear partial differential equations describing the complete system (linearized reacting-diffusing medium plus well-localized, nonlinear sites), may be transformed into a set of nonlinear ordinary integral equations among the composition variables at the sites.[13]

Taking a matrix $\mathbf{D}$ of constant diffusion coefficients, we may write the equation of motion for the set of concentrations $\mathbf{n}(r,t)$ at point r and time t as

$$\frac{\partial \mathbf{n}}{\partial t} = \mathbf{D}\nabla^2 \mathbf{n} + \mathbf{F}[n] + \sum_\alpha \mathbf{G}^{(\alpha)}[\mathbf{n}]\delta(r - r^\alpha). \qquad (3.1)$$

The bulk reactions contribute a term $\mathbf{F}[n]$. The heterogeneous reaction sites are taken to be well (delta-function) localized to the surface of the site and to have a rate $\mathbf{G}^{(\alpha)}(\mathbf{n})$ for the αth site.

We introduce the steady state $\mathbf{n}^*(r)$ of the bulk reaction system, $0 = \mathbf{D}\nabla^2 \mathbf{n}^* + \mathbf{F}[\mathbf{n}^*]$, satisfying specified boundary conditions. The matrix $\Omega(r) \equiv (\partial \mathbf{F}/\partial \mathbf{n})_{\mathbf{n}^*}$ determines the linearized chemical kinetics. We then define the matrix propagator $\Xi(r,r';t)$, which satisfies the linearized bulk dynamics:

$$\frac{\partial \Xi}{\partial t} = \mathbf{D}\nabla_r^2 \Xi + \Omega(r)\Xi \qquad (3.2)$$

with $\Xi(r,r';t=0) = \mathbf{I}\delta(r-r')$. For infinite homogeneous bulk systems Ξ depends only on $r - r'$ and may be determined straightforwardly by standard Fourier (r) and Laplace (t) transform methods. The propagator $\Xi(r,r';t)$ is zero for points r on the boundary for any r' in the system volume if the concentration is maintained fixed there. After all transients have subsided, the system obeys the set of ordinary nonlinear integral equations

$$\mathbf{n}(r^\alpha,t) = \mathbf{n}^*(r^\alpha) + \sum_\beta \int_0^t dt' \Xi(r^\alpha,r^\beta;t-t')\mathbf{G}^\beta[\mathbf{n}(r^\beta,t')]. \qquad (3.3)$$

We have neglected nonlinear terms in the bulk.

The equation for the steady states of the complete system may be attained by imposing the condition that $\mathbf{n}(r^{\alpha}, t)$ is independent of time and evaluating $\int_0^t \Xi(r^{\alpha}, r^{\beta}; t - t') dt'$ as $t \to \infty$. This leads to a set of nonlinear algebraic equations for the steady-state concentrations at the localized sites. In general we may expect multiple solutions because of feedback mechanisms (1) contained totally at the local sites and (2) involving both localized reactions and exchanges with the bulk and between sites. Simple reaction schemes have been constructed to demonstrate the existence of localized multiple steady states and the properties of the transitions between them.

Stability analysis may be carried out by linearization of (3.2), which yields a linear integral equation for the perturbations around the steady state. The evolution of small perturbations may be oscillatory. In the presence of oscillatory growth we may expect the possibility of local stable limit cycles. This has been verified by analytical solutions of the integral equations for several simple model systems.

Neighboring sites may interact chemically via the diffusing reacting media between them. In general the behavior of coupled sites may be qualitatively different from that of widely separated, essentially uncoupled sites. In particular, a pair of sites with identical localized reaction schemes may become unstable with respect to asymmetric perturbations in concentration while remaining stable with respect to symmetric ones. Such symmetry-breaking instabilities have been found in model systems and may be shown to lead to stable asymmetric structures (concentration distributions).

In a variety of biological problems we would like our localized sites to represent regions bounded by a permeable membrane (the cells in slime-mold aggregation, the mitochondria and nucleus in a cell, etc.). This may be accomplished by assuming that part of the local reaction is the permeation step reversibly taking material from the bulk medium into the localized region. Thus, all developments considered in the variable membrane systems (Section II) may be extended to include a more general interaction with the environment.

Localized undulatory spatial patterns may be induced in the region surrounding a given site.[14] Thus, the interaction between localized sites may vary in an undulatory way with their separation and, for a dispersion of reaction centers, with site density.

Aspects of cellular dynamics may be expressed in terms of the localized theory in the absence of any essential cytoplasmic nonlinearity (i.e., unstable chemical feedback). However, such formulations may not be sufficient for some systems, such as yeast cells, where under some conditions stable limit-cycle oscillations have been demonstrated in cell-free extract. Nonetheless, the existence of mitochondrial oscillations demonstrates the importance of the localized instabilities. Nuclear and other localized multiple stable steady states may provide a local basis for cell differentiation, as discussed earlier.

IV. SYMMETRY BREAKING IN MITOSIS

Many examples exist of cell division wherein a parent cell of type A, with no apparent architectural asymmetry, divides into daughters of type B and C (where C may be A). Turing[15] conjectured that the development of structure in biology (morphogenesis) may find a physical basis in the instability of certain reacting systems with respect to symmetry-breaking perturbations. This instability toward the formation of patterning in the concentration distribution signals the onset of dissipative structures sustained by nonequilibrium conditions. We have proposed that such a mechanism may underlie the well-controlled development of asymmetry in certain mitotic events in embryonic development, stem-cell division, and other bifurcations, as well as the evolutionary onset of multicellular organisms.[16]

The particular mechanism for reaction and transport leading to an asymmetric division based on chemical symmetry breaking may vary with cell type. For this instability there is the requirement that the time scale for interdaughter transport and reaction be comparable, so that the asymmetries may develop. However, as the mitosis proceeds, the interdaughter coupling is gradually decreased, passing through a large range of values. Symmetry breaking may set in as a result of cytoplasmic reaction and diffusion at the earliest stages of mitosis, possibly even before nuclear division has commenced, as in the first division in the eggs of fucus. At later stages of the mitosis, the rate-limiting transport process is not diffusion, but the flux through partially formed membranes (the phragmoplast) in plant cells, or a decreasing cross section of common cytoplasm during cleavage in animal cells.

There are a large variety of processes coupling to the interdaughter transport that may lead to the symmetry breaking. Homogeneous feedback mechanisms are exemplified by product-activated enzyme and other homogeneous biochemical cross coupling. Heterogeneous sites of reaction may lead to symmetry breaking, as mentioned in Section III. Thus, if the chromosomes split into identical pairs and then commence to travel toward opposite poles along the spindle apparatus, a critical separation may exist for which feedback associated with the activation and repression of DNA sites could constitute a cooperative repression--activation loop, leaving one daughter with a given template repressed and the same template activated in the other daughter. Similar arguments could be made for other localized sites such as the (originally symmetrically distributed) mitochondria, completed nuclei, ribosomes, and catalyzing membranes. Furthermore, mechanisms involving variable plasma membrane permeability may result in daughters in alternate states of membrane-transport-mediated metabolism (as discussed in Section II). Other processes leading to symmetry breaking involve photochemical reactions (Section VII) and Liesegang ring processes, believed to have autocatalytic precipitation mechanisms.[17]

The role of external gradients in the development of asymmetric division may be to drive the asymmetry in an otherwise stable symmetric development. In this case the external gradients in which the dividing parent cell is situated must have appreciable concentration variations over cellular dimensions. On the contrary, if the mitosis is unstable with respect to asymmetric perturbations, only a vanishingly small gradient is necessary to determine the polarity of a given mitosis. The control of the orientation of the mitosis may be required for the purpose of a particular morphology in a developing multicellular system. Furthermore, the primitive origins of such symmetry breaking could have importance in understanding the genesis of multicellular systems with specialized cell functions. Indeed, a two-cell system with dissipative structure may be able to achieve a greater overall production of necessary metabolites than two identical isolated cells, and thus the dissipative structure would have a selection advantage.

V. NONLINEAR WAVES IN OSCILLATORY CHEMICAL REACTIONS

Waves of chemical composition have been observed in continuous physicochemical systems[18-20] and in dispersions of aggregating amoeboid slime-mold cells.[12] We discuss some theoretical aspects of the waves that arise upon the imposition of heterogeneities on otherwise homogeneously oscillating systems and note possible applications to developmental biology and peristaltic motion.

An analysis of the equations of reaction and diffusion describing nonlinear chemical waves has been carried out in terms of (1) a nonsecular perturbation theory in the strength of the effect of the heterogeneity[21] and (2) numerical solutions of the nonlinear partial differential equations for model systems.[22,20] Qualitatively different modes of wave propagation are found, depending on whether the homogeneous kinetics yields a smoothly varying or a relaxation oscillation.[23]

The starting point for our analysis is the set of dynamical equations for reaction and diffusion describing the time evolution of the concentrations $\mathbf{n}(r, t)$:

$$\frac{\partial n}{\partial t} = \nabla \cdot \mathbf{D}\nabla\mathbf{n} + F[\mathbf{n}] + \gamma\mathbf{G}[r, \mathbf{n}] \tag{5.1}$$

where $\mathbf{D}$ is a matrix of diffusion coefficients, $F[\mathbf{n}]$ is the contribution due to the homogeneous reaction kinetics, and $\gamma\mathbf{G}[r, \mathbf{n}]$, with strength parameter γ, is the heterogeneous effect on the chemical kinetics. If then we assume that the homogeneous (undisturbed) system

$$\frac{d\mathbf{n}}{dt} = F[\mathbf{n}] \tag{5.2}$$

has a stable limit-cycle chemical oscillation $\mathbf{n}^c(t)$, what is the effect of the heterogeneity, and in particular can this effect be characterized by a perturbation analysis in the limit $\gamma \to 0$?

In this limit we expect the undisturbed phase t of the oscillation to be changed to a local phase $\phi(r, t; \gamma)$, and thus, neglecting terms with amplitude of order γ, that the solution is to be of the form

$$\mathbf{n}(r, t; \gamma) \underset{\gamma \to 0}{\sim} \mathbf{n}^c(\phi(r, t; \gamma)) \tag{5.3}$$

By changing independent variables from (r, t) to (r, ϕ), a perturbation theory may be conveniently developed that automatically accounts for local renormalization (local changes in frequency) and the development of phase gradients.[21] The results of the theory, stated here for the simplest case of constant diffusion coefficients, are most clearly demonstrated in terms of the phase correction $t_1(r, t)$, where $\phi(r, t) \underset{\gamma \to 0}{\sim} t - \gamma t_1(r, t)$. It may be shown that $t_1(r, t)$ obeys a *phase diffusion* equation

$$\frac{\partial t_1}{\partial t} = D_p \nabla^2 t_1 - g(r). \tag{5.4}$$

The phase diffusion coefficient D_p is an average of the diffusion coefficients weighted by the regression properties of perturbations from the stable homogeneous oscillation. The phase correction source term $g(r)$ is a measure of the average per cycle of the degree to which the heterogeneous kinetics $\mathbf{G}[r, \mathbf{n}]$ resembles the bulk (oscillatory) kinetics $F[\mathbf{n}]$.

The development of a gradient in the phase of the oscillation, $t \to \phi(r, t)$ as γ increases from zero, is observed as the presence of waves traveling with phase velocity $V_p = (\partial\phi/\partial t)/|\nabla\phi|$ at a given point r and time t.

An analysis of the structure of the terms in the γ expansion underlying the phase-wave picture discussed above reveals that the expansion must break down for relaxation oscillations. The latter are a result of the presence of widely differing time scales in the homogeneous reaction kinetics. Such homogeneous chemical relaxation oscillations have been observed experimentally[24] and theoretically.[25] In the relaxation oscillation a stage of relatively slow change is followed by a quasidiscontinuous short-time-scale kinetic process. During this latter stage, in the presence of a weak heterogeneity, we expect that a front of reaction activity, propagating toward the spatial direction in which the catastropic process is about to occur, could sustain or perhaps sharpen the concentration profile, developing a reaction-enhanced diffusional front.[20]

For a smoothly varying undisturbed cycle $n^c(t)$ (e.g., one in which the period of oscillation is the only essential time scale), the phase-wave picture holds for $\gamma \to 0$. The theory shows that waves are emitted from a well-localized pacemaker region (one with a heterogeneity that speeds up the local oscillation) to an unperturbed region, with a velocity and wavelength that increase with increasing distance from the local heterogeneity. When $n^c(t)$ is a relaxation oscillation, we expect that waves emanating from a localized heterogeneity are in the form of a train of reaction-enhanced diffusion fronts separated by regions in which the potentiality for "catastrophe" is being regenerated. The velocity, wavelength, and waveform propagate essentially undisturbed until the outmost front annihilates with the bulk oscillatory catastrophe. Dimensional analysis leads one to expect that the front velocity is of the order of $\sqrt{D_{rlx}/t_{rlx}}$, where D_{rlx} is a maximum diffusion coefficient of the species participating in the fast process, and t_{rlx} is the short time during which that process takes place. These predictions have been verified numerically.

Phase waves travel from a localized heterogeneity at constant velocity over distances of the order of $R(t) = 2\sqrt{D_p t}$ in a system that initially ($t = 0$) was homogeneous. Beyond R the velocity and wavelength increase rapidly. The catastrophe fronts can penetrate the system at constant velocity over distances far greater than the diffusion length $R(t)$.

Recent developments in the theory of chemical waves using bifurcation theory and extensions of the phase wave theory may be found in Ref. 23 and references cited.

Wave phenomena are to be found in a variety of biological systems, including aggregation in slime molds,[12] and the development of organization in embryos,[26-28] and is believed to play a role in the breaking of symmetry in the first division of the egg of fucus.[29] When coupled to mechanical (i.e., contractile) processes, chemical waves appear as peristaltic and other types of biological motion on the multi- and unicellular levels.

VI. INHOMOGENEOUS NOISE IN BIOCHEMICALLY OSCILLATING SYSTEMS

Oscillatory biological systems are subject to random external and internal fluctuations. One of the most striking effects of such noise is the fact that the phase of an otherwise homogeneous oscillation may be altered globally or locally. Even stable limit-cycle oscillations are weakly stable with respect to long-wavelength perturbations in phase. As a result of this weak stability, chemical oscillations can remain coherent only over a finite distance when subject to small-amplitude local noise.

The theory of small amplitude inhomogeneous noise in such systems proceeds as in the previous section. The starting equation is (5.1) with $\gamma G = \gamma \xi(r,t)$, where ξ is a random variable. If the root-mean-square noise is small, we may introduce a local phase and calculate the phase correlation function $C(r,t) \equiv \langle [\phi(r,t) - \phi(0,t)]^2 \rangle$, which gives the average square difference in phase of points separated by a distance r. We assume for simplicity that at $t = 0$ the system is in phase at all points. If the noise is applied quite locally in space and time (i.e., $\xi(r,t)$ is delta-correlated in space and time) then the phase-wave analysis shows that $C(r,t)$ is proportional to $t^{1/2}$ for large r. For small $r/2\sqrt{2D_p t}$ (where D_p is a phase diffusion coefficient, as discussed in Section V), $C(r,t)$ is linear in $|r|$. Furthermore, $C(r,t)$ is proportional to the mean square noise. Hence there exists a correlation length r_c, defined by $C(r_c, \infty) = T^2$ (where T is the period of the cycle), such that at r_c the rms phase difference is of the order of period itself. Thus, r_c is proportional to T^2 and inversely proportional to the rms noise.

VII. ILLUMINATED SYSTEMS

The illumination of chemical reactions provides the possibility of feedback mechanisms and consequently an interesting variety of phenomena.[30] Consider a reaction

$$A + \cdots = B + \cdots \tag{7.1}$$

Let A absorb light of a given frequency to which the other species partaking in the reaction are transparent. Furthermore, let the molecule A be of sufficient complexity such that after the absorption of a photon, the molecule undergoes a rapid radiationless transition to a vibrationally excited ground electronic state of the molecule, and this in turn is followed by rapid thermal deexcitation. The net result of the illumination in this process is a heating of the reacting mixture proportional to the concentration of A. If now we have the reacting mixture in a vessel at chemical equilibrium and subject it to a steady illumination of the selective-absorption frequency, then after some time a stationary state will be achieved. The equations describing the system are

$$\frac{dA}{dt} = -k_1 A + k_2 B \tag{7.2}$$

$$\frac{dT}{dt} = -\alpha A - \beta(T - T_e) - \lambda \frac{dA}{dt} \tag{7.3}$$

The symbols α and β represent absorption and Newton cooling coefficients, respectively, and T_e is the external temperature. The equations are coupled nonlinearly because of the exponential temperature dependence of the rate coefficients. Steady states correspond to letting the left-hand sides of both equations equal zero. For endothermic reactions we find a single stationary state, which is stable. Depending upon the parameters of the system, perturbations from this state may decay in an oscillatory way. No limit-cycle behavior is observed for this system. For exothermic reactions we find the possibility of three stationary states, of which two are stable and one is unstable. Transitions from one branch of a stable stationary state to another are accompanied by hysteresis.

A thermodynamic analysis of the system confirms the necessity for an asymmetrically maintained stationary state in order to serve the phenomena described. The asymmetry in this case is due to the dependence of the heat input on the density of one of the reactants. This allows the study of fluctuations from stationary states of a closed system that are far from chemical equilibrium.

Coupling of photochemical processes and transport (diffusion and conduction) allows for the possibility of dissipative spatial structures maintained by ambient illumination.[31] Thus, photochemical processes may be important in biomorphogenesis.[15]

The illumination of chemically reacting systems under circumstances described here allows the introduction of far-from-equilibrium conditions in closed systems (primordial pond?). Self-replicating species or closed cycles arising from error copies or fluctuations,[32] which may take part in photochemical processes, may have competitive advantage beyond critical levels of illumination. Model systems with self-replicating species. some of which use photosynthetic processes, competing in a closed isothermal system have been constructed that show (1) multiple steady states, transitions among which correspond to the domination of one species or another, and (2) instability toward the genesis of photosynthetic error copies or cycles. It may well be necessary to take photochemical mechanisms into account in a theory of the evolution of prebiotic molecules.

Acknowledgments

It is with great pleasure that we thank E. Bimpong-Bota, J. Deutch, M. Flicker, R. Gilbert, H. Hahn, S. Hudson, and A. Nitzan for their contributions to various aspects of this work.

References

1. P. Glansdorff and I. Prigogine, *Thermodynamic Theory of Structure, Stability and Fluctuations* Wiley-Interscience, New York, 1971.
2. N. Rashevsky, *Mathematical Biophysics*, Vol. I, Dover, New York, 1960.
3. A. Babloyantz and M. Sanglier, "All or None Type Transitions in β-Galactosidase Induction in E. coli" (to be published).
4. H. Hahn, P. Ortoleva, and J. Ross, *J. Theor. Biol.* **41**, 503 (1973).
5. A. Ghosh and B. Chance, *Biochem. Biophys. Res. Commun.*, **16**, 174 (1964).
6. A. Betz and E. Sel'kov, *FEBS Letters*, **3**, 5 (1965).
7. E. K. Pye, *Biochronometry*, Nat. Acad. Sci., Washington, 1971.
8. A. Braun, in *Cell Differentiation*, R. Harris, P. Allin, and D. Viza, Eds., Scandinavian University Books, Copenhagen, 1972.
9. P. Ortoleva and J. Ross, "Crown Gall Tumors. An Interpretation Based on Multiple Steady States" (unpublished).
10. A. Boiteux, H. Degn, and B. Chance, "Oscillating Respiration in Mitochondria and its Control" (to appear).
11. E. J. Harris, M. P. Hofer, and B. C. Pressman, "Multiparameter Oscillations Induced in Rat Liver Mitochondria with Valinomycin and Oleate" (to appear).
12. M. H. Cohen and A. Robertson, in *Cell Differentiation*, R. Harris, P. Allin, and D. Viza, Eds., Scandinavian University Books, Copenhagen, 1972.
13. E. Bimpong-Bota, P. Ortoleva, and J. Ross, *J. Chem. Phys.* **60**, 3124 (1974).
14. P. Ortoleva and J. Ross, *J. Chem. Phys.*, **56**, 4397 (1972).
15. A. M. Turing, *Phil. Trans. Roy. Soc., London*, Ser. B, **237**, 37 (1952).
16. P. Ortoleva and J. Ross, *Develop. Biol.* **39** F19 (1973). P. Ortoleva and J. Ross, *Biophys. Chem.* **1**, 87 (1973).
17. M. Flicker and J. Ross, *J. Chem. Phys.* **60**, 3458 (1974).
18. A. Zaikin and A. Zhabotinsky, *Nature*, **225**, 537 (1972).
19. A. Winfree, *Science*, **175**, 634 (1972).
20. J. Field and R. Noyes, *Nature*, **237**, 390 (1972).
21. P. Ortoleva and J. Ross, *J. Chem. Phys.* **58**, 5673 (1973).
22. M. Herschkowitz-Kaufman and G. Nicolis, *J. Chem. Phys.*, **56**, 1890 (1972).
23. P. Ortoleva and J. Ross, *J. Chem. Phys.* **60**, 5090 (1974).

24. R. M. Noyes, R. J. Field and E. Körös, *J. Am. Chem. Soc.*, **94**, 1395 (1972).
25. B. Lavenda, G. Nicolis, and M. Herschkowita-Kaufman, *J. Theoret. Biol.*, **32**, 283 (1971).
26. L. Wolpert, *J. Theoret. Biol.*, **25**, 1 (1969).
27. B. C. Goodwin and M. H. Cohen, *J. Theoret. Biol.*, **25**, 49 (1969).
28. M. H. Cohen, *Symp. Soc. Exp. Bio.*, **XXV**, 455 (1974).
29. B. C. Goodwin, this volume, p. 269.
30. A. Nitzan and J. Ross, *J. Chem. Phys.* **59**, 241 (1973).
31. A. Nitzan, P. Ortoleva, and J. Ross, *J. Chem. Phys.* **60**, 3134 (1974).
32. M. Eigen, *Naturwissenschaften* **58**, 465 (1971).

FINITE FLUCTUATIONS, NONLINEAR THERMODYNAMICS, AND FAR-FROM-EQUILIBRIUM TRANSITIONS BETWEEN MULTIPLE STEADY STATES

JACK S. TURNER

*Center for Statistical Mechanics and Thermodynamics,
The University of Texas at Austin, Austin, Texas*

CONTENTS

Abstract

A model biochemical system for which the usual deterministic chemical kinetics predicts a far-from-equilibrium region of multiple steady states is examined. A stochastic approach to chemical kinetics is adopted to study the effect on a predicted hysteresis in the transition point of *finite fluctuations* around the coexisting stable states. The results of a numerical solution of the stochastic master equation are found to differ *qualitatively* from predictions of the purely macroscopic theory. The expected generality of these results, together with their possible thermodynamic interpretation, is discussed.

I. INTRODUCTION

The importance of fluctuations in determining the nonequilibrium behavior of many nonlinear systems has been well established in recent years, particularly for systems that may exhibit instabilities in their steady-state solutions.[1-3] Nonequilibrium transitions occurring beyond instabilities have long been known in hydrodynamics (e.g., laminar→

63

turbulent fluid flow), and are now a subject of great interest in regard to purely dissipative systems as well.[3,4] Examples of the latter—chemically reacting mixtures under open system conditions, for example—are especially relevant to understanding the appearance of spatial, temporal, and functional order in living systems.[3,4]

At or near thermodynamic equilibrium, the macroscopic description for all such systems yields necessarily a unique stationary solution that is always stable with respect to arbitrary perturbations. Sufficiently far from equilibrium, however, the governing kinetic equations for non-linear systems may admit steady states other than the continuous extension of the equilibrium solution into the nonequilibrium domain. Beyond a critical distance from equilibrium, moreover, this *thermodynamic branch* may become unstable in response to fluctuations occurring in the medium. Such an instability may be symmetry-breaking,* in which case, beyond a critical affinity,† the response of the system to an infinitesimal disturbance leads ultimately to a new operating regime characterized by organization in space or time.[3] The resulting *dissipative structure*, an ordered state completely unlike the initial homogeneous thermodynamic branch, is then infinitesimally stable, any subsequent evolution to other possible states occurring in response to finite-amplitude fluctuations.[5,6] Examples of such dissipative structures abound in biological systems, and for a variety of phenomena specific macroscopic models have been worked out,[7] some of which are discussed in the chapters by I. Prigogine and R. Lefever (p. 1), G. Nicolis (p. 29), and Ortoleva and Ross (p. 49).

For instabilities that are not symmetry-breaking, on the other hand, the situation is less well understood. Typically, the macroscopic kinetic equations predict a region of overall affinity in which, for example, three homogeneous steady states are accessible to the system (e.g., Fig. 1). In this case, the middle branch of the resulting S-shaped steady-state curve is infinitesimally unstable, while the upper and lower branches remain stable throughout the coexistence region. The macroscopic analysis, therefore, predicts a hysteresis effect in the transition point between two stable branches of the steady-state solution. The existence of multiple steady states is a common feature of macroscopic models for many diverse biological processes: for example, membrane transport,[8] nerve excitation,[9] enzymatic reactions,[10,11] and many flip-flop-type mechanisms for biological control.[12] Unanswered in such deterministic models, however, is the important question of when the "real" system

---

*The usage here refers to temporal as well as spatial symmetry.

†In anticipation of the focus of this paper, we introduce already the terminology appropriate to chemical reactions.

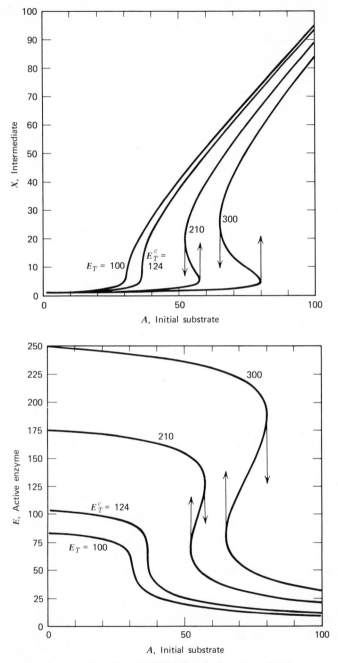

Figure 1. Steady states for the model biochemical system. Continuous curves are predictions of the macroscopic theory, parameterized by E_T, with $B = 1$; arrows indicate points of infinitesimal instability. Rate constants are $k_1 = k_{-1} = k_2 = k_{-3} = 0.2$ and $k_{-2} = k_3 = 1$.

65

will actually jump from one branch to another. As the affinity increases, for example (Fig. 1), will the system remain on the initial branch until the point of infinitesimal instability (arrows, Fig. 1), as the macroscopic theory predicts, or will it be driven from that branch at some earlier point in response to finite fluctuations in the medium?

This question of nonequilibrium transitions among multiple steady states has many features in common with equilibrium first-order phase transitions. The van der Waals theory of the liquid-vapor transition,[13] for example, predicts a range of pressure in which equilibrium isotherms (below the critical temperature) exhibit three possible volume states. The middle state is ruled out on physical grounds (negative isothermal compressibility!), while the other two are in principle accessible throughout the coexistence region. The liquid-vapor phase change should occur therefore at a lower pressure than the reverse transition. That a single equilibrium transition pressure is found regardless of the direction of the change is a direct consequence of the response to finite-amplitude fluctuations in the initial fluid state.

The problem, therefore, for both equilibrium and nonequilibrium transitions, is to understand the range of validity of the purely macroscopic description in the neighborhood of an instability. To approach this question of stability with respect to finite fluctuations for nonequilibrium states we must, as in the equilibrium case, turn to a molecular description of the phenomenon. Unfortunately, a complete microscopic treatment in terms of nonequilibrium statistical mechanics is not yet possible. Nevertheless, a significant and highly successful first step in this direction has been provided for chemical systems by the adoption of a stochastic approach to chemical kinetics.[14] Although such a description is by no means expected to provide the *final* answer, it is important, nevertheless, as an initial inquiry into the *true* nature of transitions between simultaneous nonequilibrium states, to investigate macroscopic and stochastic models of typical systems for which the macroscopic analysis predicts a region of multiple steady states. Accordingly, the remaining sections of this chapter present parallel macroscopic and stochastic steady-state analyses of a simple model biochemical system.[11]

In Section II are summarized briefly the results of a deterministic kinetic investigation, followed in Section III by a stochastic analysis of the same system. The results of a numerical solution of the stochastic master equation are reported in Section IV, with particular emphasis on unexpected features of a far-from-equilibrium transition between two branches of the macroscopic solution. The fundamental differences between the macroscopic and stochastic descriptions are then discussed, followed by a possible thermodynamic interpretation of the results in Section V. The chapter concludes with a discussion of principal results and a few remarks on computational aspects of the problem.

II. THE MODEL: MACROSCOPIC DESCRIPTION

In what follows we consider a model biochemical scheme,[11] which involves the autocatalytic production of an intermediate compound X and its subsequent enzymatic degradation:

$$A + X \underset{k_{-1}}{\overset{k_1}{\rightleftarrows}} 2X$$

$$X + E \underset{k_{-2}}{\overset{k_2}{\rightleftarrows}} C \qquad (2.1)$$

$$C \underset{k_{-3}}{\overset{k_3}{\rightleftarrows}} B + E$$

Here A and B represent the initial reactant and final product, respectively; E the free enzyme; and C its complex with the substrate X.

We assume that the reservoirs of components A and B are time independent. If we let the symbol A denote the average number of molecules* of component A, and so on, then the macroscopic kinetic equations for a spatially homogeneous system take the form

$$\frac{dX}{dt} = k_1 AX - k_{-1} X^2 - k_2 XE + k_{-2}(E_T - E)$$

$$\frac{dE}{dt} = -k_2 XE - k_{-3} BE + (k_{-2} + k_3)(E_T - E) = -\frac{dC}{dt} \qquad (2.2)$$

where the condition of conservation of total enzyme has been imposed

---

*Hence the bimolecular rate constants k_i ($i = 1, -1, 2, -3$) have dimensions of molecule^{-1} sec^{-1}, for example, and are related to more familiar constants [e.g., in (moles/liter)$^{-1}$ sec^{-1}] by a factor containing the volume of the system. Introducing the *number-of-particles notation* here avoids having to define later *new* rate constants for the stochastic model. The choice of the numerical values of the k_i's remains of course arbitrary, although the latter should be adjusted in the larger system computer experiments that are now in progress.

to eliminate the enzymatic complex C from the analysis:

$$E_T = E + C \qquad (2.3)$$

In the time-independent state, the system (2.2) may be solved directly to give the steady-state populations of X and E (and C) for fixed A, B, E_T. Setting the time derivatives equal to zero, we obtain, after a few simple manipulations, a cubic equation the solutions of which represent possible populations of X in the steady state:

$$k_{-1}k_2 X^3 + [k_{-1}(k_{-2} + k_3 + k_{-3}B) - k_1 k_2 A]X^2$$

$$+ [k_2 k_3 E_T - k_1 A(k_{-2} + k_3 + k_{-3}B)]X - k_{-2}k_{-3}E_T B = 0 \qquad (2.4a)$$

with

$$E = \frac{(k_{-2} + k_3)E_T}{(k_{-2} + k_3 + k_{-3}B) + k_2 X} \qquad (2.4b)$$

The conditions for thermodynamic equilibrium are easily seen to be

$$X_{eq} = \frac{k_1}{k_{-1}}A_{eq} = \frac{k_{-2}k_{-3}}{k_2 k_3}B_{eq} \qquad (2.5)$$

whatever the value of E_T.

To illustrate the nature of the steady states, a typical family of solutions to Equations (2.4), parameterized by E_T for $B = 1$, is displayed in Fig. 1. For E_T less than a "critical" value, $E_T^c \approx 124$ in this case, there is a unique solution for all A. Beyond E_T^c, however, there exists a range of A within which there are three possible steady states. It is easy to verify—by direct normal-mode analysis of Equations (2.2), for example —that the upper and lower branches are stable with respect to infinitesimal perturbations, while the middle branch is unstable.[3,11] For fixed $E_T > E_T^c$, therefore, a transition between the two stable branches of the solution occurs at a higher or lower value of A (as indicated by the arrows in Fig. 1), depending on whether the population of the initial reactant A (for fixed B, a measure of chemical affinity) is increasing or decreasing.

III. THE STOCHASTIC MODEL

In order to investigate the effect of fluctuations around the deterministic steady states given by Equations (2.4), particularly in the three-state region, we turn now to a stochastic description of the model system (2.1). We denote the *numbers* of molecules of type A, X, and so on in the system at time t by the discrete random variables $a(t)$, $x(t)$, and so on, respectively, and make the usual assumption that the state of the system is then prescribed by the probability function $P(a,b,c,x,e)$ of those random variables. The master equation of the birth-and-death type that gives the evolution of this function is then[14]

$$P(x,e;t+\Delta t) = P(x,e;t)$$

$$+\Delta t \{[k_1 A(x-1)P(x-1) + k_{-1}(x+1)xP(x+1)$$

$$+ k_2(x+1)(e+1)P(x+1,e+1)$$

$$+ k_{-2}(E_T - e + 1)P(x-1,e-1) + k_3(E_T - e + 1)P(e-1)$$

$$+ k_{-3}B(e+1)P(e+1)] - [k_1 Ax + k_{-1}x(x-1)$$

$$+ k_2 xe + k_{-2}(E_T - e) + k_3(E_T - e) + k_{-3}Be]P(x,e)\}$$

$$(3.1)$$

Here we have summed over the reservoir variables a and b, assuming that the boundary conditions for the system (2.1) do not depend on internal variables. Mathematically this means that the conditional mean values of a and b do not depend on x and e explicitly:

$$\sum_{a,b} \binom{a}{b} P(a,b,x,e;t) = \binom{A}{B} P(x,e;t) \qquad (3.2)$$

In the last two equations, A and B represent the appropriate boundary values for components A and B, component C has again been eliminated by the enzyme conservation condition (2.3), and in addition, where no ambiguity is possible, unchanged variables have not been included as arguments of $P(x,e;t)$.

Once an initial distribution $P(x,e;0)$ and appropriate values for A, B, E_T are specified, the stationary distribution corresponding to the stochastic model (3.1) is obtained asymptotically by direct numerical integration. Briefly, we consider the two-dimensional probability $P(x,e;t)$ having elements $p_{ij}(t) = P(x=i,e=j;t)$, where i and j represent the N_x and N_e possible integer values assumed by the random variables x and e, respectively. Given an initial distribution $P(x,e;0) = \{p_{ij}(0)\}$, we then solve approximately $N_x N_e$ coupled difference equations of the form (3.1) until a stationary distribution is reached. The upper and lower bounds on the grid coordinates (i,j) are allowed to vary dynamically as the calculation proceeds so that at each instant of time the distribution $P(x,e;t)$ lies entirely within these bounds.

It is clear at the outset that computer storage and time limitations severely restrict the size of the region of xe-space to be examined. In order to get reasonable resolution of the final distribution, and yet to do so with not unreasonable expenditure of computer time, we must treat a system containing at most a few hundred molecules of each species. The consequences of this restriction, and other questions concerning the computations, are discussed in Section VI.

Once we select the boundary conditions for the stochastic model to correspond to those used in the macroscopic analysis, there remains only the choice of E_T and the initial distribution. The macroscopic steady-state curve for $E_T = 300$ is plotted in Fig. 2. This rather large value (relative to E_T^c) is chosen to provide an appreciable number of integral A values in the anticipated multiple-steady-state region. Finally, we note that the stochastic master equation is *linear* in $P(x, e; t)$, implying a unique steady state solution, so that the choice of initial

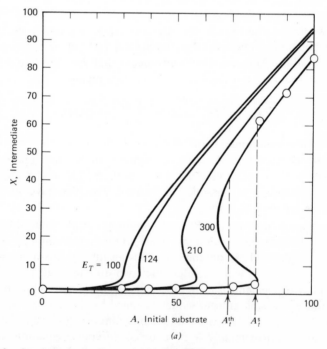

Figure 2. Steady states for the model biochemical system. Continuous curves represent the macroscopic predictions, while open circles denote stochastic mean values. The dashed lines represent the unique transition paths predicted, at A_f^s, by the stochastic "experiments," and at A_f^{th}, by the thermodynamic analysis.

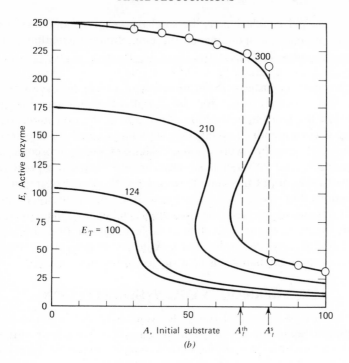

(b)

distribution is arbitrary. For a particular value of A, therefore, we take a convenient uniform initial distribution over a region of xe-space appropriate to the stochastic boundary conditions. Such a distribution rapidly assumes a form peaked near various candidates (x,e) for the stochastic steady state, and evolves in the long-time limit to a stationary distribution, the mean values of which are evaluated for comparison with the predictions of the macroscopic theory.

IV. STEADY STATES FOR THE STOCHASTIC MODEL

The results of the stochastic model "experiments" are displayed in Fig. 2, where the open circles denote the stochastic mean values $X = \bar{x}$ and $E = \bar{e}$ at selected numbers A of the initial reactant. The continuous curve representing the macroscopic steady states is included for comparison.

The most striking feature of the stochastic results is the absence of any hysteresis effect in the transition point.[16]

The principal conclusion of the stochastic experiments, therefore, is that the presence of spontaneous finite-amplitude fluctuations in this system precludes the appearance of multiple steady states, and that the nonequilibrium transition between the stable states will always occur at a single point, regardless of the direction in which the overall affinity (here, the boundary value for the initial substrate) may be varied. Moreover, the loops appearing on either side of the transition point A_t in Fig. 2 correspond to metastable stationary states for the reaction scheme (2.1). In other words, those regions of the macroscopic curve continuing *beyond* the stochastic "experimental" transition point represent states that are at the same time *stable* with respect to infinitesimal fluctuations but *unstable* with respect to finite disturbances in the system. In this sense, then, the transition exhibited by the system (2.1) already bears a strong resemblance to equilibrium first-order phase transitions, as suggested in Section I. A deeper analogy, from the point of view of thermodynamics, is also suggested by these results.

V. THERMODYNAMIC INTERPRETATION

In the case of a nonideal fluid, the problem of locating precisely the equilibrium liquid-vapor transition pressure (for fixed temperature) is solved once and for all when we have a complete theory of the equation of state (e.g., from statistical mechanics). Were it feasible at this time, a microscopic analysis of the scheme (2.1), using nonequilibrium statistical mechanics, would answer in the same manner all questions regarding the nature and location of the nonequilibrium transition predicted by the macroscopic theory of Section II. A first step in this direction has been taken via the stochastic model discussed in preceding sections of this paper.

On the other hand, if we consider the van der Waals equation of state, for example, as a model (admittedly incomplete) for a nonideal fluid, then the appropriate transition point may be obtained in a heuristic fashion from a purely macroscopic point of view. Classical thermodynamics requires for phases in equilibrium that the chemical potential of each component be the same in every phase. This condition leads directly to the well-known Maxwell equal-area construction for determining the equilibrium transition pressure on a p-v coexistence isotherm.[13] In a remarkable paper[8] (see also his chapter in this volume, p. 319), Kobatake has suggested that the corresponding procedure for nonequilibrium transitions be based on recent advances in nonequilibrium thermodynamics, due to Glansdorff and Prigogine,[3] by which the scope of thermodynamics has been broadened enormously to include situations in which nonlinear irreversible processes play a dominant role.

For a macroscopic model of electrical transport through a porous charged membrane, Kobatake found, at transmembrane pressure differences above a critical value, a discontinuous transition between two states corresponding to high and low electrical resistance of the membrane. Just as for the van der Waals model of a nonideal gas, the theoretical (macroscopic) analysis predicted a hysteresis effect in the transition point which was not verified experimentally. Pursuing the phase-transition analogy further, Kobatake sought a thermodynamic criterion for predicting the single transmembrane electrical potential difference at which the flip-flop transition would occur.[8]

According to thermodynamics,[15] the change in entropy for an open system during a time interval dt takes the schematic form

$$dS = d_e S + d_i S \tag{5.1}$$

where $d_e S$ is the *entropy flow* due to exchange with the surroundings and $d_i S \geqslant 0$ the *entropy production* due to irreversible processes within the system. Moreover, the entropy production may be written as a bilinear form,

$$P[S] \equiv \frac{d_i S}{dt} = \int \sum_\alpha J_\alpha X_\alpha \equiv \int \sigma \, dV \geqslant 0 \tag{5.2}$$

explicitly in terms of macroscopic quantities of direct physical interest: J_α, the fluxes or rates of irreversible processes (e.g., chemical reaction rates, heat flow), and X_α, the corresponding forces (e.g., differences in chemical potentials, temperature gradient).

Sufficiently near equilibrium that the rates and forces may be related by linear laws, the steady state (for time-independent boundary conditions) satisfies the principle of minimum entropy production[15]:

$$\frac{dP}{dt} \leqslant 0 \tag{5.3}$$

This extremum condition on the thermodynamic potential P at the same time guarantees the stability of the steady state which satisfies the equality in Equation (5.3).

Far from thermodynamic equilibrium, however, linear phenomenological laws are no longer valid, and the inequality (5.3) breaks down. Nevertheless, Glansdorff and Prigogine[3] have shown that if we decompose dP into separate contributions $d_J P$ due to changes in the flows or rates and $d_x P$ due to changes in the forces [according to Equation (5.2)], then even in the nonlinear regime we obtain a general inequality (again, for fixed boundary conditions) which prescribes the natural evolution of the system,[3]

$$\frac{d_x \sigma}{dt} = \sum_\alpha J_\alpha \frac{dX_\alpha}{dt} \leqslant 0 \qquad (5.4)$$

Near equilibrium this inequality reduces to the principle of minimum entropy production. For nonlinear situations, however, $d_x \sigma$ is *not* in general the differential of a function of the state of the system, so that the inequality (5.4) does not imply stability for a steady state. It is possible, however, to derive from Equation (5.4) a separate inequality which then serves as a stability criterion for nonequilibrium steady states[3] (see also the Chapter by Prigogine and Lefever in ths volume, p. 1),

$$\sigma[\delta S] = \sum_\alpha \delta J_\alpha \delta X_\alpha \geqslant 0 \qquad (5.5)$$

Here the symbol δ refers to *excess* quantities (entropy, flow, force) due to infinitesimal deviation of the system from the steady reference state.

Within the framework of the generalized thermodynamics mentioned above, therefore, the role of the free energy change in providing evolution and stability criteria for *equilibrium* states is taken for *nonequilibrium* states by the quantity $d_x \sigma$ and the *excess entropy production* $\sigma[\delta S]$, respectively. Moreover, although in general not an exact differential outside the linear domain of irreversible thermodynamics, $d_x \sigma$ may in some instances be shown to yield a new thermodynamic potential which remains valid for arbitrary deviations from equilibrium. In particular, if the state of a system is specified by giving the value of a single independent variable (ξ, for example), then we may always write

$$d\Phi \equiv d_x \sigma = \sum_\alpha J_\alpha \frac{dX_\alpha}{d\xi} d\xi \qquad (5.6)$$

where Φ is a *kinetic potential*, which for a steady state will be an extremum with respect to ξ.

Having now a potential, Φ, given by Equation (5.6) in terms of a single internal variable ξ [e.g., the concentration of the intermediate X in the scheme (2.1)], we assert, following Kobatake,[8] that the nonequilibrium transition between two steady states a and b in their region of coexistence will occur when

$$\Phi(\xi_a) = \Phi(\xi_b) \qquad (5.7)$$

In other words, the kinetic potential Φ plays here a role analogous to that of the chemical potential for first-order transitions between equilibrium phases. It is precisely the condition (5.7) which permits in the membrane transport model the explicit construction of an equal-area rule for locating the transition point on a plot, similar to that in Fig. 2, of transmembrane electric current against electric potential.[8]

It is important to realize that the conjecture implicit in Equation (5.7), unlike its equilibrium counterpart, has at present no clear justification even from a purely macroscopic point of view. On the other hand, we recognize the qualitative agreement with experiment obtained by Kobatake for the porous charged membrane.[8] For this reason, therefore, and from a desire to resolve a similar disagreement between the macroscopic and stochastic models of the system (2.1), we adopt without further comment the criterion (5.7) for transitions between multiple steady states. Deferring until Section VI additional discussion of this question, we turn now to a thermodynamic analysis of our model biochemical scheme (2.1).

If chemical reactions constitute the only irreversible processes occurring in the system, then the entropy production (5.2) becomes

$$\sigma[S] = \sum_\alpha v_\alpha \frac{\mathcal{Q}_\alpha}{T} \qquad (5.8)$$

Here the reaction rates for the system (2.1) are

$$v_1 = k_1 A X - k_{-1} X^2, \qquad v_2 = k_2 X E - k_{-2} C, \qquad v_3 = k_3 C - k_{-3} B E \qquad (5.9)$$

and the corresponding chemical affinities (for an ideal mixture) are given in terms of constituent populations by

$$\mathcal{Q}_1 = kT \ln\left(\frac{k_1 A}{k_{-1} X}\right), \qquad \mathcal{Q}_2 = kT \ln\left(\frac{k_2 X E}{k_{-2} C}\right), \qquad \mathcal{Q}_3 = kT \ln\left(\frac{k_3 C}{k_{-3} B E}\right) \qquad (5.10)$$

where k is Boltzmann's constant.

Applying the prescription (5.4) and using the last three equations, we obtain, after a few manipulations,

$$d_x\sigma = -\frac{1}{X}\frac{dX}{dt}dX - \frac{E_T}{EC}\frac{dE}{dt}dE \tag{5.11}$$

where A and B are fixed, C is given by Equation (2.3), and for convenience entropy is measured in units of Boltzmann's constant. It is easy to verify from Equations (2.2) and (2.3) that $d_x\sigma$ is *not* a total differential in the independent variables X and E, so that a kinetic potential is not available directly from Equation (5.11). In order to illustrate the thermodynamic approach, however, we now *assume* that even away from the steady state, E is given approximately by its steady-state value in terms of X. Inasmuch as this assumption is clearly incompatible (except at the steady state) with the numerical values used in the analyses of previous sections, we do not anticipate quantitative agreement between earlier results and those of the present thermodynamic calculations.

According to our approximation, therefore, we take the population of intermediate X as the single independent variable for fixed boundary values A, B, and E_T, and find from Equation (5.11)

$$\frac{d\Phi}{dX} = -k_1 A + k_{-1}X + \frac{(k_2 k_3 X - k_{-2}k_{-3}B)E_T}{X[k_2 X + (k_{-2}+k_3+k_{-3}B)]} \tag{5.12}$$

where C and E have been eliminated in favor of X via Equations (2.3) and (2.4). Finally we integrate this expression to obtain the desired kinetic potential

$$\Phi(X) = \Phi_0 - k_1 AX + \tfrac{1}{2}k_{-1}X^2 + \frac{E_T}{k_{-2}+k_3+k_{-3}B}[(k_{-2}+k_3)(k_3+k_{-3}B)$$

$$\times \ln|k_2 X + (k_{-2}+k_3+k_{-3}B)| - k_{-2}k_{-3}B\ln|X|] \tag{5.13}$$

where Φ_0 is a constant of integration.

The form of the kinetic potential for a typical value of A in the coexistence region is displayed in Fig. 3. The minima correspond to the stable steady states predicted by the macroscopic theory, and the maximum to the unstable one. According to our analogy with the equilibrium case, we expect that the lower minimum represents the *preferred* operating regime for the system (2.1), the higher one being a metastable situation. Moreover, Equation (5.7) implies that as A is changed the system will shift to the other regime at a value $A = A_t^{th}$ for

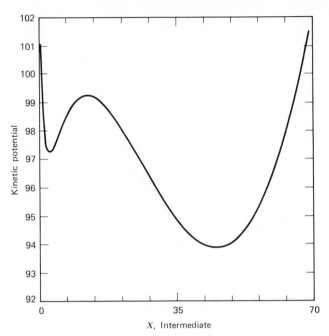

Figure 3. A typical plot of the kinetic potential Φ in the deterministic region of multiple steady states. The minima and maximum correspond to infinitesimally stable and unstable steady states, respectively. In the thermodynamic interpretation the upper minimum is in fact a *metastable* state.

which the kinetic potential is the same for both minima. In the present case this degeneracy of the kinetic potential is found to occur at $A_t^{th} \approx 69.2$.

Our understanding of the interpretation suggested here is facilitated by a plot of the kinetic potential for the steady states against the deviation from equilibrium. Figure 4 shows Φ against A for various values of the total enzyme population E_T. Below $E_T^c \approx 124$, the system possesses a unique steady state, and accordingly $\Phi(X)$ has a single minimum whatever the value of A. Above the critical point, however, Φ becomes multivalued in the coexistence region, with typical stable branches RP and PQ in Fig. 4, and unstable branch RQ.

As the system is driven from equilibrium by increasing numbers of the initial reactant A, its steady states lie along the thermodynamic branch OPQ. According to the macroscopic theory, the system remains on that branch until the point Q, where it becomes unstable with respect to infinitesimal fluctuations and jumps to the other branch RPS. Simi-

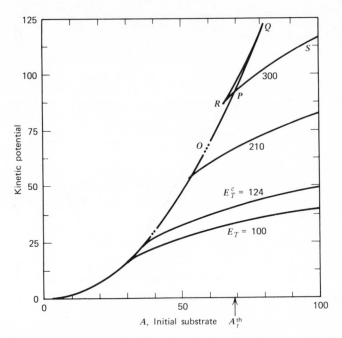

Figure 4. Kinetic potential Φ as a function of the deviation from thermodynamic equilibrium, as measured here by the population of initial substrate A. For E_T beyond the critical value $E_T^c \approx 124$, the kinetic potential is multivalued. According to the thermodynamic analysis, the system experiences a transition at the point P for $E_T = 300$), with stable branches OP and PS, unstable branch RQ, and metastable branches RP and PQ.

larly, as A is decreased, states on the latter branch remain stable until the point R.

Our thermodynamic interpretation of the phenomenon, on the other hand, implies that states along the branches PQ and RP are metastable and in fact will not be observed because of finite-amplitude fluctuations occurring in the system—a result that is consistent with the stochastic experiments reported in Section IV. Thus as A increases and the system reaches the point P at which Φ is degenerate, finite fluctuations cause the system to take the branch (PS) having the lower kinetic potential. In other words, the system appears to *prefer* the branch along which the increase $d_x\sigma$ in dissipation is smaller as the overall affinity increases. Moreover, the steady reaction rate for the global process $A \rightarrow B$ is much larger for the new branch PS than for the continuation of the original branch OPQ. This remarkable result implies that the system at each point *selects* for its steady state the one in which its operation will be most efficient with respect to the changing thermodynamic forces.

It cannot be overemphasized that such behavior has no explanation in terms of response to infinitesimal fluctuations, which will always be damped in either steady state of the coexistence region. It is precisely the presence of finite-amplitude fluctuations that permits the system in one steady state to be constantly testing for other, possibly more favorable, steady states. Now in a real chemical system the probability of spontaneous *homogeneous* fluctuations of finite size is vanishingly small. What we expect, rather, is that once a *local* fluctuation from a metastable state (Fig. 3) causes the system locally to pass the potential barrier into the deeper well (i.e., to become locally unstable), that fluctuation will grow until the entire system has passed to the more stable regime.

In concluding this section it is important to realize the deep difference between transitions involving (homogeneous) multiple steady states and those resulting from symmetry-breaking instabilities. In the model considered here, for example, it is easy to verify that *both* infinitesimally stable branches (*OPQ* and *RPS* of Fig. 4) belong to the *thermodynamic branch* in that each is attainable by continuous extension from a (different) state of thermodynamic equilibrium.[3,11] Thus by exploiting an additional degree of freedom (B or E_T in this example, temperature for the liquid-vapor system) it is possible to connect any two points on the different branches by *continuous* paths of steady (plus possibly equilibrium) states. Hence the curve of Fig. 1 corresponding to the critical value E_T^c is indeed analogous to the "critical isotherm" in the equilibrium liquid-vapor system. This striking feature of multiple steady states is in sharp contrast with *qualitatively new* states which appear beyond symmetry-breaking instabilities. States such as stable limit cycles or spatial structures are *always* separated from the thermodynamic branch by a discontinuous transition originating in an instability due to infinitesimal fluctuations.

VI. DISCUSSION

In our analysis of the model biochemical system (2.1), we have presented a possible thermodynamic interpretation of the stochastic model calculations[16] reported in this paper, and have pointed out a number of similarities between far-from-equilibrium transitions involving multiple steady states and first-order phase transitions.[17] Within the context of that analysis the usual macroscopic or deterministic approach to chemical kinetics is seen to give a picture that is even *qualitatively* incorrect in the coexistence region. In order to get a feeling for the possible generality of this result, we turn now to a discussion of the assumptions and computer-experimental "data" on which the conclusion is based.

Quantitatively the location of the transition point predicted by nonlinear thermodynamics ($A_t^{th} \approx 69.2$) differs markedly from the stochastic result ($A_t^s \approx 79 \leftrightarrow 80$). (This is certainly not surprising in view of the approximation introduced in Section V.) Moreover, it is clear from Fig. 2 that the stochastic transition, far from appearing near the *middle* of the deterministic multiple-steady-state region, occurs only slightly before the point of infinitesimal instability for the lower branch [Fig. 2(a)]. If the stochastic approach is taken as a valid test of the macroscopic predictions, then both discrepancies may well be partly due to the limited size of the xe-sample space available to the stochastic analysis. A first look at the numbers involved makes the asymmetry in the transition point all the more surprising: Since the lower branch [Fig. 2(a)] has mean values $\bar{x}$ of order unity, one expects that fluctuations should be *more destabilizing* on that branch than on the higher one. The combined features of Fig. 2(a) and (b), however, suggest a complex interplay of fluctuations in *both* internal variables, resulting in the ($\bar{x} \approx 1$, $\bar{e} \approx 210$) branch having greater stability in the transition region than the ($\bar{x} \approx 60$, $\bar{e} \approx 40$) branch. In this view, calculations on larger systems should lessen or remove the asymmetry and produce successively better agreement with thermodynamic predictions as constituent populations are increased.

Another question concerning the small sample space is more difficult. Clearly mean populations of order 1 to 100 are hardly comparable to the roughly molar quantities assumed in a macroscopic description. In addition, it is obvious that discrete fluctuations about the smaller mean values in this analysis can in no way be considered *small*. One may be tempted to conclude, therefore, that even the main *qualitative* feature of the stochastic result is due to analysis of an untenably small system, and that the hysteresis loop will reappear (as evidenced by a bimodal final distribution) if a suitably large system is studied. This is of course a possibility which must be considered. Nevertheless, the *excellent* agreement (Fig. 2) between the stochastic mean values and the deterministic steady state curve *outside the transition region* can hardly be accidental, and therefore must be weighed as well in this connection. On the other hand, stochastic models of purely chemical systems sufficiently nonlinear to exhibit dissipative instabilities have so far eluded *analytical* solution.[18] Moreover, in model studies of biochemical schemes important to understanding life processes on the *cellular* level, which is undeniably a principal aim of the study of chemical instabilities, one must acknowledge the relatively small populations of various important chemical species (e.g., of order 10^4 per cell for typical enzymes). It is not at all clear that a purely macroscopic description remains valid in such cases, particularly near instabilities, and this point is indeed a prime motivation for the present study. As to the reappearance of the hysteresis loop for larger constituent populations, that is still an open question, and awaits as a next step the conclusion of current extensions of the stochastic computations. In this regard the method discussed in the paper by Kitahara in this volume may prove useful in attempts to understand this type of question.

Concerning the computations themselves, two remarks are appropriate here. First, in the formulation of the stochastic approach used here,[14] the master equation that results is *by definition* a *finite-difference equation* in both the population variables and the time variable. This means that the model system's "phase space" and time dimension are already discrete, so that (unlike the case of *differential* equations) *no approximation* is involved in adapting the stochastic master equation (3.1) for numerical computation. Second, the *decoupling procedure* (3.2), used to reduce the full master equation to the simpler one of Equation (3.1), implies a wide separation of time scales characteristic of changes in composition: a large time scale for the reservoir variables A, B, E_T, and a smaller one for the internal system variables X, E. This condition is nothing more than the approximation made in defining what is meant by a steady nonequilibrium state.[19] The computational scheme is therefore a faithful model, within the framework of the stochastic approach, of the finite chemical system examined in this paper. Additional details of computational aspects of the work are presented elsewhere.[20]

In summary, a principal conclusion of this work is that highly nonequilibrium transitions between infinitesimally stable steady states correspond to situations in which *fluctuations are no less important in determining the evolution of the system than are the macroscopic state variables* (mean values). In other words, the decoupling of macroscopic behavior from fluctuations, which is guaranteed at or near thermodynamic equilibrium *except near phase transitions or critical points*, is no longer possible in general far from equilibrium, and is especially doubtful for systems that may experience macroscopic instabilities.[1,2] In such cases it is essential to study simultaneously, in a self-consistent fashion, the evolution of the fluctuations along with the evolution of the macroscopic state. Because it is the simplest attempt in that direction, the birth-and-death type of stochastic model for chemical kinetics is viewed as a first step toward adequate treatment of fluctuations near macroscopic instabilities.[21]

Note Added In Proof

Recent computer experiments involving the present model with larger constituent populations indicate that the numerical evolution of the distributions is extremely sensitive both to boundary conditions imposed on the distribution itself and to the accuracy of the individual steps in the numerical integration of the master equation. (See Ref. 20 for details of the computational procedure used in the work discussed in this chapter.) Indeed, in these refined calculations bimodal distributions are seen to persist for long times,[22] in agreement with analytical studies of several simpler (single-intermediate) chemical models yielding such distributions as steady state solutions for finite systems.[22] As was suggested at the end of Section V, therefore, it appears that an adequate stochastic theory of states which may be metastable must incorporate explicitly the existence of fluctuations which are spatially localized rather than globally homogeneous. A new procedure which appears promising in this regard has been proposed recently by Nicolis and coworkers.[23]

Acknowledgments

The author wishes to acknowledge the collaboration of Dr. David P. Chock during an early stage of this work, fruitful discussions with Professors R. S. Schechter, I. Prigogine, and R. M. Noyes, and helpful suggestions following a critical reading of the manuscript by Professor G. Nicolis.

References

1. G. Nicolis, *J. Stat. Phys.*, **6**, 195 (1972), and references therein.
2. G. Nicolis and I. Prigogine, *PNAS (USA)*, **68**, 2102 (1971).
3. P. Glansdorff and I. Prigogine, *Thermodynamic Theory of Structure, Stability and Fluctuations*, Wiley-Interscience, London, 1971.
4. For an overview and recent results related especially to living systems see I. Prigogine, G. Nicolis, and A. Babloyantz, *Physics Today*, **25**, (No. 11), 23 and (No. 12), 38 (1972). See also the following review articles: G. Nicolis, *Adv. Chem. Phys.*, **19**, 209 (1971); G. Nicolis and J. Portnow, *Chem. Reviews*, **73**, 365 (1973); J. S. Turner, in *Lectures in Statistical Physics*, W. C. Schieve and J. S. Turner, Eds., Vol. 23 of *Lecture Notes in Physics*, Springer-Verlag, Berlin, 1974, p. 248.
5. Reference 3, Chapters XIV and XV, and references therein.
6. R. Lefever, G. Nicolis, and I. Prigogine, *J. Chem. Phys.*, **47**, 1045 (1967); I. Prigogine and R. Lefever, *J. Chem. Phys.*, **48**, 1695 (1968); M. Herschkowitz-Kaufman and G. Nicolis, *J. Chem. Phys.*, **56**, 1890 (1972).
7. See especially Refs. 3 and 4 for details and additional citations.
8. Y. Kobatake, *Physica*, **48**, 301 (1970).
9. R. Blumenthal, J.-P. Changeux, and R. Lefever, *J. Membrane Biol.*, **2**, 351 (1970).

10. R. A. Spangler and F. M. Snell, *Nature*, **191**, 457 (1961); R. A. Spangler and F. M. Snell, *J. Theor. Biol.*, **16**, 381 (1967).

11. B. B. Edelstein, *J. Theor. Biol.*, **29**, 57 (1970).

12. J. Monod and F. Jacob, *Cold Spring Harb. Symp. Quant. Biol.*, **26**, 389 (1961); A. Babloyantz and G. Nicolis, *J. Theor. Biol.*, **34**, 185 (1972).

13. See, for example, I. Prigogine and R. Defay, *Chemical Thermodynamics*, Longmans, Green, London, 1954, Chapter XVI.

14. For an excellent review of theory and applications, see D. A. McQuarrie, *Stochastic Approach to Chemical Kinetics*, Vol. 8 of Suppl. Rev. Series in Applied Probability, Methuen, London, 1967.

15. S. R. deGroot and P. Mazur, *Non-equilibrium Thermodynamics*, North Holland, Amsterdam, 1962; I. Prigogine, *Introduction to Thermodynamics of Irreversible Processes*, 3rd ed., Wiley-Interscience, New York, 1967.

16. J. S. Turner, *Phys. Letters*, **44A**, 395 (1963).

17. Phase-transition analogies for nonequilibrium instabilities are certainly not new, and are by no means unique to any one discipline. For a discussion of electronic analogies involving multiple steady states, see R. Landauer and J. W. F. Woo, in *Statistical Mechanics: New Concepts, New Problems, New Applications*, S. A. Rice, K. F. Freed, and J. C. Light, Eds., Univ. of Chicago Press, Chicago, 1972, p. 299. More relevant to the context of this paper are Refs. 1 and 2 and the comments in Chapter XVI of Ref. 3.

18. Approximate methods, valid in the limit of *small* fluctuations, are available [e.g., N. G. van Kampen, *Adv. Chem. Phys.*, **15**, 65 (1969)], and in fact have been applied successfully to systems exhibiting chemical dissipative instabilities (e.g., Ref. 1).

19. The decoupling assumption is discussed in G. Nicolis and A. Babloyantz, *J. Chem. Phys.*, **51**, 2632 (1969).

20. J. S. Turner, *Bull. Math. Biology*, **36**, 205 (1974).

21. Indeed, Nicolis and Prigogine[1,2] have shown that the birth-and-death type of stochastic model is itself inadequate to give a complete description of fluctuations in nonequilibrium systems. Within the framework of macroscopic thermodynamics,[15] the description of nonequilibrium systems is based on the concept of *local equilibrium*. In the current context this means that the velocity distribution for each component of the chemical mixture deviates little from a local Maxwellian. On a molecular level continuous restoration to a local Maxwellian is accomplished by frequent elastic collisions (with inert solvent molecules) between relatively infrequent inelastic (reactive) collisions, which tend to distort the velocity distribution. Because it does not take into account the two widely separated time scales characterizing these competing processes, a description in which the mass-balance equations are assumed to define a Markov process in the number-of-particles space cannot ensure that local equilibrium is maintained, and therefore cannot be expected to give the *total* picture of fluctuations in a thermodynamic regime. A detailed study of fluctuations in the present model is in progress.

22. J. S. Turner, to be published.

23. G. Nicolis, M. Malek-Mansour, K. Kitahara, and A. Van Nypelseer, *Phys. Letters*, **48A**, 217 (1974).

THE
HAMILTON-JACOBI-EQUATION
APPROACH
TO FLUCTUATION PHENOMENA

KAZUO KITAHARA

Chimie Physique II, Faculté des Sciences,
Université Libre de Bruxelles, Bruxelles, Belgium

CONTENTS

Abstract

The stochastic master equation, which describes phenomena of fluctuation and of relaxation in a large system, is reduced to an equation of Hamilton-Jacobi type. The properties of the reduced equation are shown for simplified models of a bistable system and oscillating systems.

I. INTRODUCTION

Recently a new formulation of the theory of stochastic processes in a large system was reported.[1] One introduced as a small parameter the inverse of the size of the system, which plays the same role as $\hbar/i$ in quantum mechanics. Therefore, in analogy with quantum mechanics, one may reduce the original stochastic master equation into the form of a Hamilton-Jacobi equation, which has been quite well studied by mathematicians and physicists.[2]

Since the main assumption here is only that the system is large and the process is Markoffian, the range of application seems to be quite wide, whether the system is closed or open, near equilibrium or far from equilibrium.

The phenomena of fluctuation become important when the macroscopic state of the system approaches the instability. If one changes some external parameters, the system goes over from a stable regime to an unstable regime. The boundary of the stable regime is usually called a marginal situation. In a marginal situation the system loses the restoring force to the stationary state. This is found in a closed system near equilibrium as "a critical slowing down." The fact that the restoring force vanishes in a marginal situation is related to large fluctuations around the stationary state. For, if the system is driven off from the stationary state by random perturbations, it takes an extremely long time for the system to come back to the stationary state. Thus the distribution of the macroscopic states around the stationary state is broadened.

The method of Hamilton-Jacobi equation gives a transparent way of looking into the relation between the stability and the fluctuation.

In this short note, first we explain briefly the formulation in Section II, and then in Section III show some simple examples in order to see how the theory works. In Sections IV and V we discuss simple birth-and-death processes that may describe systems whose macroscopic states behave like a limit cycle and a Volterra-Lotka cycle. Section VI is devoted to some concluding remarks.

II. FORMULATION

One considers a random variable X, which is macroscopic in the sense that it is an extensive variable. Therefore, one may scale X as

$$X = \Omega x \tag{2.1}$$

where Ω is the size of the system and x is a variable of order Ω^0.

We study the evolution of the probability $P(X,t)$ that the system variable takes on the value X at time t. Under the assumption that the process is Markoffian, the stochastic master equation, which drives the probability, is written as

$$\frac{\partial}{\partial t} P(X,t) = - \sum_r W(X \to X + r) P(X,t)$$

$$+ \sum_r W(X - r \to X) P(X - r, t) \tag{2.2}$$

where $W(X \to X + r)$ is the transition probability per unit time and r is the amount of the transition.

If one expresses the probability $P(X,t)$ in terms of x as

$$P(X,t) = \psi(x,t) \qquad (2.3)$$

the equation for $\psi(x,t)$ becomes a "Schrödinger equation,"

$$\frac{1}{\Omega} \frac{\partial}{\partial t} \psi(x,t) = -H\left(x, \frac{1}{\Omega} \frac{\partial}{\partial x}\right) \psi(x,t) \qquad (2.4)$$

The "Hamiltonian" H in the above is given by

$$H(x,p) = \sum_r (1 - e^{-rp}) w(x;r) \qquad (2.5)$$

where we have introduced a new notation $w(x;r)$ defined by

$$\Omega w(x;r) = W(X \rightarrow X + r) \qquad (2.6)$$

In Equation (2.4), $1/\Omega$ corresponds to $\hbar/i$ in quantum mechanics. Therefore, in analogy with the WKB method in quantum mechanics, one may expect that Equation (2.4) has a solution of the form

$$\psi(x,t) \sim \exp[\Omega J(x,t)] \qquad (2.7)$$

in the limit of $\Omega \rightarrow \infty$. Rigorous proof of Equation (2.7) is given in K.M.K.[1] using Feynman's path-integral method.

If one substitutes Equation (2.7) into Equation (2.4) and takes the dominant terms, one easily finds a Hamilton-Jacobi equation for $J(x,t)$,

$$\frac{\partial J(x,t)}{\partial t} + H\left(x, \frac{\partial J(x,t)}{\partial x}\right) = 0 \qquad (2.8)$$

Thus the original master equation (2.2) is now reduced to a Hamilton-Jacobi equation.

In the case that there exist several random variables, $X_1, X_2, \ldots, X_N$, one may scale them as

$$X_i = \Omega x_i, \qquad i = 1, 2, \ldots, N, \qquad (2.9)$$

and the Hamiltonian becomes

$$H(x_1, x_2, \ldots, x_N, p_1, p_2, \ldots, p_N)$$

$$= \sum_{r_1, r_2, \ldots, r_N} (1 - e^{-\sum_i r_i p_i}) w(x_1, x_2, \ldots, x_N; r_1, r_2, \ldots, r_N) \qquad (2.10)$$

where the transition probability is scaled as

$$W(X_1 \to X_1 + r_1, X_2 \to X_2 + r_2, \ldots, X_N \to X_N + r_N)$$

$$= \Omega w(x_1, x_2, \ldots, x_N; r_1, r_2, \ldots, r_N) \quad (2.11)$$

The problem in this case is reduced to solving a multidimensional Hamilton-Jacobi equation,

$$\frac{\partial}{\partial t} J(x_1, x_2, \ldots, x_N, t) + H\left(x_1, x_2, \ldots, x_N, \frac{\partial J}{\partial x_1}, \frac{\partial J}{\partial x_2}, \ldots, \frac{\partial J}{\partial x_N}\right) = 0 \quad (2.12)$$

The method of solving the Hamilton-Jacobi equation (2.12) is well established.[2] Suppose the initial condition

$$J(x_1, x_2, \ldots, x_N, 0) = f(x_1, x_2, \ldots, x_N) \quad (2.13)$$

is given. In order to solve Equation (2.13) with this initial condition, one introduces a characteristic flow in $(x_1, x_2, \ldots, x_N, p_1, p_2, \ldots, p_N)$-space. The flow starts from a point $(x_1^0, x_2^0, \ldots, x_N^0, p_1^0, p_2^0, \ldots, p_N^0)$, where

$$P_i^0 = \frac{\partial f}{\partial x_i}(x_1^0, x_2^0, \ldots, x_N^0), \qquad i = 1, 2, \ldots, N \quad (2.14)$$

and it is driven by the following characteristic equations

$$\frac{dx_i(s)}{ds} = \frac{\partial H}{\partial p_i}(x_1(s), x_2(s), \ldots, x_N(s), p_1(s), p_2(s), \ldots, p_N(s)) \quad (2.15)$$

and

$$\frac{dp_i(s)}{ds} = -\frac{\partial H}{\partial x_i}(x_1(s), x_2(s), \ldots, x_N(s), p_1(s), p_2(s), \ldots, p_N(s)) \quad (2.16)$$

where $i = 1, 2, \ldots, N$. Therefore the characteristic flow is expressed as

$$x_i(s) = \xi_i(s; x_1^0, x_2^0, \ldots, x_N^0), \quad (2.17)$$

and

$$p_i(s) = \pi_i(s; x_1^0, x_2^0, \ldots, x_N^0), \quad (2.18)$$

where $i = 1, 2, \ldots, N$. Along the flow, one integrates the Langrangian up to $s = t$,

$$\tilde{J}\left(t; x_1^0, x_2^0, \ldots, x_N^0\right)$$

$$= \int_0^t ds \left[\sum_i p_i(s)\dot{x}_i(s) - H(x_1(s), x_2(s), \ldots, x_N(s), p_1(s), p_2(s), \ldots, p_N(s)) \right]$$

$$+ f(x_1^0, x_2^0, \ldots, x_N^0). \tag{2.19}$$

Suppose at time t one arrives at a point $(x_1, x_2, \ldots, x_N, p_1, p_2, \ldots, p_N)$, that is,

$$x_i = \xi_i\left(t; x_1^0, x_2^0, \ldots, x_N^0\right) \tag{2.20}$$

and

$$p_i = \pi_i\left(t; x_1^0, x_2^0, \ldots, x_N^0\right) \tag{2.21}$$

where $i = 1, 2, \ldots, N$. Then one may write the x_i^0's as functions of t and the x_i's,

$$x_i^0 = \eta_i(t; x_1, x_2, \ldots, x_N), \qquad i = 1, 2, \ldots, N \tag{2.22}$$

Using Equation (2.22), the function $\tilde{J}(t; x_1^0, x_2^0, \ldots, x_N^0)$ is transformed into a function of $x_1, x_2, \ldots, x_N$, and t,

$$\tilde{J}\left(t; x_1^0, x_2^0, \ldots, x_N^0\right) = J(x_1, x_2, \ldots, x_N, t) \tag{2.23}$$

It is known that the function $J(x_1, x_2, \ldots, x_N, t)$, obtained above, satisfies Equation (2.12). Therefore, in order to solve Equation (2.12), one solves the ordinary differential equations, (2.15) and (2.16), and estimates the integral in Equation (2.19).

There is another useful property of the Hamilton-Jacobi equation. Using Equation (2.21) and (2.22), one may express the p_i's in terms of the x_i's,

$$p_i = \gamma_t^i(x_1, x_2, \ldots, x_N), \qquad i = 1, 2, \ldots, N \tag{2.24}$$

Then one has

$$\frac{\partial J}{\partial x_i}(x_1, x_2, \ldots, x_N, t) = \gamma_t^i(x_1, x_2, \ldots, x_N), \qquad i = 1, 2, \ldots, N \tag{2.25}$$

Suppose $J(x_1, x_2,\ldots,x_N, t)$ has a maximum at a point $(\alpha_t^{(1)}, \alpha_t^{(2)},\ldots,$ $\alpha_t^{(N)})$ in $(x_1, x_2,\ldots,x_N)$-space. Then one has

$$\frac{\partial J}{\partial x_i}\left(\alpha_t^{(1)}, \alpha_t^{(2)},\ldots,\alpha_t^{(N)}, t\right)=0, \qquad i=1, 2,\ldots,N \qquad (2.26)$$

Equation (2.26) implies that the motion of the peak of $J(x_1, x_2,\ldots,x_N, t)$ is identical with that of a particular characteristic flow with $p_1 = p_2 = \cdots$ $= p_N = 0$. Therefore, putting $p_1 = p_2 = \cdots = p_N = 0$ in Equation (2.15), one gets equations for the $\alpha_t^{(i)}$s,

$$\frac{d\alpha_t^{(i)}}{dt} = \frac{\partial H}{\partial p_i}\left(\alpha_t^{(1)}, \alpha_t^{(2)},\ldots,\alpha_t^{(N)}, 0, 0,\ldots,0\right)$$

$$= \sum_{r_1,\ldots,r_N} r_i w\left(\alpha_t^{(1)}, \alpha_t^{(2)},\ldots,\alpha_t^{(N)}; r_1, r_2,\ldots,r_N\right) \qquad (2.27)$$

where one has used Equation (2.10).

Equation (2.25) is especially useful in the case of $N=1$. In this case one may write Equation (2.24) as

$$p = \gamma_t(x) \qquad (2.28)$$

The set of points (x,p) that satisfy Eq. (2.28) forms a curve γ_t in (x,p)-space. Let us call it a front curve. The solution of the Hamilton-Jacobi equation (2.8) is given by

$$\frac{\partial J}{\partial x}(x,t) = \gamma_t(x) \qquad (2.29)$$

or

$$J(x,t) = \int^x dx' \gamma_t(x') \qquad (2.30)$$

Suppose the front curve γ_t crosses the x-axis at $x = \alpha_t$; then $J(x,t)$ has a maximum or minimum for $x = \alpha_t$ at time t. In fact, at $x = \alpha_t$ on the x-axis one has

$$p = \gamma_t(\alpha_t) = 0 \qquad (2.31)$$

Therefore

$$\frac{\partial J}{\partial x}(\alpha_t, t) = 0 \qquad (2.32)$$

It is easily seen that if one has

$$\gamma_t'(\alpha_t) > 0$$

then

$$\frac{\partial^2 J}{\partial x^2}(\alpha_t, t) > 0$$

and accordingly $J(x,t)$ has a minimum for $x = \alpha_t$. In the case of $\gamma_t'(\alpha_t) < 0$, $J(x,t)$ has a maximum for $x = \alpha_t$.

III. ONE-VARIABLE CASES

The simplest model of a Brownian motion was discussed in K.M.K. They found that the front curve always tends to a line

$$p = \gamma_\infty(x) = -\frac{x}{\sigma_e}.$$

Therefore, starting from any initial condition except unnormalizable ones, the probability tends to a Gaussian form.

Here we discuss a simple model, which may describe a system with two stable stationary macroscopic states and an unstable one. A spin system below the critical temperature and an enzymatic reaction with multiple steady states are typical examples.

The model may be given in the form of a "Hamiltonian"

$$H(x,p) = px(a^2 - x^2) - Dp^2 \tag{3.1}$$

where $a > 0$ and $D > 0$, or one may express the model in terms of the transition probability as

$$\sum_r rw(x;r) = x(a^2 - x^2) \tag{3.2}$$

$$\sum_r r^2 w(x;r) = 2D \tag{3.3}$$

and

$$\sum_r r^n w(x;r) = 0 \tag{3.4}$$

for $n \geqslant 3$.

From Equations (2.27) and (3.2), one finds an equation of the motion of the peak of $J(x,t)$,

$$\frac{d\alpha_t}{dt} = \alpha_t(a^2 - \alpha_t^2) \tag{3.5}$$

The motion of the peak given above is supposed to correspond to the macroscopic behavior of the system. Obviously Equation (3.5) has three stationary solutions: $\alpha_t = a$, $\alpha_t = -a$, and $\alpha_t = 0$. The first two solutions are stable with respect to a small deviation from them, while the last one is unstable.

The characteristic flows are driven by the following equations:

$$\frac{dx(s)}{ds} = \frac{\partial H}{\partial p}(x(s),p(s)) = x(s)\left[a^2 - x(s)^2\right] - 2Dp(s) \qquad (3.6)$$

and

$$\frac{dp(s)}{ds} = -\frac{\partial H}{\partial x}(x(s),p(s)) = -\left[a^2 - 3x(s)^2\right]p(s) \qquad (3.7)$$

Using the fact that on each characteristic flow one has a constant of motion,

$$H(x(s),p(s)) = E, \qquad (3.8)$$

one may draw characteristic flows with various E in the (x,p)-plane. In Fig. 1 these characteristic flows are shown. The arrows stand for the direction of flow.

Suppose the initial condition is given by

$$J(x,0) = -\tfrac{1}{2}(x - \alpha_0)^2 \qquad (3.9)$$

which corresponds to a Gaussian distribution; then the initial front curve γ_0 is a straight line, that is,

$$\gamma_0: \quad p = \gamma_0(x) = -(x - \alpha_0) \qquad (3.10)$$

Moreover, one assumes that $\alpha_0 > a$ and the initial front curve γ_0 does not cross the curve

$$p = \frac{x(a^2 - x^2)}{D}$$

for $0 < x < a$. Then from the pattern of flow in Fig. 1 one easily sees that the front curve γ_t tends to a final front curve γ_∞, which is given by

$$\gamma_\infty(x) = \begin{cases} \dfrac{x}{D}(a^2 - x^2), & x < -a \\ 0, & -a \leqslant x \leqslant 0 \\ \dfrac{x}{D}(a^2 - x^2), & x > 0 \end{cases} \qquad (3.11)$$

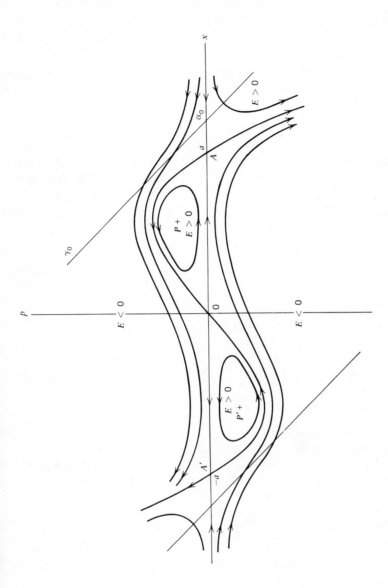

Figure 1. The pattern of characteristic flows for the model given in Equation (3.1). Here one puts $a = D = 1.0$. P and P' are the centers of the rotating flows; γ_0, the initial front curve given in Equation (3.10).

Accordingly one gets

$$J(x,\infty) = \begin{cases} \dfrac{a^2}{2D}x^2 - \dfrac{x^4}{4D} + C, & x < -a \\[2mm] \dfrac{a^4}{4D} + C, & -a \leqslant x \leqslant 0 \\[2mm] \dfrac{a^2}{2D}x^2 - \dfrac{x^4}{4D} + \dfrac{a^4}{4D} + C, & x > 0 \end{cases} \qquad (3.12)$$

where C is a constant.

Suppose, on the other hand, $\alpha_0 < -a$ in Equation (3.10); then one gets a different final front curve,

$$\gamma_\infty(x) = \begin{cases} \dfrac{x}{D}(a^2 - x^2), & x < 0 \\[2mm] 0, & 0 \leqslant x \leqslant a \\[2mm] \dfrac{x}{D}(a^2 - x^2), & x > a \end{cases} \qquad (3.13)$$

Accordingly,

$$J(x,\infty) = \begin{cases} \dfrac{a^2}{2D}x^2 - \dfrac{x^4}{4D} + \dfrac{a^4}{4D} + C', & x < 0 \\[2mm] \dfrac{a^4}{4D} + C', & 0 \leqslant x \leqslant a \\[2mm] \dfrac{a^2}{2D}x^2 - \dfrac{x^4}{4D} + C', & x > a \end{cases} \qquad (3.14)$$

where C' is a constant.

Equations (3.12) and (3.14) show that the position of the peak of $J(x,\infty)$ is dependent on the position of the initial peak. It should be noted here that the stationary solution of the Hamilton-Jacobi equation (2.8) is given by

$$x(a^2 - x^2)J'_{st}(x) - D(J'_{st}(x))^2 = 0 \qquad (3.15)$$

Eq. (3.15) has two solutions: a trivial solution

$$J_{st}^{(1)}(x) = \text{const} \qquad (3.16)$$

and a nontrivial one

$$J_{st}^{(2)}(x) = \frac{a^2}{2D}x^2 - \frac{1}{4D}x^4 + \text{const} \qquad (3.17)$$

which has two maxima at $x = a$ and $x = -a$ and a minimum at $x = 0$. As shown in Equations (3.12) and (3.14), the asymptotic solution in the long-time limit is one of the two stationary solutions, $J_{st}^{(1)}(x)$ and $J_{st}^{(2)}(x)$, in one interval of x, and the other in the other interval. According to the initial condition, one of the peaks of $J_{st}^{(2)}(x)$ is suppressed in the asymptotic solution. This may correspond to the phenomena of symmetry breaking.

As shown in Fig. 1, there are two regions in (x,p)-plane where the characteristic flows rotate. The two points P and P' are the centers of the rotation. If the initial front curve γ_0 crosses these regions, the behavior of the front curve γ_t for $t > 0$ is complicated.

Suppose the initial condition is given by

$$J(x,0) = -\tfrac{1}{2}(x - \alpha_0)^2 \qquad (3.18)$$

with

$$\alpha_0 = \frac{a}{\sqrt{3}}\left(1 + \frac{a^2}{3D}\right) \qquad (3.19)$$

Then the initial front curve γ_0 is a straight line, which passes through the point P.

It is shown in Appendix A that the period of the rotation of the flow is a continuous function of the constant of motion E along the flow. The point P corresponds to

$$E = E_0 \equiv \frac{1}{27}\frac{a^6}{D} \qquad (3.20)$$

and the period of the rotation with infinitesimal amplitude around P is given by

$$T(E_0) = \frac{\sqrt{3}\,\pi}{a^2} \qquad (3.21)$$

As E decreases, the period $T(E)$ of the rotation becomes longer, and finally it becomes infinite:

$$T(0) = \infty \qquad (3.22)$$

Therefore flows with different values of E gradually become out of phase. Actually, the front curve γ_t deforms into an S-shaped curve, as shown in Fig. 2.

In such situations, to some values of x (for example, $x = b$ in Fig. 2) there correspond three points l, m, and n on the front curve γ_t. These

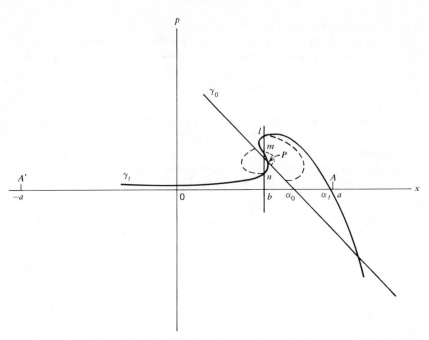

Figure 2.   The front curve γ_t in the rotating flow field ($t = 2.5$). γ_0 is the initial front curve given in Equation (3.18). The dotted lines are the charactersitic flows.

three points are on different characteristic flows. Therefore the one-to-one correspondence [(2.20) and (2.22)] between the initial value x^0 and the final value x in the characteristic equations breaks down. This implies that the solution $J(x,t)$ becomes a multivalued function of x. In fact, if one calculates $J(x,t)$ using Equation (2.19), one obtains an explicit form of $J(x,t)$ as shown in Fig. 3.

As time proceeds, the front curve γ_t turns around the point P many times. Accordingly $J(x,t)$ becomes an infinitely multivalued function of x.

However, the probability $P(X,t)$, which obeys the difference-differential equation (2.2), is a one-valued function. Therefore the function $J(x,t)$ defined by Equations (2.3) and (2.7) is also a one-valued function of x.

In order to avoid this difficulty, one should return to the justification of Equation (2.7). As is discussed in Appendix B, as far as one is concerned with the asymptotic behavior of the probability in the limit of

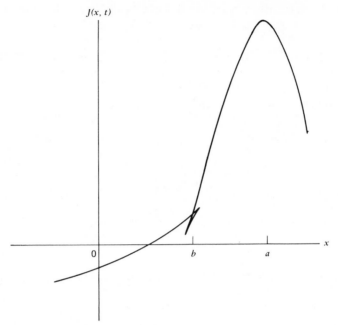

Figure 3. The form of $J(x,t)$ corresponding to the front curve γ_t in Fig. 2.

$\Omega \to \infty$, or more precisely, with the function $\phi(x,t)$ defined by

$$\phi(x,t) = \lim_{\Omega \to \infty} \frac{1}{\Omega} \log \psi(x,t) \tag{3.23}$$

one should take the maximal branch of the multivalued function $J(x,t)$ as the asymptotic solution $\phi(x,t)$ of Equation (2.4). The function $\phi(x,t)$ thus obtained is a one-valued function.

Starting from the initial condition (3.18) with (3.19), $\phi(x,t)$ tends to a definite function that is peaked at $x = a$.

It should be noted here that the final $\phi(x, \infty)$ has a peak at $x = a$ ($x = -a$) if initially $\phi(x,0)$ has a peak on the positive (negative) side of x.

IV. LIMIT-CYCLE CASE

In many cases of oscillatory phenomena in chemical and biological systems, it is found that the oscillations are of limit-cycle type rather than of Lotka-Volterra type.

The limit cycle has a restoring force in the sense that the deviation from the asymptotic orbit (i.e., the limit cycle) vanishes as time proceeds. Therefore one may expect that a proper model of stochastic processes in such systems will give, in the long-time limit, an asymptotic form of the probability whose peak is along the asymptotic orbit and with finite width.

R. Lefever and G. Nicolis[3] studied a simple reaction scheme that shows limit-cycle behavior. The model contains two species, 1 and 2, and the kinetic equations for the concentrations of these species are

$$\frac{dx_1}{dt} = 1 + x_1^2 x_2 - Bx_1 - x_1 \tag{4.1}$$

and

$$\frac{dx_2}{dt} = Bx_1 - x_1^2 x_2 \tag{4.2}$$

The coupled equations (4.1) and (4.2) have a stationary solution

$$x_1 = 1 \quad \text{and} \quad x_2 = B$$

The stationary solution is stable if $B < 2$, and it is unstable if $B > 2$. For each value of $B > 2$, one definite limit cycle around the stationary solution exists.

In order to construct a stochastic model, one looks into the kinetic equations (4.1) and (4.2). The first term on the right-hand side of Equation (4.1) corresponds to the input of species 1 from external sources; therefore one may write the transition probability for this process as

$$w(x_1, x_2; +1, 0) = 1$$

The second term on the right-hand side of Equation (4.1) together with the second term on the right-hand side of Equation (4.2), corresponds to the creation of species 1 and the annihilation of species 2 by the three-body collision; therefore the transition probability corresponding to this process is

$$w(x_1, x_2; +1, -1) = x_1^2 x_2$$

In a similar way one gets the transition probability corresponding to other terms in Equations (4.1) and (4.2). The complete form of the transition probability is as follows:

$$
w(x_1, x_2; r_1, r_2) =
\begin{cases}
1, & r_1 = +1, & r_2 = 0 \\
x_1^2 x_2, & r_1 = +1, & r_2 = -1 \\
Bx_1, & r_1 = -1, & r_2 = +1 \\
x_1, & r_1 = -1, & r_2 = 0 \\
0, & \text{otherwise}
\end{cases}
\qquad (4.3)
$$

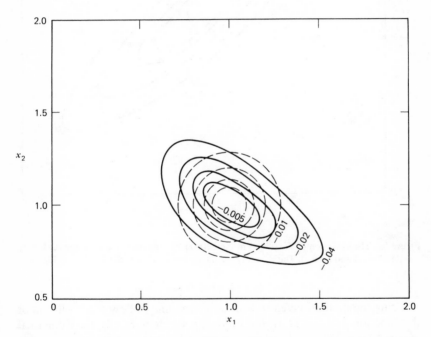

Figure 4. The evolution of $J(x_1, x_2, t)$ of Lefever-Nicolis system around the stationary state ($B = 1.0$ and $t = 0.5$).

$$
J(x_1, x_2, t) =
\begin{cases}
-0.005 & \text{on the innermost closed curve} \\
-0.01 & \text{on the second inner closed curve} \\
-0.02 & \text{on the third inner closed curve} \\
-0.04 & \text{on the outermost closed curve}
\end{cases}
$$

The dashed closed curves correspond to the initial condition $J(x_1, x_2, 0)$.

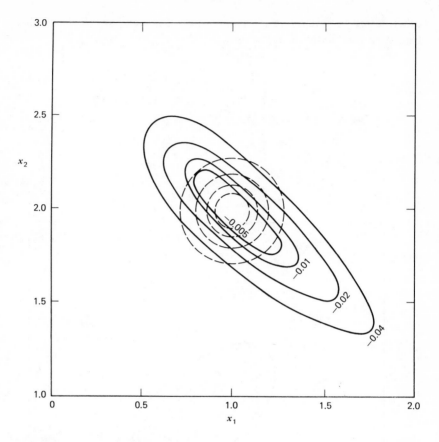

Figure 5. The evolution of $J(x_1, x_2, t)$ in the Lefever-Nicolis system around the stationary state ($B = 2.0$ and $t = 0.5$). The meanings of the closed curves are the same as in Fig. 4.

Using the method given in Section 2, one may follow the evolution of the function $J(x_1, x_2, t)$. In the following we show briefly the numerical results.[4]

First, we study the evolution of the probability around the stationary solution, $x_1 = 1$ and $x_2 = B$. We put the initial condition as

$$J(x_1, x_2, 0) = -\tfrac{1}{2}\left[(x_1 - 1)^2 + (x_2 - B)^2\right] \tag{4.4}$$

which corresponds to a Gaussian distribution around the stationary solution. After some time ($t = 0.5$), the function $J(x_1, x_2, t)$ deforms as shown in Figs. 4, 5, and 6. Fig. 4 corresponds to the stable regime

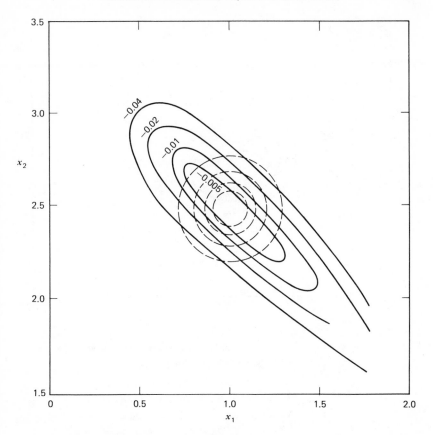

Figure 6.   The evolution of $J(x_1, x_2, t)$ in the Lefever-Nicolis system around the stationary state ($B = 2.5$ and $t = 0.5$). The meanings of the closed curves are the same as in Fig. 4.

($B = 1.0$), Fig. 5 to the marginal one ($B = 2.0$), and Fig. 6 to the unstable one ($B = 2.5$). On each closed curve, $J(x_1, x_2, t)$ takes on the same value. The innermost closed curve corresponds to $J(x, x, t) = -0.005$; the second one, the third one, and the outermost one correspond to $J(x_1, x_2, t) = -0.01$, -0.02, and -0.04, respectively. The dashed closed curves show the initial form of $J(x_1, x_2, 0)$.

It is clearly seen that in the unstable regime (Fig. 6), the probability is rapidly spreading out, with strong correlation between x_1 and x_2.

Second, we study the evolution of the probability around the transient macroscopic motion, starting from the initial condition

$$J(x_1, x_2, 0) = -\tfrac{1}{2}\left[(x_1 - 1)^2 + (x_2 - B + 0.5)^2\right] \tag{4.5}$$

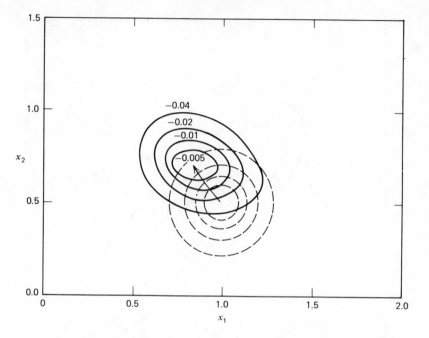

Figure 7.   The evolution of $J(x_1, x_2, t)$ in the Lefever-Nicolis system around the macro-scopic motion ($B = 1.0$ and $t = 0.5$).

$$J(x_1, x_2, t) = \begin{cases} -0.005 & \text{on the innermost closed curve} \\ -0.01 & \text{on the second inner closed curve} \\ -0.02 & \text{on the third inner closed curve} \\ -0.04 & \text{on the outermost closed curve} \end{cases}$$

The dashed closed curves correspond to the initial condition $J(x_1, x_2, 0)$. The arrow stands for the macroscopic motion.

The initial peak is at $x_1 = 1$, $x_2 = B - 0.5$, and it begins to move according to Equations (4.1) and (4.2). After some time ($t = 0.5$), $J(x_1, x_2, t)$ deforms as shown in Figs. 7, 8, and 9.

Figures 7, 8, and 9 correspond to $B = 1.0$ (the stable regime), $B = 2.0$ (the marginal situation), and $B = 3.0$ (the unstable regime), respectively. Here one sees that in the case of $B = 3.0$ the probability becomes broader in the direction of the macroscopic motion. This implies that the phase is much more randomized than the orbit itself.

The calculations above are only for small t. For small t, the shape of the probability changes rather continuously from the stable regime to

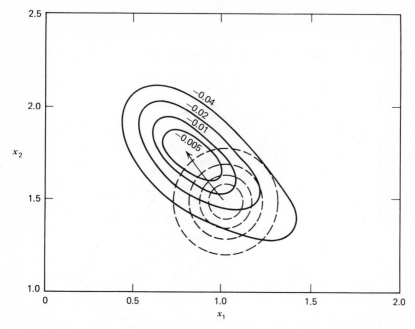

Figure 8. The evolution of $J(x_1, x_2, t)$ in the Lefever-Nicolis system around the macro-scopic motion ($B = 2.0$ and $t = 0.5$). The closed curves and the arrow have the same meanings as in Fig. 7.

the unstable regime. For large t, the shape will change drastically from the stable regime to the unstable regime, and the shape in the marginal situation is of particular interest. It is now under investigation.

V. LOTKA-VOLTERRA MODEL

An ecological model given by Lotka has attracted attention recently.[5] Since this model has an infinite number of orbits neighboring one another, each orbit is unstable with respect to a perturbation. In other words, no orbit has a restoring force. The system on one orbit can jump to another orbit, and it is not necessary for the system to come back to the initial orbit.

Therefore one may expect that if the system is initially in a macroscopic state in a coherent way, the macroscopic state will become less coherent and the distribution of states around the macroscopic state will be broadened.

The Lotka-Volterra system contains two species, 1 and 2, and the

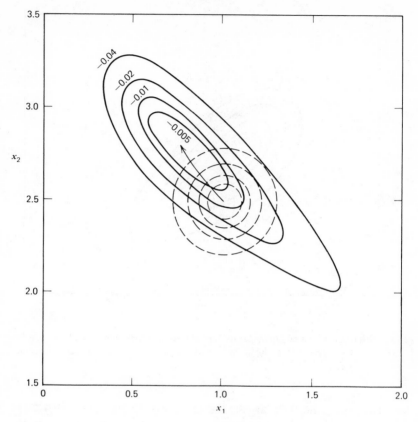

Figure 9.   The evolution of $J(x_1, x_2, t)$ in the Lefever-Nicolis system around the macro-scopic motion ($B = 3.0$ and $t = 0.5$).

kinetic equations for the concentrations of the two species are

$$\frac{dx_1}{dt} = x_1 - x_1 x_2 \tag{5.1}$$

and

$$\frac{dx_2}{dt} = x_1 x_2 - x_2 \tag{5.2}$$

Equations (5.1) and (5.2) have a trivial stationary solution: $x_1 = x_2 = 0$, and a nontrivial one: $x_1 = x_2 = 1$.

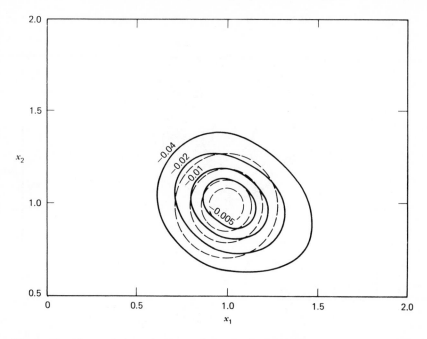

Figure 10. The evolution of $J(x_1,x_2,t)$ in the Lotka-Volterra system around the stationary state ($t=0.5$).

$$J(x_1,x_2,t) = \begin{cases} -0.005 & \text{on the innermost closed curve} \\ -0.01 & \text{on the second inner closed curve} \\ -0.02 & \text{on the third inner closed curve} \\ -0.04 & \text{on the outermost closed curve} \end{cases}$$

The dashed closed curves correspond to the initial condition $J(x_1,x_2,0)$.

Following the same procedure as in Section IV, one may construct the transition probability as

$$w(x_1,x_2;r_1,r_2) = \begin{cases} x_1 & r_1=1, & r_2=0 \\ x_1x_2 & r_1=-1, & r_2=+1 \\ x_2 & r_1=0, & r_2=-1 \\ 0 & \text{otherwise} \end{cases} \qquad (5.3)$$

If one substitutes Equation (5.3) into Equation (2.27), one finds equations of the motion of the peak of $J(x_1,x_2,t)$ that are identical with Equations (5.1) and (5.2).

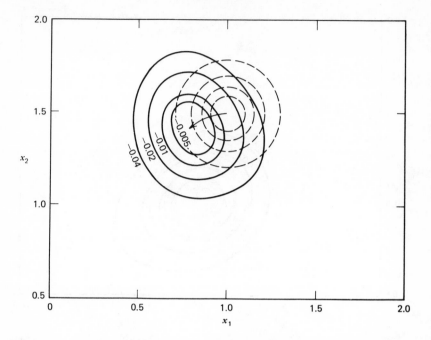

Figure 11. The evolution of $J(x_1, x_2, t)$ in the Lotka-Volterra system around the macroscopic motion ($t = 0.5$). The closed curves have the same meanings as in Fig. 10. The arrow stands for the macroscopic motion.

First, we study the evolution of $J(x_1, x_2, t)$ around the stationary solution $x_1 = x_2 = 1$ of the kinetic equations (5.1) and (5.2). One puts the initial condition as

$$J(x_1, x_2, 0) = -\tfrac{1}{2}\left[(x_1 - 1)^2 + (x_2 - 1)^2\right] \qquad (5.4)$$

which corresponds to a Gaussian distribution around $x_1 = x_2 = 1$. At time $t = 0.5$, $J(x_1, x_2, t)$ becomes broader isotropically as shown in Fig. 10.

Second, if one puts the initial condition as

$$J(x_1, x_2, 0) = -\tfrac{1}{2}\left[(x_1 - 1)^2 + (x_2 - 1.5)^2\right] \qquad (5.5)$$

then the peak, which is located at $x_1 = 1$, $x_2 = 1.5$, begins to move according to the kinetic equations (5.1) and (5.2). At time $t = 0.5$, the distribution of states around the moving peak becomes broader isotropically, as shown in Fig. 11.

The manner of broadening in both cases of the Lotka-Volterra system is in great contrast with that of the limit-cycle system. The former is isotropic, while the latter shows strong correlation between the fluctuations of the two species.

VI. CONCLUDING REMARKS

In Section III, we discussed the evolution of the probability in a bistable system. Sometimes such a system has a hysteresis with respect to an external parameter. In fact, if one changes the model in Section III slightly to

$$H(x,p,h) = [x(a^2 - x^2) + h]p - Dp^2 \tag{6.1}$$

where h is a parameter, then the corresponding macroscopic behavior of the system is given by

$$\frac{d\alpha_t}{dt} = \alpha_t(a^2 - \alpha_t^2) + h \tag{6.2}$$

The stationary solution of Equation (6.2) is a multivalued function of h, and it forms an S-shaped hysteresis curve. It should be interesting to see how the probability or $J(x,t)$ evolves when one changes h slowly enough to keep the peak on the hysteresis curve. This analysis may give us some information about the sudden jump of the macroscopic state as found in membranes.[6]

In Sections IV and V we studied the short-time behavior of the probability in oscillating systems and found a contrast in the manner of broadening of the probability; in the case of the limit cycle the broadening is strongly anisotropic, while in the case of the Lotka-Volterra system it is fairly isotropic.

In order to relate the stability of the macroscopic orbit (i.e., the macroscopic response of the system to the external force) with the fluctuation in the unperturbed system, we should study the long-time behavior of the probability. It is difficult to calculate the long-time behavior numerically because the motion of the characteristic flows becomes divergent as time proceeds. Of course, this does not exclude the possibility of finding the stationary form of the probability.

Acknowledgment

The author is thankful to Professor I. Prigogine of the Université Libre de Bruxelles (U.L.B.) for his interest in this work and for his hospitality during the author's stay in Brussels.

This work was initiated by enlightening discussions with Professor R. Kubo and Dr. K. Matsuo at the University of Tokyo.

He expresses his gratitude to Professor G. Nicolis of the U.L.B. for his stimulating advice and for his careful reading of this manuscript.

Fruitful discussions with Professor R. Balescu and Dr. J. Vardalas of the U.L.B. and Professor D. H. Kobe of North Texas State University are also gratefully acknowledged.

APPENDIX A. PERIOD OF THE ROTATING FLOW

One rotating flow is characterized by the constant of motion E, that is,

$$H(x,p) = x(a^2 - x^2)p - Dp^2 = E \tag{A.1}$$

From Equation (3.6), one gets

$$p = \frac{1}{2D}[x(a^2 - x^2) - \dot{x}] \tag{A.2}$$

Equations (A.1) and (A.2) lead to a differential equation for x,

$$\dot{x} = \pm\sqrt{x^2(a^2 - x^2)^2 - 4ED} \tag{A.3}$$

From Equation (A.3), the period of the rotation is calculated as

$$T(E) = 2\int_{x_-}^{x_+} dx \frac{1}{\sqrt{x^2(a^2 - x^2)^2 - 4ED}} \tag{A.4}$$

where x_+ and x_- are the positive solutions of an equation for x,

$$x^3 - a^2 x + 2\sqrt{ED} = 0 \tag{A.5}$$

Equation (A.4) may be transformed into an elliptic integral,

$$T(E) = \frac{2}{\sqrt{(\xi_m - \xi_-)\xi_+}} \int_0^{\pi/2} dO \frac{1}{\sqrt{1 - k^2 \sin^2 O}} \tag{A.6}$$

where ξ_m, ξ_+, and ξ_- ($\xi_m \geqslant \xi_+ \geqslant \xi_- > 0$) are the solutions of an equation for ξ,

$$(\xi - a^2)^2\xi - 4ED = 0 \tag{A.7}$$

and

$$k^2 = \frac{(\xi_+ - \xi_-)\xi_m}{(\xi_m - \xi_-)\xi_+} \tag{A.8}$$

Near the point P, which corresponds to

$$E = E_0 = \frac{a^6}{27D}$$

one may write

$$E = E_0 - \frac{a^6}{D}\Delta$$

and expand (A.4) in powers of Δ. Then the period is estimated as

$$T(E) = \frac{\sqrt{3}}{a^2}\pi\left[1 + \tfrac{176}{63}\Delta + O(\Delta\sqrt{\Delta}\,)\right] \qquad (A.9)$$

APPENDIX B. DERIVATION OF EQUATION (2.7)

In K.M.K., the Green's function of the Schrödinger equation (2.4) is derived in the form of a path integral,

$$G(x,t|x_0) \equiv \langle x|e^{-\Omega H(x,(1/\Omega)\,\partial/\partial x)t}|x_0\rangle$$

$$= \int \delta\sigma(x,p)\exp\left[-\Omega\int_0^t ds\left\{H(x(s,\sigma),p(s,\sigma)) - p(s,\sigma)\dot{x}(s,\sigma)\right\}\right] \qquad (B.1)$$

where the paths in (x,p)-space are under the constraint

$$x(0,\sigma) = x_0 \qquad (B.2)$$

and

$$x(t,\sigma) = x \qquad (B.3)$$

Using the saddle-point approximation, the Green's function is estimated in the limit of $\Omega \to \infty$ as

$$G(x,t|x_0) \approx \int \delta\sigma(x)\exp\left\{\Omega J_\sigma(x,t|x_0)\right\} \qquad (B.4)$$

where

$$J_\sigma(x,t|x_0) = \int_0^t ds\,\mathcal{L}(x(s,\sigma),\dot{x}(s,\sigma)) \qquad (B.5)$$

$$\mathcal{L}(x,\dot{x}) = p(x,\dot{x})\dot{x} - H(x,p(x,\dot{x})) \qquad (B.6)$$

and $p(x,\dot{x})$ is given by the Legendre transformation

$$\dot{x} = \frac{\partial H}{\partial p}(x, p(x,\dot{x})) \tag{B.7}$$

Suppose a path σ_0 gives the maximal value of $J_\sigma(x,t|x_0)$. Then in the path integral in Equation (B.4), contributions of other paths than σ_0 are negligible in comparison with the contribution of σ_0 in the limit of $\Omega \to \infty$. Therefore one may write the Green's function simply as

$$G(xt|x_0) \approx \exp\left\{\Omega J_{\sigma_0}(x,t|x_0)\right\} \tag{B.8}$$

The path σ_0 satisfies the Euler equation

$$\frac{d}{ds}\frac{\partial \mathcal{L}}{\partial \dot{x}} - \frac{\partial \mathcal{L}}{\partial x} = 0 \qquad (0 \leqslant s \leqslant t) \tag{B.9}$$

which is equivalent to the Hamilton equation

$$\frac{dp}{ds} = -\frac{\partial H}{\partial x}(x,p) \qquad (0 \leqslant s \leqslant t) \tag{B.10}$$

However, the condition (B.9) or (B.10) is not sufficient. There may exist several paths that satisfy Equation (B.9) or (B.10). In fact, these equations are derived from the stationarity conditions on the action integral $J_\sigma(x,t|x_0)$ with respect to the variation of the path. Therefore one must choose one path among these paths, which gives the maximal value of the action integral.[7]

The function $\phi(x,t)$, defined in Equation (3.23), evolves through the Green's function; therefore $\phi(x,t)$ should also take the maximal value of all possible values of $J(x,t)$, which obeys the Hamilton-Jacobi equation (2.8).

It should be noted here that the value of x at which $\phi(x,t)$ changes from one branch to another of $J(x,t)$ corresponds to the value that obeys "Maxwell's rule" for the S-shaped front curve in the (x,p)-plane.

References

1. R. Kubo, K. Matsuo, and K. Kitahara, to be published; referred to as K.M.K. in this chapter.
2. D. Courant and R. Hilbert, *Methods of Mathematical Physics*, Vol. 2, Wiley-Interscience, London, 1962.
3. R. Lefever and G. Nicolis, *J. Theor. Biol.*, **30**, 267 (1971); P. Glansdorff and I. Prigogine, *Structure, Stabilité et Fluctuations*, Masson, Paris, 1971.
4. The procedure of the numerical analysis is as follows. One takes a point in the (x_1,x_2)-plane and defines at this point the initial values of p_1 and p_2 and $\tilde{J}(0; x_1,x_2)$,

using Equations (2.14) and (2.13). Then one starts with the characteristic equations (2.15) and (2.16) together with the integration (2.19). At time t, one arrives at a point (x_1, x_2), and one associates with this point the integral (2.19). Following the same procedures from other points in the (x_1, x_2)-plane, one may obtain the values of $J(x_1, x_2, t)$ for several points in the (x_1, x_2)-plane. One connects by a curve the points with the same value of $J(x_1, x_2, t)$.

5. I. Prigogine and G. Nicolis, *Proc. Nat. Acad. Sci. U.S.A.*, **68**, 2102 (1971); I. Prigogine and G. Nicolis, presented at the 3rd Int. Symp. "Theoretical Physics and Biology," Versailles, 1971; E. W. Montroll et al., *Rev. Mod. Phys.*, **43**, 231 (1971).

6. Y. Kobatake, in this volume (p. 319).

7. The fact that in the case of the master equation only the maximal path σ_0 may contribute significantly to the Green's function in the limit of $\Omega \to \infty$ is in a clear contrast with the usual mechanics. In mechanics, since the small parameter $\hbar/i$ is purely imaginary, all paths that give extrema of the action integral contribute equally in magnitude to the Green's function, and these extremum paths correspond to the classical motions.

FUNCTIONAL ORGANIZATION IN ARTIFICIAL ENZYME MEMBRANES— ACCOMPLISHMENTS AND PROSPECTS

DANIEL THOMAS*

E.R.A. $n = 338$ *du* C.N.R.S., *Laboratoire de Technologie Enzymatique Université de Technologie de Compiègne, France.*

CONTENTS

I. INTRODUCTION

Within the living cell, the great majority of the enzymes are attached to membrane structures or contained in cell organelles. When enzymes are isolated, they are removed from their natural state and quite often are highly unstable. The artificial binding of enzymes into membranes makes possible a study of the interaction between diffusion and enzyme reaction within a well-defined context. This approach requires the insolubilization of enzymes by techniques that preserve a major part of their initial activities. In this way Goldman et al.[1-4] incorporated papain and phosphatase in collodion membranes, and other authors (Selegny et al.,[5] Broun et al.[6] have done likewise with several enzymes in cellophane

*This paper was written during a stay on leave of the author in Harvard Medical School, Biophysical Laboratory, Boston, Mass.

membranes. These methods produce active artificial membranes, but the active site distribution is not well defined, and the theoretical treatment is difficult.

In order to give a homogeneous distribution of enzyme molecules inside the membrane, it was necessary to synthesize the membrane and to incorporate the enzymes at the same time. The co-cross-linking of enzyme molecules with an inert protein appears to be a proper solution. Purely active proteic films were created by using this procedure.[7–9] These artificial enzyme membranes can be used in the study of heterogenous enzymes kinetics and for modeling biological membranes. In particular, it is possible to achieve facilitated or active transport with these membranes.

The phenomena in the enzyme membranes can be classified in two parts.

1. *The effect of the composition and structure of the membrane itself on the enzyme membrane behavior.* The binding of enzyme into an insoluble phase allows the creation of structural models that include diffusion constraints; but at the same time it allows the production of an asymmetrical distribution of active sites in the structures. For example, a structural multilayer bienzyme membrane, composed of a metabolically and spatially sequential enzyme system, results in active glucose transport.[9, 10] This system, presented here, demonstrates a specific effect of spatial transport that is closely related to an oriented structure, that is to say, with the arrangement of functional elements in a logical sequence in space.

2. *The effect of the local concentration distribution of the reactants on the enzyme membrane behavior.* First of all, the behavior of the enzymes in the membrane differs markedly from the behavior of the unbound enzymes in solution. *It is pertinent to note that the milieu in which the enzyme bound to a membrane acts might be determined not only by the composition and structure of the membrane itself, but also by the local concentration distribution of substrate and products.* The microenvironment in the membranes is the result of a balance between the flow of matter and enzyme reactions. The substrate and product concentrations in the membrane differ from point to point across the membrane and also differ from those at the outer solution. This may be demonstrated for simple and stable stationary states of the products and substrate concentration profiles inside the membrane.[2] But it is still more important when nonlinearity of the enzyme reactions can produce a pattern or structure due to an instability in the homogeneous state.

The stable stationary states of the concentration profiles inside the membrane modulate only the kinetic properties of the enzymes. The presence of instabilities introduces totally new properties in the system, such as asymmetrical behavior. This new inhomogeneous distribution can be called a "dissipative structure," according to Prigogine.[11] Until now, experimental evidence of instabilities in biochemical reactions has been practically limited to the observation of sustained oscillations.

Experimentally, sustained oscillations have been observed and established beyond doubt for glycolysis. Most of the experiments have been carried out by Hess[12] and by Betz and Chance.[13] Theoretical studies attribute these oscillations to the enzyme phosphofructokinase, which is an allosteric enzyme. Sel'kov[14] and Higgins[15] have worked out models to represent the observed oscillations. Recently Goldbeter and Lefever[16] have constructed a model explicitly taking into account the allosteric effects that is free from phenomenological factors. According to Prigogine and Nicolis,[17] these systems are also likely to occur, under different experimental conditions, in spatially ordered states. In this way, artificial enzyme membranes, or at least artificial immobilization of the enzymes, could be a means to get these structures experimentally. That is because, according to the next section, these systems are ruled by the diffusion-reaction equations without convection, and the active sites are truly immobilized. Moreover, from the experimental point of view it is easily possible to get those changes of the boundary conditions so critical for these kinds of systems.

II. METHODS OF PREPARATION OF ARTIFICIAL ENZYME MEMBRANES[7,8]

In the co-cross-linking method used, a bifunctional agent, such as glutaraldehyde, and a bulk protein, such as plasma protein, are mixed with one or more enzymes. This solution is spread on a plane glass surface. After the chemical reaction, a complete insolubilization occurs, and active enzyme membranes are obtained. This process gives a homogeneous distribution of the active sites within the membrane. The films thus achieved are transparent and have good mechanical resistance. They show mean diffusion coefficients of the same magnitude as collodion films.

Variation of enzyme activity by volume unit of membrane is easily obtained by introducing different amounts of enzyme.

It is possible to immobilize, with a high enzyme activity yield (80%), enzymes with different catalytic activities, molecular weights, and isoelectric points.[8] Fragile enzymes also give good activity ratios after immobilization, if cofactors, inhibitors, or substrates are used as protectors.

The resistance to thermal denaturation increases after immobilization in the membrane. The enzymes are stable for several days at 4, 20, and 37°C. Chemical bonding and proteic environment give better stability. The specificity of tested enzymes is not modified after immobilization, and the activation energy is also the same.

The membranes are studied in diffusion cells. In these classical cells, the active membrane separates two compartments containing metabolite solutions, and it is very easy, from the experimental viewpoint, to change the boundary conditions of these systems. The solutions are stirred and thermostatted. The fluxes entering and leaving the membrane can be measured in these diffusion cells, and the symmetry properties of the membrane tested.

III. BASIC EQUATIONS

The enzyme activity expressions (Michaelian or not) are by definition valid only for homogeneous media and are not applicable to the membrane system as a whole. It is not possible to give one value to the substrate or other reactant concentrations inside the whole matrix. These concentrations vary with position in the system.

But the activity expression remains valid for an elementary volume, small enough to be considered as homogeneous with regard to substrate concentration. This elementary volume contains a large enough number of molecules to permit the use of these expressions.

The time evolution of the substrate concentration in this one dimensional system is given by the well known conservation-of-mass equation.

$$\frac{\partial S}{\partial t} = D_S \frac{\partial^2 S}{\partial x^2} - V(s)$$

$V(s)$ describes the effect of the enzyme reaction (Michaelian or more complicated expressions). Similar equations can be written for the products and for H^+ or OH^-.

This partial differential equation is generally nonlinear. No general theory is available for these systems, but it is always possible to do a numerical simulation on a computer for a particular case.[18] Enzyme membranes are governed by equations very close to the equations written by Prigogine and his school[17] for hypothetical systems. These equations could be brought closer to reality by writing for the reaction term a plausible sequence of multistep enzyme reactions.

The boundary conditions, especially the substrate concentration on both sides, can be symmetrical, fixed or mobile. One of the boundaries can be a wall, impermeable to all the reactants; such is the case when an enzyme coating is attached to a sensitive glass electrode.

IV. THE EFFECT OF THE COMPOSITION AND STRUCTURE OF THE MEMBRANE ITSELF ON THE ENZYME MEMBRANE BEHAVIOR

In reference to this, it is possible to describe examples of facilitated transport and active transport.

Facilitated transport of CO_2 with a membrane bearing carbonic anhydrase was described by Broun et al.[19] In this system a hydrophobic membrane separates two compartments containing buffer solutions. This membrane is permeable to gases like CO_2, but impermeable to water and electrolytes. The CO_2 diffusion velocity was measured for a membrane without enzyme, with grafted enzyme on one side, and with grafted enzyme on both sides. With enzyme the apparent permeability increased 2 times and 4 times, respectively. The concentration profiles of CO_2 with and without enzyme can explain this phenomenon.

An active glucose transport effect occurs with a bienzyme membrane composed of two active protein layers and two selective films.[9, 10] The active enzyme films carry, respectively, hexokinase and phosphate co-cross-linked with an inert protein. Both are impregnated with ATP and covered on their external sides by two selective films permeable to glucose, but impermeable to glucose-6-phosphate. In this asymmetrical membrane, glucose is temporarily phosphorylated, and the system behaves chemically as a simple ATP-ase. In the first layer glucose is a substrate, and glucose-6-phosphate diffuses along its own concentration gradient into the second layer, where glucose is a product. The glucose and glucose-6-phosphate concentration profiles explain how this diffusion reaction is converted into a pumping phenomenon (Fig. 1).

When the concentration in the donor compartment remains constant during the experiment (Fig. 2), the concentration in the receptor compartment increases regularly at first, and then attains a plateau. For an ideal system with the thickness of a biological membrane a maximum, an increase in concentration of 130 times can be predicted by computer simulation.[18] This system gives an experimental physicochemical example of a transformation of scalar chemical energy into a "vectorial catalysis effect."

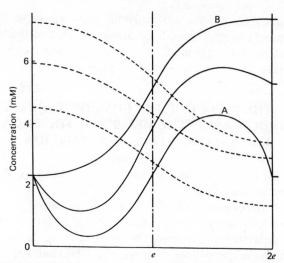

Figure 1. Substrate and product concentration profiles in a bienzyme membrane with a hexokinase-phosphatase double-layer system. Glucose concentration profiles are given in solid lines; glucose-6-phosphate concentration profiles, in dashed lines. When $S_1 = S_2$, the glucose profile is given by the curve A. When $S_1 < S_2$ and $J_1 = J_2 = 0$, the glucose profile is given by B. The third glucose profile shows an active-transport effect.

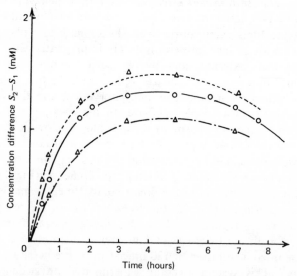

Figure 2. Glucose concentration difference between the two compartments as a function of time (hours) in the described system. Each curve corresponds to one value of initial glucose concentration.

V. THE EFFECT OF THE REACTANT
CONCENTRATION DISTRIBUTIONS ON THE
ENZYME MEMBRANE BEHAVIOR

The existence of concentration profiles due to the diffusion limitations inside the membrane greatly modulates the enzyme behavior. The effects of these profiles were studied with monoenzyme membranes by Goldman et al.[1-4] and Thomas et al.[8] A study of activity as a function of inhibitor concentration for enzyme membrane with different Thiele moduli[20] was done. Enzyme in a membrane is less sensitive to the effect of inhibitor. In contrast, the effect of the reaction reversibility on an enzyme reaction is increased by the diffusion limitations. These problems were studied experimentally and by simulation on a computer.[21]

We can now present a system in which the nonlinearity of the enzyme reactions can spontaneously produce an inhomogeneous distribution of one of the products, beyond an instability.[22]

An artificial membrane bearing two different enzymes (glucose-oxidase and urease) in a spatially homogeneous fashion, is produced by using the method previously described. The glucose-oxidase reaction increases the pH, and the urease reaction decreases the pH. The pH activity profiles show an autocatalytic effect for the glucose-oxidase in the range of pH values greater than the optimum; for the urease, smaller than the optimum (Figs. 3 and 4). When the two enzymes are mixed together, the global pH variation is zero for one well-defined pH value.

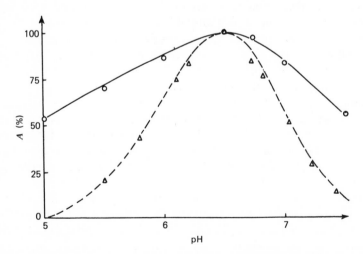

Figure 3. Enzyme activity (in precent) as a function of pH. Glucose-oxidase in solution (solid line, ○) and cross-linked into a membrane (dashed line, △).

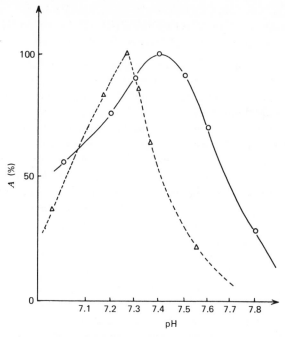

Figure 4.   Enzyme activity (in percent) as a function of pH. Urease solution (solid line, O), and cross-linked into a matrix (dashed line, △).

The active membrane separates two compartments, and it is possible to get this pH value throughout the system, in presence of the two substrates, by the transient use of a buffer. The pH values outside are controlled and H^+ fluxes measured by pH-stat systems. After small asymmetrical perturbations of the pH values at the boundaries (0.05), an inhomogeneous pH distribution arises spontaneously inside the membrane. The initial perturbations are amplified, and the pH values in the compartments tend to evolve in opposite directions. The H^+ fluxes entering and leaving the membrane can be determined by pH-stat measurements. If the boundary pH values are not maintained constant by a pH-stat, the system evolves to a new stationary characterized by a pH gradient of 2 pH units across the membrane. It seems possible to explain this behavior by the diffusion-reaction equations. (The pH distribution inside the membrane after perturbation would be similar to the shape of profile A, Fig. 1.) Paradoxically, this experimental system is similar to the absolutely fictitious system described by Turing.[23] In this system, the simplest case consists of two cells connected by a permeable wall. In each cell, a chemical reaction takes place in which one substance is converted into an another one by an autocatalytic process. With one well-defined homogeneous distribution of matter, the concentrations do not change with time. If, however, a fluctuation of a few percent occurs, a rather important change will take place, as in the experimental system previously presented.

In both systems, a positive fluctuation in the concentration in one compartment does not result in a negative value of the derivatives with time, but in a positive one. In these systems the initial homogeneous steady state is unstable, and the system will develop spontaneously toward a nonhomogeneous concentration distribution. The experimental system can illustrate the potential of artificial enzyme membranes. *Moreover, oscillations in time of the profile of one of the products can also occur in an artificial enzyme membrane.*[24]

A coating, bearing one enzyme (papain), is produced on the surface of a glass pH electrode by the method previously described (co-cross-linking). The papain reaction decreases the pH, and the pH-activity variation gives an autocatalytic effect for pH values greater than the optimum, as in the case of glucose-oxidase. When the enzyme electrode is introduced in a solution with a high pH and a well-defined substrate (benzoyl arginin ethyl ester) concentration, an oscillation in time of the measured pH inside the membrane occurs spontaneously. It is important to note that this enzyme, which has been extensively studied, does not give oscillations in solution for any conditions of pH and substrate concentration. The period of oscillation is around $\frac{1}{2}$ min, and the oscillation is abolished by introducing an enzyme inhibitor. The phenomenon can be explained by the autocatalytic effect and by a feedback action of OH^- diffusing in from the outside solution. The diffusion of this ion is quicker than the diffusion of the substrate. There is qualitative agreement between the computer simulation and the experimental results.

Quite recently, memory in artificial enzyme membranes was described.[25]

VI. CONCLUSION

Enzymes were classically studied in a homogeneous and isotropic medium—the stirred solution. A stage is necessary between this classical enzymology and the study of the properties of enzymes in very complex biological systems. This stage is achieved by using well-defined preparations of membrane-supported enzymes. In this way, it is possible to show in very simple systems the influence of structure and diffusional constraints on global enzyme-membrane properties.

There is experimental evidence for the existence of local concentrations inside enzyme membranes.[26] The local concentrations and/or distributions of product can not only modulate enzyme kinetics but also introduce totally new properties such as asymmetrical behavior and active transport. Naturally, with enzymes in stirred solutions, and even without stirring, it is impossible to get this kind of behavior.

Acknowledgments

The author wants to thank A. Owen for his help in writing this paper in English. He also wants to thank Professor Prigogine's group and Professor Caplan's group for helpful discussions.

References

1. R. Goldman, H. I. Silman, S. R. Caplan, O. Kedem, and E. Katchalski, Papain membrane on a collodion matrix: preparation and enzyme behavior, *Science*, **150**, 758 (1965).

2. R. Goldman, O. Kedem, H. I. Silman, S. R. Caplan, and E. Katchalski, Papain-collodion membranes—preparation and properties, *Biochemistry*, **7**, 486 (1968).

3. R. Goldman, O. Kedem, and E. Katchalski, Papain-collodion menbranes. Analysis of the kinetic behavior of enzymes immobilized in artificial membranes, *Biochemistry*, **7**, 4518 (1968).

4. R. Goldman, O. Kedem, and E. Katchalski, Kinetic behavior of alkaline phosphatase-collodion membranes, *Biochemistry*, **10**, 165 (1971).

5. E. Selegny, S. Avrameas, G. Broun, and D. Thomas, Membranes à activité enzymatique, *C. R. Acad. Sci., C*, **266**, 1931 (1968).

6. G. Broun, S. Avrameas, E. Selegny, and D. Thomas, Enzymatically active membranes: some properties of cellophane membranes supported by cross-linked enzymes, *Biochim. Biophys. Acta*, **185**, 260 (1969).

7. G. Broun, E. Selegny, and D. Thomas, New membranes regulation and transport models based on film bearing cross-linked enzymes, in *Protides of Biological Fluids*, Peeters, Ed., Vol. 54, Pergamon, New York, 1970, p. 75.

8. D. Thomas, G. Broun, and E. Selegny, Monoenzymatical model membranes: diffusion-reaction kinetics and phenomena, *Biochimie*, **54**, 229 (1972).

9. D. Thomas, C. Tran Minh, G. Gellf, D. Domurado, B. Paillot, R. Jacobsen, and G. Broun, Films bearing cross-linked enzymes: applications to biological models and to membrane biotechnology, *Biotech. Bioeng. Symp.*, **3**, 299 (1972).

10. G. Broun, D. Thomas, and E. Selegny, Structured bienzymatical models formed by sequential enzymes bound into artificial supports: active glucose transport effect, *J. Membrane Biol.*, **8**, 373 (1972).

11. I. Prigogine, Structure, dissipation and life, in *Theoretical Physics and Biology*: *Proceedings*, M. Marois, Ed., North-Holland, Amsterdam, 1969, pp. 23–52.

12. B. Hess, *In Funktionelle und morphologische Organisation der Zelle*, Springer, Berlin, 1962.

13. A. Betz, and B. Chance, Phase relationship of glycolytic intermediates in yeast cells with oscillatory metabolic control, *Arch. Biochem. Biophys.*, **109**, 585 (1965).

14. E. E. Sel'kov, Self-oscillations in glycolysis, *Eur. J. Biochem.*, **4**, 79 (1968).
15. J. Higgins, A chemical mechanism for oscillation of glycolytic intermediates in yeast cells, *Proc. Nat. Acad. Sci. U.S.A.*, **51**, 989 (1964).
16. A. Goldbeter and R. Lefever, Dissipative structures for an allosteric model. Application to glycolytic oscillations, *Biophys. J.*, **12**, 1302 (1972).
17. I. Prigogine and G. Nicholis, Biological order, structure and instabilities, *Quarterly Rev. of Biophys.*, **4**, 107 (1971).
18. J. P. Kernevez and D. Thomas, Evolution and control of membrane systems bearing cross-linked enzymes, *Journées d'Informatique Medicale* (I.R.I.A.), **2**, 240 (1972).
19. G. Broun, E. Selegny, C. Tran Minh, and D. Thomas, Facilitated transport of CO_2 across a membrane bearing carbonic anhydrase, *F.E.B.S. Letters*, **7**, 223 (1970).
20. E.W. Thiele, *Ind. Engng. Chem.*, **31**, 916 (1939).
21. D. Thomas and G. Broun, Regulation and transport in enzyme membrane models (lecture) in Symposium on Chemically Grafted and Cross-linked Proteins (1972), *Biochimie*, **55**, 975 (1973).
22. D. Thomas, A. Goldbeter, and R. Lefever, Spontaneous inhomogeneous distribution of one product in an artificial enzyme membrane, to be published.
23. A. M. Turing, The chemical basis of morphogenesis, *Phil. Trans. Roy. Soc.*, **B237**, 37 (1952).
24. A. Naparstek, S. R. Caplan, and D. Thomas, *Biochem. Biophys. Acta*, **323**, 643 (1973).
25. A. Naparstek, J. L. Romette, J. P. Kernevez, and D. Thomas, Memory in enzyme memberane, *Nature* **249**, 490 (1974).
26. J. N. Barbotin and D. Thomas, Electron microscopic and kinetic studies dealing with artificial enzyme membrane, *J. Histochem. Cytochem.* **22**, (1974) in press.

THE GLOBAL ASYMPTOTIC STABILITY OF PREY–PREDATOR SYSTEMS WITH SECOND-ORDER DISSIPATION

CHARLES WALTER

*Departments of Biomathematics and Biochemistry,
University of Texas M. D. Anderson Hospital, Houston,
Texas*

Abstract

Models of biological development, evolution, and control should take into account that very small numbers of cells or chemicals or individuals eventually grow into stable, large populations. The simplified two-component model used in these studies includes the following: (1) first-order decay, (2) first-order autocatalysis, (3) negative feedback, (4) positive feedback, (5) second-order decay, (6) second-order autocatalysis. A positive definite Lyapunov function is constructed and shown to have a negative definite total derivative. The stationary state $x > 0$, $y > 0$ therefore possesses global asymptotic stability. This means that sustained oscillations cannot occur. Another stationary state, $x = y = 0$, is shown to be unstable. This means that infinitesimally small perturbations of $x = y = 0$ will result in the evolution of the variables to the stable stationary state. This result contrasts with that obtained with the Lotka-Volterra model in that small perturbations of $x = y = 0$ for that model result in sustained, oscillating excursions; the smaller the initial perturbation, the larger these excursions will be.

A simulation illustrates that stable populations of 10^{20} x's and y's can arise from a single x and y. x grows more or less continuously, but y remains extremely small for 80% of the time interval required for the variables to approach their stable populations.

A large number of macroscopic biological events are characterized by one or more of the following:

1. First-order decay.

2. First-order autocatalysis: "self-reproduction" or reproduction by a single unit.

3. Negative feedback: decay due to interactions between units of different species.

4. Positive feedback: "reproduction" due to interactions between units of different species.

5. Second-order decay: "cannibalism" or decay due to interactions between units of the same species; decay due to crowding.

6. Second-order autocatalysis: "reproduction" by two units of the same species.

Mathematically, these events can be described by

$$\dot{x} = -ax \tag{1}$$

$$\dot{x} = \alpha x \tag{2}$$

$$\dot{x} = -bxy \tag{3}$$

$$\dot{x} = \beta xy \tag{4}$$

$$\dot{x} = -gx^2 \tag{5}$$

$$\dot{x} = \zeta x^2 \tag{6}$$

where all the coefficients are positive.

We now consider a two-component system wherein the following events occur:

1. Initially there is no x or y present.

2. A very small amount of y is formed "spontaneously."

3. The y "replicates" itself via first- and second-order autocatalysis, but:

4. It also experiences first- and second-order decay. Assume that the rate constants for decay are greater than the corresponding constants for "replication" ($a > \alpha$ and $g > \zeta$). Thus, nothing much "happens."

5. A very small amount of x is formed "spontaneously."

6. The x experiences first- and second-order decay, but

7. It also "replicates" itself. Assume that the rate constant for "self-replication" is greater than the constant for first-order decay ($\alpha > a$), but the second-order constant for "replication" is less than the second-order constant for decay ($g > \zeta$).

8. When x and y "encounter" one another, y increases in number but x decreases.

9. y continues, as in 4, to experience first- and second-order decay.

The model described above could refer to molecular evolution, species selection, differentiation and/or oncogenesis, a genetic and/or a metabolic control circuit, a hormonal control mechanism, cell division and growth, or other macroscopic biological phenomena. The differential equations for x and y are

$$\dot{x} = (\alpha_1 - a_1)x - bxy + (\zeta_1 - g_1)x^2 \tag{7}$$

$$\dot{y} = (\alpha_2 - a_2)y + \beta xy + (\zeta_2 - g_2)y^2 \tag{8}$$

In what follows we consider the case wherein $\alpha_1 > a_1$ (the first-order constant for "self-reproduction" of x is greater than the first-order constant for decay), $g_1 > \zeta_1$ and $g_2 > \zeta_2$ (the second-order constants for decay are greater than the constants for second-order autocatalysis), and $a_2 > \alpha_2$ (the first-order constant for decay of y is greater than the constant for "self-replication"). Equation (7) includes negative-feedback control of x by y, and Equation (8) describes a positive feedback by x on y.

There are two singular points for Equations (7) and (8):

$$x = y = 0 \tag{9}$$

and

$$x^* = \frac{(\alpha_1 - a_1)(g_2 - \zeta_2) + (a_2 - \alpha_2)b}{(g_2 - \zeta_2)(g_1 - \zeta_1) + b\beta}$$

$$y^* = \frac{(\alpha_1 - a_1)\beta + (a_2 - a_2)(g_1 - \zeta_1)}{(g_2 - \zeta_2)(g_1 - \zeta_1) + b\beta} \tag{10}$$

In the neighborhood of the singularity $x = y = 0$, the abridged differential equations are

$$\dot{x} = (\alpha_1 - a_1)x \tag{11}$$

$$\dot{y} = (\alpha_2 - a_2)y \tag{12}$$

For small initial amounts of $x(0)$ and $y(0)$,

$$x(t) = x(0)e^{(\alpha_1 - a_1)t} \tag{13}$$

$$y(t) = y(0)e^{(\alpha_2 - a_2)t} \tag{14}$$

Equations (13) and (14) describe $x(t)$ and $y(t)$ in the neighborhood of small $x(0)$ and $y(0)$. If $\alpha_1 > a_1$ or $\alpha_2 > a_2$, the initial movement of $x(t)$ or $y(t)$ is away from $x = y = 0$: The motion is thus unstable in the sense that it is away from the singular point. This type of instability near the origin may be important in biology because it is precisely this sort of "evolution" of large populations from relatively small initial amounts that is so widespread in biology. If $a_1 > \alpha_1$ and $a_2 > \alpha_2$, the initial movement of $x(t)$ and $y(t)$ is toward $x = y = 0$. Under these conditions the motion is stable in the neighborhood of the origin. The origin is thus stable if the rate constants for the first-order decay of *both* components exceed the respective rate constants for first-order autocatalysis. On the other hand, "self-reproduction" by at least one of the components is a necessary condition for instability near the origin.

In order to ascertain the stability properties of the other singular point (x^*, y^*), consider the positive definite Lyapunov function*

$$V = \left(\frac{x^* e^{x/x^*}}{x} \right)^\eta \left(\frac{y^* e^{y/y^*}}{y} \right)^\mu - e^{\eta + \mu} \tag{15}$$

This Lyapunov function has the property $\lim x, y \to \infty \ V = \infty$,[1] which Brashin and Krasovsky[2] have shown guarantees closure of the Lyapunov surfaces. This point is critical for studies of *global* (rather than local) stability. The derivative of V is

$$-\dot{V} = \left(\frac{x^* e^{x/x^*}}{x} \right)^\eta \left(\frac{y^* e^{y/y^*}}{y} \right)^\mu$$
$$\times \left(\frac{\eta(g_1 - \zeta_1)}{x^*}(x - x^*)^2 + \frac{\mu(g_2 - \zeta_2)}{y^*}(y - y^*)^2 \right) \tag{16}$$

If we choose $\eta = x^*/b$ and $\mu = y^*/\beta$, then $\dot{V}$ is negative definite provided $g_1 > \zeta_1$ and $g_2 > \zeta_2$. These are the same conditions that guarantee nonnegative x and y; whenever they are met, the singular point (x^*, y^*) possesses global asymptotic stability.

For real systems involving nonnegative x and y, we have $g_1 > \zeta_1$, $g_2 > \zeta_2$, and in addition $\alpha_1 > a_1$. We therefore expect the singularity at the origin to be unstable for real biological systems described by Equations (7) and (8). Thus, even if $x(0)$ and $y(0)$ are initially present in very small quantities, we expect the motion to be away from $x = y = 0$. Since the singularity (x^*, y^*) possesses global asymptotic stability, $x(t)$ and $y(t)$ must always move toward $x = x^*$ and $y = y^*$. The approach

*I am indebted to Gregory Dunkel who first suggested this Lyapunov function to me in 1967.

toward this singular point may be oscillatory or it may not, but *sustained* oscillations about (x^*, y^*) cannot occur.

In Fig. 1 appear typical $x(t)$ and $y(t)$ for relatively low $x(0)$ and $y(0)$. $x^* = y^* = 1$, and $x(0) = y(0) = 10^{-10}$ in this figure. As expected, there is an initial interval (about 167 time units for x and about 193 time units for y) during which x and y remain below 1% of x^* and y^*. Subsequently, there is a shorter time period during which $x(t)$ and $y(t)$ approach x^* and y^*.

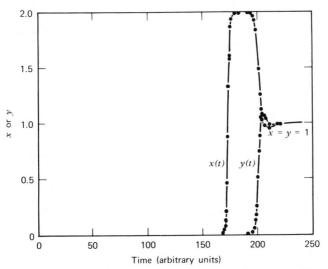

Figure 1. $x(t)$ and $y(t)$ for Equations (7) and (8). $x(0) = 10^{-10}$, $y(0) = 10^{-10}$, $\alpha_1 - a_1 = 1$, $b = 0.5$, $\zeta_1 - g_1 = -0.5$, $\alpha_2 - a_2 = -0.25$, $\beta = 0.5$, and $\zeta_2 - g_2 = -0.25$. $x^* = 1$ and $y^* = 1$.

In Fig. 2 appears the relationship between $x(0)$ and the time period during which x remains below $0.01 x^*$ (denoted by $T_{0.01x^*}$). The empirical relationship between $x(0)$ and $T_{0.01x^*}$ is $\log T_{0.01x^*} = -[0.194 + 3.93 \log x(0)]$ for $x(0) = 10^{-2}$ to 10^{-20}.

The behavior of Equations (7) and (8) is quite different from the behavior reported by Lotka[3] and Volterra[4] for the "prey–predator" equations without second-order dissipation. In the Lotka-Volterra equations the singular point $(x = y = 0)$ is unstable, but the other singular point (x^*, y^*) is not sufficiently stable to attract trajectories from everywhere in the phase space. If the initial conditions are in the neighborhood of (x^*, y^*), then $x(t)$ and $y(t)$ remain near (x^*, y^*), but

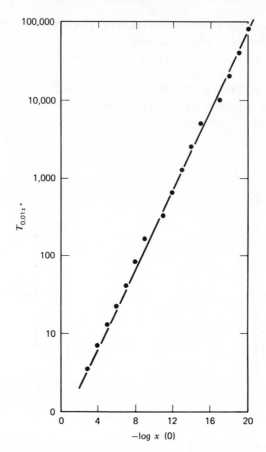

Figure 2. Relationship between the time necessary for $x(t)$ to reach 1% of x^* (denoted by $T_{0.01x^*}$) and $x(0)$ for Equations (7) and (8). $\alpha_1 - \alpha_1 = 1$, $b = 0.5$, $\zeta_1 - g_1 = -0.5$, $\alpha_2 - \alpha_2 = -0.25$, $\beta = 0.5$, and $\zeta_2 - g_2 = -0.25$.

for initial conditions more remote from (x^*, y^*), and especially for initial conditions near $x = y = 0$, $x(t)$ and $y(t)$ rotate in large excursions above and below x^* and y^*.

In both the Lotka-Volterra model and this model, there is a question whether or not it is appropriate to use a continuous description for small populations. Analog-computer simulations of Equations (7) and (8) were conducted to investigate this question. Since when $x(0)$ and $y(0)$ are small, $x(t)$ may become large compared to $y(t)$ (see Fig. 1), fluctuations of $y(t)$ in the analog simulations sometimes caused the whole system to become unstable. The result of this instability was usually the negation of a variable which in the real world is nonnegative. If a real system evolved into stable populations according to Equations 7 and 8, perhaps it was necessary that the initial formation of small amounts of x and y during "noisy" periods was followed by relatively "quiet" periods during which the stable dynamical behavior was achieved.

References

1. C. Walter, in *Biochemical Regulatory Mechanisms in Eukaryotic Cells*, E. Kun and S. Grisola, Eds., Wiley-Interscience, New York, 1972, Chapter 11, p. 394.
2. E. A. Barashin and N. N. Krasovsky, *Dokl. A.N. SSSR*, **36**, 3 (1953).
3. A. Lotka, *J. Phys. Chem.*, **14**, 271 (1910), and *Proc. Nat. Acad. Sci.* (U.S.) **6**, 410 (1920).
4. V. Volterra, *Leçons sur la Theorie Mathematique de la Lutte pour la Vie*, Gauthiers-Villars, Paris, 1931.

A SHORT REMARK ABOUT VARIOUS DISSIPATIVE STRUCTURES

E. KAHRIG AND H. BESSERDICH

Akademie der Wissenschaften der DDR,
Zentralinstitut für Molekularbiologie, 1115 Berlin-Buch,
D. D. R.

Doubtless, the number of dissipative structures in chemical networks is very large and of special interest for modern biology because of the different possibilities for nonlinearities.[1,2] Nevertheless it is useful and informative to have a look at the dissipative phenomena characterized by the presence of hydrodynamics.

For the dissipative phenomena collected in Table I the fluctuations of the generating generalized forces are given in accordance to the excess entropy production. The special phenomenon of a dissipative structure depends on the boundary conditions. For instance, the relative orientation of the gradients and the existence of external forces are important. The latter was verified in the case of a magnetic field by Pamfilov et al.[3] Benard convection cells with chemical reaction[4] and the Marangoni effect[5,6] are of importance for chemical technology, especially for mass and heat transfer processes. Schaaffs concentration zones[7,8] are instructive and impressive dissipative structures, as was mentioned during the panel discussion at the meeting reported in this volume. The analysis and understanding of all these dissipative structures in physical and physicochemical systems is of importance for modeling biological systems.

Furthermore, there are physical examples of dissipative structures without hydrodynamics—for instance, the well-known diffusion zones within a discharge valve[9] and the stationary movement of a charged particle in a special four-pole condenser.[10] We hope that this short comment will direct attention to dissipative structures that are not in the center of the main activities in the field of nonlinear thermodynamics.

133

TABLE I

Dissipative phenomena	Fluctuations of the generalized forces[a]							
	$\delta\nabla p$	$\delta\nabla\mu$	$\delta\nabla T$	$\delta\nabla\varphi$	$\delta\nabla\sigma$	δH	Gradv	δA
Benard convection cells[11,12]	√		√	√				
Benard convection cells with thermodiffusion[1]	√	√	√	√				
Benard convection cells of a binary solution in a magnetic field[3]	√	√	√			√		
Benard convection cells with chemical reaction[4]	√	√	√					√
Gambale-Gliozzi diffusion cells[13]	√	√						
Concentration zones in the melt during crystal growth[14]	√	√	√					
Schaaffs concentration zones[7,8]	√	√	√					
Baranowski convection cells[15]	√	√		√				
Laminar-turbulent transition[1]	√						√	
Taylor instability of Couette flow[16]	√						√	
Teorell's oscillator[17,18]	√	√		√				
Marangoni effect[5,6]	√	√			√			
Marangoni effect[5,6]	√		√		√			

[a]Symbols: p, pressure; μ, chemical potential; T, temperature; φ, electrical potential; σ, surface tension; H, magnetic field intensity; v, stream velocity; A, affinity.

References

1. P. Glansdorff and I. Prigogine, *Thermodynamic Theory of Structure, Stability and Fluctuations*, Wiley-Interscience, New York, 1971.
2. I. Prigogine, G. Nicolis, and A. Babloyantz, *Physics Today*, November, 1972, p. 23.
3. A. V. Pamfilov, V. N. Kapranov, and V. V. Nečiporuk, paper presented at the 1st All-Union Conference on Thermodynamics of Irreversible Processes, Černovcy (1972).
4. V. V. Nečiporuk, A. I. Lopušanskaja, and P. M. Grigorišin, paper presented at the 1st All-Union Conference on Thermodynamics of Irreversible Processes, Černovcy (1972).
5. C. V. Sternling and L. E. Scriven, *A.I.Ch.E. J.*, **5**, 514 (1959).
6. H. Linde and B. Sehrt, *Chem. Techn.*, **16**, 583 (1964).
7. W. Schaaffs, *Kolloid-Zeitschrift*, **227**, 131 (1968).
8. L. Haun and W. Schaaffs, *Kolloid-Zeitschrift*, **239**, 616 (1970).
9. E. Grimsehl, *Lehrbuch der Physik*, Bd. II, Leipzig (1960).
10. Gradewald, after a private communication by Ebeling.
11. S. Chandrasekhar, *Hydrodynamic and Hydromagnetic Stability*, Oxford U.P., Oxford, England, 1961.

12. J. A. Whitehead, *American Scientist*, **59**, 444 (1971).
13. F. Gambale and A. Gliozzi, paper presented at the 1st European Biophysical Congress, Baden bei Wien (1971).
14. B. Chalmers, *Principles of Solidification*, Wiley-Interscience, New York, 1964.
15. B. Baranowski and A. L. Kawczynski, *Roczniki Chemii, Ann. Soc. Chim. Polonorum*, **44**, 2447 (1970).
16. L. Prandtl, *Strömungslehre*, Braunschweig, 1956.
17. T. Teorell, *J. Gen. Physiol.*, **42**, 847 (1959).
18. Y. Kobatake, *Physica*, **48**, 301 (1970).

SPATIOTEMPORAL ORGANIZATION IN CHEMICAL AND CELLULAR SYSTEMS*†

B. HESS, A. BOITEUX, H. G. BUSSE

*Max-Planck-Institut für Ernährungsphysiologie,
46 Dortmund, Rheinlanddamm 201, Germany*

and

G. GERISCH

*Friedrich-Miescher-Laboratorium der
Max-Planck-Gesellschaft,
74 Tübingen, Spemannstrasse 37-39, Germany*

DEDICATED TO PROFESSOR A. BUTENANDT ON THE
OCCASION OF HIS 70TH BIRTHDAY

CONTENTS

I. INTRODUCTION

Spatiotemporal organization is a fundamental property of living systems. The temporal organization is clearly displayed in the oscillatory

*Based on a lecture given by B. H. on the occasion of the EMBO Conference on Membranes, Dissipative Structures and Evolution, Bruxelles, November 1972.
†Part of this work (G. G.'s) is supported by the Deutsche Forschungsgemeinschaft and the Stiftung Volkswagenwerk.

137

and rhythmic phenomena covering a bandwidth of over ten orders of magnitude in the living world. The spatial organization of matter is given in the structural order and complexity of living systems. Both types of organizations include what in recent years has been defined as "dissipative structures" by I. Prigogine and his colleagues[1,2]: states that evolve in open systems distant from equilibrium.

In the context of this volume, it is fitting to present experimental evidence. Whereas a recent summary covers the broad field of oscillatory phenomena in biochemistry,[3] here we would like to discuss examples of oscillatory systems demonstrating typical properties of some dissipative structures, which we have been expecially interested in: the *soluble* multienzyme system of glycolysis, the *membrane*-bound multienzyme system of mitochondrial respiration, the *cellular* system of Dictyostelium discoideum, and finally the *chemical* system of the malonic acid oxidation.

II. GLYCOLYSIS

In our laboratory, glycolytic oscillations have been studied in intact cells and cell-free extracts of yeast (*S. carlsbergensis*) (for a summary see Ref. 3) with the help of an experimental design as outlined in Fig. 1,

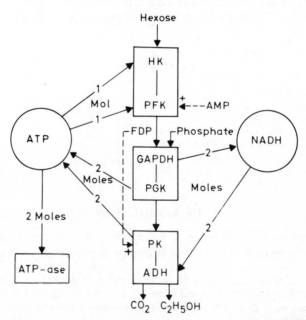

Figure 1.   Control diagram of glycolysis (courtesy of Dechema-Monographic).

demonstrating the minimum functions of enzymes (in blocks) and recycling metabolites (in circles) involved in the generation of glycolytic oscillations. An injection technique was used to allow a continuous and controlled input of glycolytic substrates and intermediates for the induction of the steady states of the pathway. Simultaneously a continuous record of some indicator parameters of the system—namely pH, NADH concentration, and CO_2 pressure—was obtained by respective recording techniques. A pH electrode reflects mainly the dynamics of the ATP-system and indicates the activity state of the phosphofructokinase reaction. The NADH concentration (free and bound) is measured fluorometrically or spectrophotometrically and indicates the activity of the "lower" glycolytic pathway relative to the activity of the "upper" glycolytic pathway. Finally, a CO_2 electrode measures the CO_2 pressure. Thus, macroscopically, input and output parameters and two internal parameters of the system can be recorded continuously. Also, manual sampling allows the analysis of glycolytic intermediates, usually assayed by enzymatic methods.

A typical experiment carried out in a suspension of glycolysing yeast

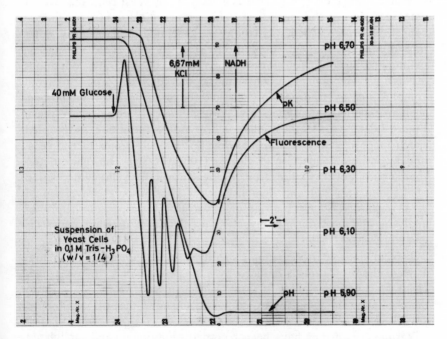

Figure 2.   Oscillatory transient of pyridinnucleotide fluorescence, pH and $p[K^+]$ in a suspension of yeast cells.

cells is presented in Fig. 2, which demonstrates a record of the kinetics of fluorescent pyridine nucleotide, of pH and p[K$^+$]. Initially, when no glycolytic substrate is added to the system, a relatively high "resting level" of NADH is observed under conditions of negligible flux through the glycolytic system. After addition of a single dose of glucose, a brief reduction of NADH is observed. This reduction is followed by a strong oxidation and later a highly damped oscillatory transient. Finally, the trace returns asymptotically to the original level of NADH fluorescence, indicating that the resting state of the cells is regained. The rate of glycolysis as indicated by the pH electrode shows that on return of the fluorescence trace to its initial state, the rate of acidification is halted also. Thus, because all glucose added initially is used up, glycolysis finally comes to a standstill.

We here observe a transient of the whole glycolytic pathway in response to a single dose of glucose. During this transient the system passes through many states. The transient is also indicated by the trace of the potassium-sensitive electrode, demonstrating the uptake of potassium by the cells during the process of glycolysis, and its release when glycolysis ceases.

When a single dose of glucose is glycolysed by yeast cells, the level of glucose drops continuously. As soon as a critical range of glucose concentration is reached, oscillations are generated spontaneously. After this domain is passed, oscillations stop. If the critical range is maintained by continuous injection of a glycolytic substrate, continuous oscillations are generated and can be recorded. In order to avoid complication by transport processes through the cellular membrane, such an experiment has been carried out in the cell-free extract of yeast. Figure 3 demonstrates the oscillation of pyridine nucleotide, as recorded spectrophotometrically at two wavelengths in a cell-free extract of yeast under continuous injection of glucose (100 mM/hr). The figure represents a cutout of four cycles from a series of over 30 cycles with a period of 4.5 min ($\pm 4.2\%$) and demonstrates a nonsinusoidal waveform of great regularity. Such a nonsinusoidal waveform is especially helpful in analyzing the mechanism of glycolytic oscillation. A detailed analysis is given elsewhere.[4,5,6]

Typical sinusoidal oscillations, slightly damped, are recorded in Fig. 4. Here the oscillation is induced by a single dose of trehalose, which yields glucose-6-phosphate at a rate sufficient to drive the system into the oscillating domain. The experiment also shows that pH oscillations are observed.

Various waveforms have been recorded: continuous and modulated sinusoids, single and double pulses, strongly asymmetric cycles,

Range of glycolytic oscillation in yeast extracts				
Input rate[+] mM/h	period min	amplitude in mM NADH	Damping	waveform
< 20	−	steady high level of NADH	−	−
20	∼ 8.6	0.2 − 0.4	−	double periodicities, nonsinusoidal
40	∼ 6.5	0.6	−	nonsinus-sinus
60 − 80	∼ 5.0	0.3	−	stabil sinus
120	∼ 3.5	0.2	−	stabil sinus
> 160	−	steady low level of NADH	+ + +	

[+]fructose or glucose serve as substrates. cellfree extract of ∼ 60 mgr / ml

Table I.

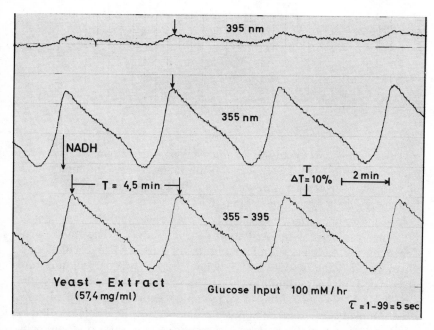

Figure 3. Oscillations of NADH recorded by absorbance measurements at 395, 355, and
355 − 395 nm. For explanation see text.

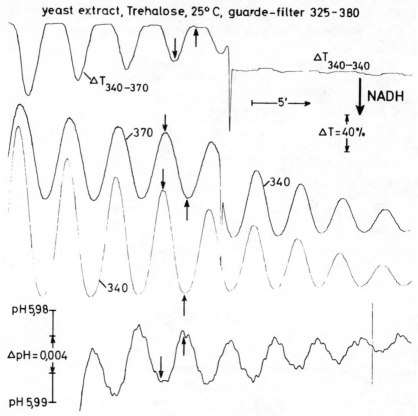

Figure 4. Oscillations of pyridinnucleotide and pH obtained in a yeast extract in the presence of trehalose (Ref. 7).

approximations of squarewaves, and spikes with frequencies between 0.05 and 0.5 min^{-1} and well over 160 continuous cycles of a high degree of stability. The temperature-dependent frequency indicates an activation energy of 19 kcal/mole from 6 to 33°C.

By continuous variation of the rate of substrate input the character of the domain of glycolytic oscillation can be analyzed. The flux dependence of the frequency and amplitude of the NADH oscillation is illustrated in Table I. With a flux below 20 mM/hr no oscillations are observed. With increasing input rates, the period is shortened with the amplitude of NADH rising and falling until—with an input rate of more than 160 mM/hr—the oscillations are damped. In the whole oscillatory range characteristic waveforms are observed. Here it should be mentioned that in a model study of phosphofructokinase Goldbeter and

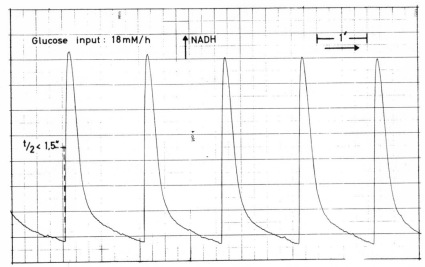

Figure 5.   Oscillations of NADH recorded by fluorometry (Ref. 9).

Lefever[8] observed the generation of oscillations within the same domain of input rates and within the same frequency range as experimentally recorded and summarized in Table 1.

The generation of self-oscillations of glycolysis can be observed not only in cell-free extracts, but also in reconstructed systems. Earlier, we described the observation of sinusoidal and square-wave motions of pyridine nucleotide in a reconstructed system.[4] Furthermore, pulses were observed. Figure 5 demonstrates the production of NADH pulses in a reconstructed glycolytic system. The record is part of a series of 30 pulses of NADH of approximately 0.2 mM with a half-rise time of less than 1.5 sec. By computation an apparent half-rise time of the order of 5 to 600 msec is obtained. A simultaneous record of the pH kinetics indicates a pulselike production of protons initiated in coincidence with the onset of the NADH rise during each pulse, generating approximately a ∆pH of 0.02.[9]

A detailed study of the kinetics of all glycolytic intermediates indicate that the concentrations of all metabolites change during the oscillatory state with equal frequency but differing phase angles relative to each other. Also, the recovery of the amount of substrate turned over into glycolytic intermediates and into the end products of glycolysis shows that the system runs with a fully balanced metabolite and energy turnover and that no side pathway needs to be considered. Furthermore, it should be stressed that the analysis of the glycolytic metabolites as well as of the flux direction clearly shows that glycolysis is a unidirectional process, running far from the thermodynamic equilibrium under conditions of oscillation.

The identification of the enzymes responsible for generation and propagation of oscillations was achieved by the following techniques (for summary see Ref. 3):

1. A test of the ability of various glycolytic substrates to initiate oscillations.

2. Analysis of the phase-angle shift of glycolytic intermediate concentrations during oscillations according to the crossover theory.

3. Analysis of the relative concentration change of metabolites in nonsinusoidal oscillations.

4. Phase-titration experiments.

The phase shift of the oscillating concentrations of the intermediates in a cell-free extract of yeast allows one to classify the glycolytic

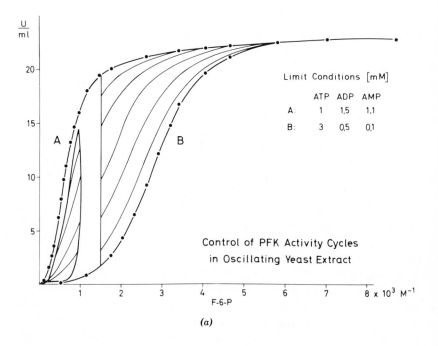

(a)

Figure 6. (a) Activity of phosphofructokinase as a function of the concentration of fructose-6-phosphate in the presence of ADP, ATP, AMP. (b) Hill plot of the experiments presented in (a).

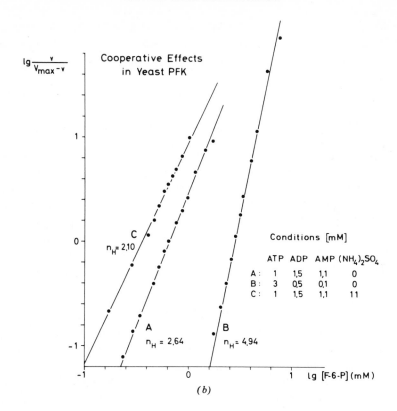

(b)

intermediates in two groups.[5] The maxima and the minima of the two groups differ by an angle $\Delta\alpha$. The analysis of these relationships identifies the enzyme phosphofructokinase as the "primary" oscillophor of the system, changing its activity up to a factor of 80 to 90 during one oscillatory period in a cell-free extract of yeast under conditions of sinusoidal oscillations. Glyceraldehyde-3-phosphate dehydrogenase/ phosphoglycerate kinase, as well as pyruvate kinase are the enzymes responsible for the propagation of the oscillatory motion via the adenine nucleotide system.[5]

Phosphofructokinase, glyceraldehyde-3-phosphate dehydrogenase, and pyruvate kinase are well-known allosteric enzymes. Work on the molecular nature of phosphofructokinase of yeast is in progress in a number of laboratories. Figure 6 demonstrates the cooperative behavior of the reaction kinetics of phosphofructokinase under ligand conditions as observed during an oscillatory cycle. The Hill coefficient of this enzyme varies depending on the ligand concentrations of fructose-6-

phosphate, ATP, ADP, and AMP between 2 and 4.9. The area on the left side of the graph (Fig. 6a) indicates the range over which the enzyme changes its activity. Also, pyruvate kinase of yeast is a cooperative enzyme.[10] Under conditions of sinusoidal oscillations the enzyme is saturated with fructose-1, 6-diphosphate[10] and controlled mainly by the ADP/ATP ratio.[5] Whereas phosphofructokinase and pyruvate kinase influence the concentration of glycolytic intermediates by a phase shift of 180°, the control function of glyceraldehyde-3-phosphate de-hydrogenase is less pronounced and more complex. Its influence on the oscillation results from the fact that the equilibrium constant of the redox reaction is near unity, which makes the reaction rate extremely sensitive to the NADH/NAD ratio as well as to the concentration of 1,3-diphosphoglycerate. The enzyme activity of glyceraldehyde-3-phosphate dehydrogenase is cross-coupled to both the adenine nucleotide system and the redox system of the cell.

The question has been asked whether glycolytic oscillations might generate spatial inhomogeneities and consequently form a regular pattern of matter. Glansdorff and Prigogine[1] analyzed the model of Selkov,[11] and pointed out that the model predicts inhomogeneities with critical wavelengths within 10^{-4} to 10^{-2} cm. Since the intracellular molarities of glycolytic enzymes are high[6] and in the range between 10^{-4} and $10^{-6}M$ per binding site, the mean molecular distance between enzymes is only 40 to 50 Å, resulting in a transit time for metabolites to travel from one enzyme to the next of approx. 1 μsec. The distance and time are small compared to the predicted range of spatial propagation. A spatial inhomogeneity of low concentration molecular intermediates is not likely for intracellular conditions.

Yet inhomogeneities might well occur. Indeed, Goldbeter and Levefer[8] recently predicted such inhomogeneities on the basis of their phosphofructokinase model of glycolytic oscillations. While their theoretical investigations were under way, we observed experimentally the spatial propagation of glycolytic oscillations in cell-free extracts of yeast, which could be explained on the basis of the relatively slow changes of states of the critical enzymes, mainly phosphofructokinase.

Qualitatively, glycolytic oscillations have been analyzed by model studies. The reduced Selkov model based on the results of our laboratory[11] gives a reasonable description of the limit-cycle behavior and predicts a frequency in the range of 2.4 min^{-1}, in agreement with our experimental results.[1] An interesting feature of this model is the relationship between the enzyme concentration, its kinetic constants, and the period being generated. As shown in Fig. 7, in the range of 10^{-4}

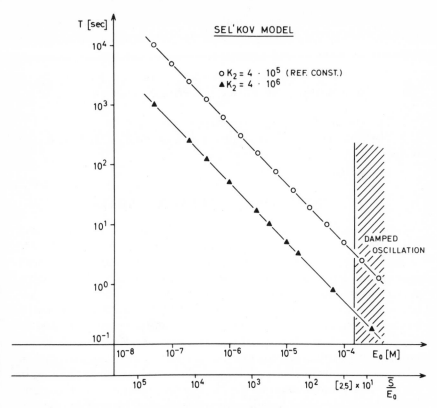

Figure 7. Relation between the period (in seconds) and enzyme concentration in the Selkov model (Ref. 12).

to 10^{-8} M, a linear relationship exists between molarity and period, the latter variable covering about five decades. Thus, such a single-enzyme system with appropriate couplings for feedback and control can generate periods over the whole range observed in biological and physiological systems, depending on concentration of enzyme and kinetic constants. The pitfall of this model, however, is its lack of response to controlling ligands other than the substrate, and also its high sensitivity towards perturbations.[12] The latter property is in clear contradiction to the properties of the experimental oscillations observed in glycolysis. A recent model of Goldbeter and Lefever fits the experimental data far better. Their allosteric model of phosphofructokinase is qualitatively and quantitatively in good agreement with the experimental observations during glycolytic self-excitation. Also, it displays a high degree of stability in response to external perturbations.[12]

Summarizing, glycolytic oscillations occur under conditions which have been predicted as being necessary for the generation of oscillatory behavior:

1. The "primary oscillophore," phosphofructokinase, has a nonlinear kinetic characteristic based on the property of cooperativity in oligomeric enzyme systems.
2. The kinetic structure has feedback properties.
3. The Onsager reciprocity relationship does not hold, and the system is open.
4. The system moves on a limit cycle.

III. OSCILLATING CELLULAR RESPIRATION

Cellular respiration, located in the mitochondria, is a complex and membrane-bound process involving in its minimum structure electron-transfer systems coupled to phosphorylation sites and ion transport phenomena. Up to now, the number of enzymes involved has not been evaluated. In 1965 two surprising reports appeared, describing oscillations of light scattering and ion fluxes in suspensions of mitochondria. Later, the oscillation of states of electron transfer components and of oxygen uptake was detected (for a summary see Refs. 3 and 13). A typical experiment is recorded in Fig. 8, demonstrating the dynamics of some components of the respiratory chain as well as light scattering in pigeon heart mitochondria, supplied with substrates and oxygen. Valinomycin, an ionophoretic antibiotic, is used to activate the potassium transport, being part of the oscillating system. However, valinomycin is not an obligatory component for the induction of the oscillation. Also, in the absence of ionophoretic antibiotics, oscillations can be induced. The oscillation of electron-transfer components is accompanied by oscillations of protons and potassium ions, which can be observed virtually undamped over a long series of cycles. Figure 9a is an example of such a record. The fluxes of potassium ions and of protons always point in opposite directions. When potassium ions accumulate inside the mitochondria, protons are released, and vice versa. The amplitude of the proton movement is in the order of 5 to 10 μeq per gram of mitochondrial protein, while the amplitude of the potassium movement can be several times higher, depending on the experimental conditions. Obviously, the oscillations of respiratory components, ion movements, and mitochondrial volume changes are not in phase with each other. Complete synchronism actually cannot be expected, since these oscillations represent different processes in the

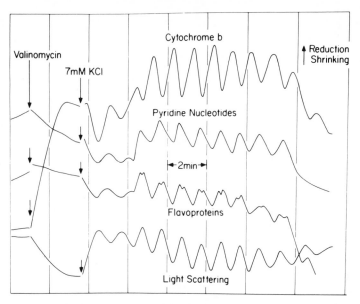

Figure 8.   Record of cytochrome *b*, pyridinnucleotide, flavoproteins, and mitochondrial-volume changes (monitored by light scattering) of pigeon heart mitochondria in suspension (Ref. 13).

mitochondrial membrane, and a detailed analysis of their phase relations yielded valuable information concerning the mechanism involved.[13]

Not only do components of the respiratory chain and the transport of ions fall into well-controlled oscillations, but also the uptake of oxygen from the solution proceeds in cycles.[14] This observation demonstrates an internal regulation of mitochondrial respiration in the oscillatory state. The respiratory control obviously is imposed by the energy transforming system.[15] In Fig. 9*b* the internal ATP level of mitochondria during oscillations clearly exhibits periodic cycles with a fixed phase relation to the swelling cycles.

Consequently, mitochondrial oscillations can be influenced by titration with adenine nucleotides. Additions of ADP at any phase angle of the oscillation result in a new synchronization of the mitochondrial oscillation at a position 50 to 70° after the maximum of the volume cycle. The phase relationship between volume change, respiration rate, the oxygen tension, and the adenine nucleotide system are given in Fig.

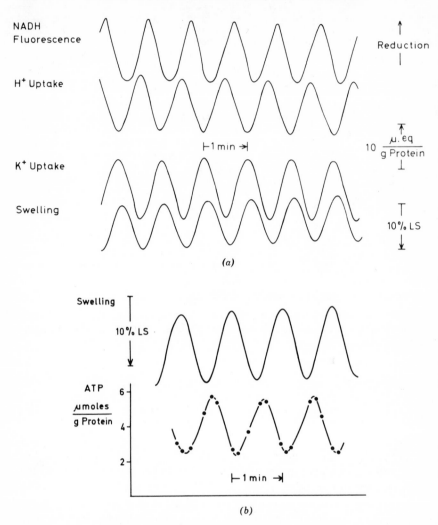

Figure 9.   Oscillations of mitochondrial components.

10. The diagram shows the position to which the system is moved on addition of ADP and ATP, respectively. The addition of adenine nucleotides leads to a phase shift within fractions of a second being followed by a continuation of the oscillation with the same frequency without time delay from the newly obtained position.

From these experiments it is obvious that the ADP/ATP ratio in the

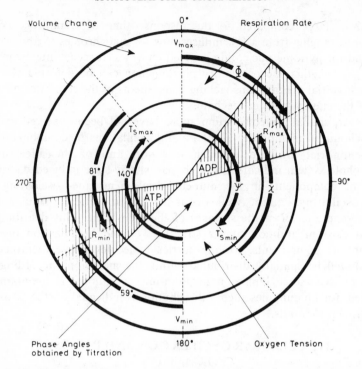

Figure 10. Phase-angle diagram of mitochondrial oscillation. For explanation see text (Ref. 13).

mitochondrial matrix is a control factor of the oscillations. However, it is not the only one. Experiments with valinomycin clearly demonstrate that period and amplitude of the oscillations are dependent on the activity of the potassium carrier.[15] The rate of ion transport to and from the matrix space is obviously another control factor of respiratory oscillations in addition to the endogeneous adenosine phosphate system. It has been shown, that ion transport and respiratory rate are mutually coupled, a proton gradient serving as external synchronizer and as a positive positive feedback component.[15]

Although the mechanism of mitochondrial oscillations seems to be rather complex, the properties 1 and 3–4 given at the end of last chapter also hold for this case. The "nonlinearity" condition is given by the involvment of vectorial transport processes. Due to the control of these processes the period of the oscillations is in the range of one minute, which is large compared to the response time of respiratory components, ranging from a few milliseconds for cytochromes to about one second for pyridine nucleotide coupled enzymes. It seems interesting to find out whether the oscillatory phenomenon can be detected also in mitochondrial preparations lacking components of the respiratory chain or of the energy-transfer mechanism.

So far, spatial nonequilibrium states have not been observed in the mitochondrial membrane. The temporal oscillations of components of mitochondrial ion and electron transport and of the adenosine-phosphate system indicate that a spatial, steady-state induced distribution of components in the mitochondrial membrane—oscillating or non-oscillating—might well occur in the direction through the membrane and/or at the surfaces of the membrane. This distribution could exist in addition to the structurally fixed organization of the membrane components. Mobile carrier molecules like cytochrome c could well be spatially distributed within the membrane according to their redox state, as long as nonequilibrium conditions are maintained. Indeed, for ubiquinones such a distribution is indicated on the basis of experimental results.

IV. CELLULAR OSCILLATION AND PATTERN FORMATION

Recently, oscillations in cellular systems have been observed (for a summary see Refs. 3 and 16). The best studied example for the function of temporal oscillations in spatial pattern formation is the development of the cellular slime mold *Dictyostelium discoideum*.[17]

After the end of their growth phase, single ameboid cells of the slime mold aggregate in response to chemotactic stimuli into multicellular masses, which eventually undergo transformation into fruiting bodies. The chemotactic system, which is believed to function by utilizing cyclic AMP as a chemotactic signal,[18] shows marked rhythmic behavior.[16,19,20] When a homogeneous population of cells is spread on an agar surface, just before full aggregation competence is reached, the cell layer organizes itself into systems of concentric or spiral-shaped moving waves (Fig. 11). Depending on the state of the cells, the frequency of wave formation varies between 0.2 and 0.3 min^{-1}, the speed of wave propagation being 40 μ/min or more. Each wavefront connects cells that at a given time are in the same phase of the chemotactic reaction cycle. The cycle is supposed to consist of (*1*) a phase of chemotactic responsiveness, (*2*) a phase of chemotactic activity, at which a pulse of chemotactic factor is released from the cells, and (*3*) a refractory

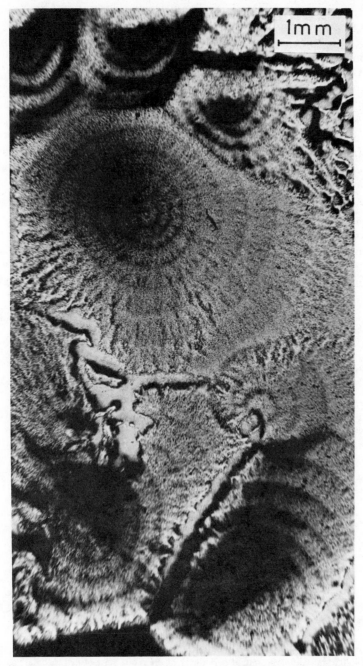

Figure 11. Wave pattern of chemotactic activity in an aggregating cell layer of *Dictyos-telium discoideum*. The dark lines represent zones of chemotactically responding cells (Ref. 17).

phase.[16,21] The latter would provide for mutual extinction of the waves along the boundaries between adjacent aggregation territories.

The question was raised whether this system, like the glycolytic system in yeast, can be manipulated to produce coupled oscillations in cell suspensions. We have found, in suspended cells of early aggregation phase, a periodic phenomenon that manifests itself by oscillations of the oxidative state of cytochrome b, and synchronously by changes in light scattering (Fig. 12). Further experiments have shown that this phenomenon is related to the periodic activity in aggregating cell layers.[22]

Another periodic phenomenon, and possibly another example of dissipative structures in biological systems, has been observed in agar-plate cultures of aggregation-deficient mutants. If a mutant colony is growing on agar plates on a homogeneous layer of food bacteria, the outer growth zone, represented by those Dictyostelium cells that are in immediate contact with the bacteria, appears as a smooth, structureless

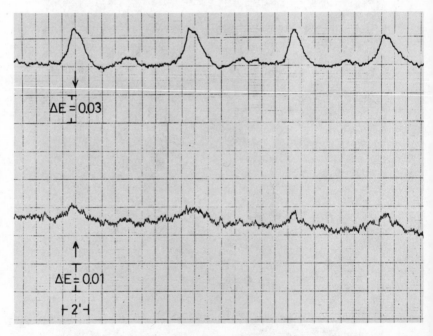

Figure 12. Oscillation of light scattering in *D. discoideum* cell suspensions (top, recorded at $\lambda = 405$ nm), and concomitant periodicity of the redox state of cytochrome b (bottom, recorded as the extinction difference measured at $430 - 405$ mn). An upward deflection demonstrates a reduction of cytochrome b and of light scattering, respectively.

line. However, the inner boundary of the wall of cells marking the growth zone is organized into projecting cell groups that represent attraction centers, into which the cells between accumulate by chemotactic response. Together with the continuous penetration of growth zone into the bacterial layer, established attraction centers stop their activity, and new centers originate from the growth zone exactly at the midlines between the earlier ones. The time-space pattern of chemotactic activity can be preserved in the form of crossing lines of accumulated cells, if conditions are imposed to prevent diffusion of the cells

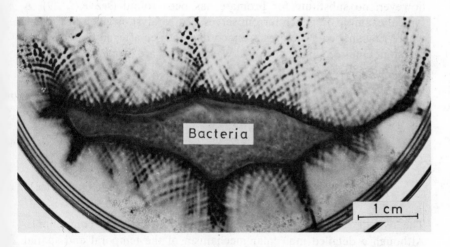

Figure 13. Pattern of cell accumulations in a mutant of *D. discoideum*. Dictyostelium cells are growing into an area occupied by bacteria. Behind the growth zone, cells are laid down in a cross-striated pattern as the result of chemotactic attraction by centers, which are lined up in a periodic array along the inner edge of the growth zone.

from the positions established at the end of chemotactic activity (Fig. 13). The regularity of the pattern indicates that the distance between attraction centers is rigidly controlled.

These examples of pattern formation illustrate what types of mechanisms can be visualized in the process of organization of even more complex systems, such as the supercellular phenomena studied by Wolpert et al.,[23] Gierer and Meinhardt,[24] and Goodwin and Cohen.[25]

V. CHEMICAL OSCILLATION AND PATTERN FORMATION

In 1958 Belovsov[26] described oscillations in a reaction system consisting of inorganic and organic compounds. Later, it was shown by Busse[27] and Zhabotinsky and Zaikin[28] that under suitable conditions the system displays spatial patterns. It was found that it is possible to generate a spatial time-dependent distribution of the system and to form spiral, ring-shaped, and more complex structures.

In an acidic medium malonic acid is not oxidized by bromate to a measurable degree at room temperature. However, in the presence of cerium ions, serving as catalyst, periodic concentration changes of the yellow ceric ions can be observed. The reaction can be recorded by measurements of light absorption or redox potentials. Malonic acid, cerium ions, and sulphuric acid may be replaced by other compounds; however, no substitute for bromate has been found (see Ref. 29). A typical example, in which ferroin serves as a catalyst, is given in Fig. 14, where ring-shaped structures are recorded 150 sec after initiation of the reaction. One recognizes chemical waves, which are propagated at a rate of 1.3 mm/30 sec.

On the basis of an investigation of the reaction components, the following overall reaction scheme has been formulated if cerium ions are used as a catalyst [31-33]:

$$5H_2C(COOH)_2 + 4BrO_3^- + 4H^+$$

$$\xrightarrow{\text{Ce}^{IV}/\text{Ce}^{III}} 2BrHC(COOH)_2 + Br_2HCCOOH + 8H_2O + 7CO_2 \quad (5.1)$$

The overall reaction drives the oscillation of the redox components. Although a detailed molecular mechanism of the temporal and spatial oscillations is not available, on the basis of the studies of Zhabotinsky and Zaikin[28] and of Noyes,[34] various reaction mechanisms have been discussed. A critical nonlinear step in the reaction system is suggested to

Figure 14. Waves of chemical activity of the malonic acid–bromate system. Ferroin was used as indicator and catalyst, and triton X-100 as a spreading agent. Conditions were used similar to the one described by Winfree (Ref. 30).

be the autocatalytic reaction in which cerous ions react with bromate ions and malonic acid, producing bromomalonic acid and ceric ions. This reaction may be inhibited by bromide ions. Also, ion complexes have been considered as intermediates. Recently, a qualitative explanation of propagation of the spatial bands has been discussed by Field and Noyes.[35]

We have asked the question whether diffusion-coupled oscillatory systems in general can be used to transmit information over a longer distance. Using the bromate–malonic acid system, we indeed have been

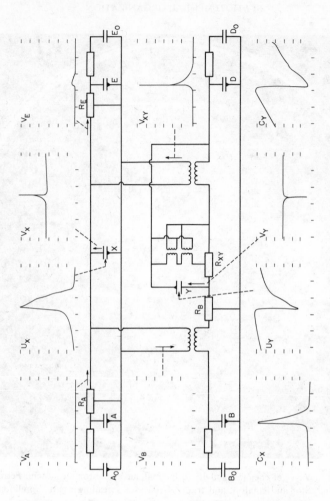

Figure 15. Network of the chemical reaction scheme: The elements of the reactions are labeled according to the equation set (5.2). A, B, D, E, X, and Y are chemical compounds, represented as charges on the respective chemical-energy-storing capacitors. A_0, B_0, D_0, and E_0 are chemical compounds serving as pools for the maintenance of constant concentrations of A, B, D, and E. The diagrams illustrate the chemical dynamics of the scheme. The oscillation proceeds by charging the capacitor Y by a current through R_B up to a threshold value. Then reaction R_{XY} fires and discharges the capacitor Y. v stands for the rate of reaction, and its subscript denotes the reaction compound it refers to. c_X and c_Y designate the variations in the concentration of X and Y, respectively. U_X and U_Y are the normalized chemical potentials of X and Y, respectively. All diagrams are drawn with the same time scale (full scale: 12 units). The scaling unit starting from zero for v_A, v_B, v_E, and v_{XY} is 35. The scaling unit for v_X and v_Y is 52.5 for the full scale, with zero being marked on the diagram. Most of the time v_X and v_Y are close to zero. The scale unit for c_X and c_Y is 2.5, starting from zero. The scaling unit for U_X and U_Y is 1.0. (Ref.37)

158

able to demonstrate a signal transmission experimentally. Under conditions far from equilibrium, neglecting reverse reactions, a simple model for such a system is (see also the chapter by Nicolis, p. 29).

$$\begin{array}{ll} A \rightarrow X & R_A \\ Y + 2X \rightarrow 3X & R_{XY} \\ B + X \rightarrow Y + D & R_B \\ X \rightarrow E & R_E \end{array} \qquad (5.2)$$

The corresponding network of this system is given in Fig. 15, which is derived from the chemical scheme as given by Hess et al.[36] For the set $A = 1$, $B = 5$, lying within the region where the system moves on the limit cycle, the time dependence of some variables is illustrated in the figure. The oscillatory behavior of this system is the one observed in case of a relaxation oscillator. A concentration (charge) is accumulated up to its threshold value; then the capacitor is discharged and the process starts anew.

The limit cycle of the essential components (X and Y) is given in Fig. 16 (following Busse and Hess[37]), showing the form of a triangle with three different phases. The motion on the limit cycle can be described by the following equations:

$$\frac{dX}{dt} = A + X^2 Y - (B+1)X \qquad (5.3a)$$

$$\frac{dY}{dt} = BX - X^2 Y \qquad (5.3b)$$

where the rate constants are choosen equal to unity. If X remains small and constant over the time of interest, the term in X^2 as well as dX/dt may be neglected, and the equation may be approximated by

$$A \approx (B+1)X \quad \text{and} \quad \frac{dY}{dt} = BX$$

Under these conditions the system moves in portion I of the limit cycle diagram, and the accumulation of Y determines the motion on the cycle. As soon as enough Y is accumulated, reaction R_{XY} is activated and dominates because of its autocatalytic character, corresponding to portion II of the limit cycle diagram. This process autocatalytically

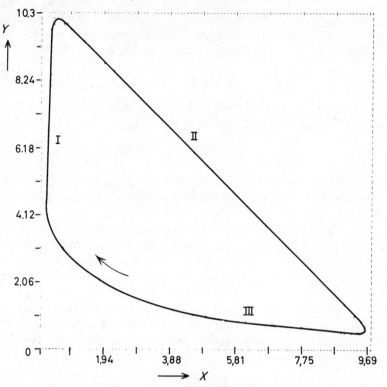

Figure 16.   Limit cycle of the reaction scheme of Fig. 15. The scaling of the concentration of X and Y is given in dimensionless units.

converts Y into X. The transition from portion I to portion II is reached if, in equation $A \approx [-XY + (B+1)]X$, the concentration of Y has grown to an extent that influences the steady state. In portion III of the limit cycle, Y is small, and X disappears at the rate $(B+1)X$ until the state of portion I is reached again.

The occurrence of diffusion can be treated by an extension of the network with an additional resistive element describing the linear relation between the diffusion flux I_D and its concentration gradient:

$$I_D = D \frac{\Delta c}{\Delta x} = D \frac{c^*}{\Delta x} (e^{U_{out}} - e^{U_{in}})$$

Here c and c^* are the concentration and the reference concentration, respectively, x is a space coordinate, D is the diffusion constant, and U is the normalized chemical potential.[28]

For small concentration differences, when $U_{in} \approx U_{out}$, rearrangement leads to

$$I_D = 2\frac{Dc^*}{\Delta x}\exp\left(\frac{U_{in}+U_{out}}{2}\right)\sinh\left(\frac{U_{out}-U_{in}}{2}\right)$$

which, on expansion of the sinh in a series of terms in $\Delta u = U_{out} - U_{in}$, leads to the first approximation

$$I_D \approx \frac{D\bar{c}}{\Delta x}\Delta u$$

where $\bar{c}$ is the geometric mean of the concentration at the ports. This constitutive relation can be used in order to approximate a continuum satisfactorily by a lumped network of resistive elements representing diffusing processes.[38]

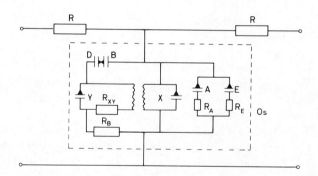

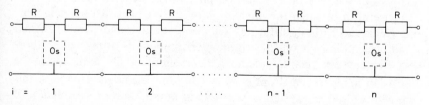

Figure 17.   The upper network is a symmetrical **T**-type four-terminal network comprising the two resistors originating from the diffusion process and the cross-linking network enclosed in the dashed box, labelled Os. The Os network is a rearranged drawing of the oscillatory network as shown in Fig. 15. In the lower graph a series of these two-ports are arranged in a chain in order to form a transmission line, composed of equal oscillatory units.

Under the assumption that except for X all other diffusion processes are negligible for a volume element in space, the two-port network shown in Fig. 17 (upper graph) is obtained. This two-port is constructed in the following way: The network given in Fig. 15 is rearranged; furthermore, the diffusive resistors R are added. Finally, all connections not carrying a net current are omitted, and the pools are eliminated. A chain of such a one-dimensional two-ports (see Fig. 17, lower graph) represents a transmission line, composed of symmetrical T-type two-ports. Since any element of the two-ports is passive, the whole two-port is also passive.

In order to analyze the properties of the transmission line, the diffusion process is included in the system of equations (5.3), giving

$$\frac{dX_i}{dt} = A + X_i^2 Y_i - (B+1)X_i + q_x(X_{i+1} + X_{i-1} - 2X_i) \quad (5.4a)$$

$$\frac{dY_i}{dt} = BX_i - X_i^2 Y_i \quad (5.4b)$$

where $q_x = D_x/\Delta x$; here D_x is the diffusion coefficient of the compound X, and Δx the approximate length of the line represented by this element. Subscript i denotes the number of the oscillatory element in the chain (see Fig. 17, lower graph).

At the boundaries $i = 1$ and $i = n$, the concentrations X_0 and X_{n+1} are zero. In the limit of an infinite number of oscillatory elements in the line, the differential equation system (5.4) approaches the partial differential equations known from the diffusion laws. The system of difference and differential equations is solved on a computer. For an analysis of the type of transmission phenomena occurring on such lines it is useful to consider the interaction of one element with its neighboring elements. Without diffusion each element moves independently on a limit cycle. Since the outflow (inflow) of compounds from the neighboring element can retard (enhance) the motion on the limit cycle, the diffusion helps to synchronize the neighboring elements. However, this happens only if the system is sensitive to an addition or subtraction of the chemical compound.

Thus there may be parts of the limit cycle where the motion is nearly independent of the state of the neighboring elements, and other parts where the motion is strongly dependent on them. Furthermore, during the sensitive part the diffusion leads to an equilibration of the phase differences between neighboring limit cycles.

In the reaction scheme (5.2) the sensitive part of the limit cycle is exemplified at the end of portion I in Fig. 16. If, in this portion of the limit cycle, compound X is added to an oscillatory element, the autocatalytic process (portion II) starts considerably earlier than without that addition. Therefore, all elements that show a high Y and a low X concentration will immediately start their autocatalytic process when the neighboring element has a sufficiently higher concentration in X, which means that as soon as an element moves into the sensitive part of its limit cycle, the interaction with its neighbor determines its future motion. If one of its neighboring elements has a higher X concentration, the given element will start its autocatalytic process and soon also have a high X concentration. Thus, in a chain of oscillatory elements all having a high Y and a low X concentration, the element that first starts, its autocatalytic process will force its neighbor to follow. If in such a chain the autocatalytic process of the first element is externally triggered, then step by step, the following elements increase their concentration in X, and a triggered pulse moves along this chain (or line) of elements.

In order to achieve an optimized transmission, the concentration of Y should be maintained for a longer time below its critical value. Thus the system stays in the sensitive portion of the limit cycle for a long time, and the pulse might travel a long distance before it reaches the limit where it interferes with the natural frequency of the oscillatory elements.

There are several ways to realize the condition of a sensitive portion of long duration. Here, a chemical trimerization process for $Y(3Y \rightarrow Z)$ is added, which competes with the production of Y, especially in the range close to the threshold value. This competition is strongly dependent on the rate constant for the reaction. The general behavior of the system remains nearly unchanged by this additional reaction. The dynamics of the oscillatory time dependence as given in Fig. 15 are almost identical with the dynamics given by this modified system, except that the part near the extreme value of Y is extended for a longer time. An analysis of a transmission line obtained by this modification is represented in Fig. 18. Three trigger pulses at different time intervals are given to the first oscillatory element in the line. As shown, the signals now run along the line with nearly constant velocity and are almost undistorted, which can be seen by comparing the states of the transmission line at different time intervals.

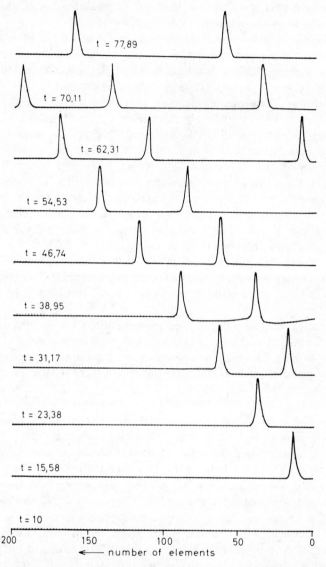

t = 77,89

t = 70,11

t = 62,31

t = 54,53

t = 46,74

t = 38,95

t = 31,17

t = 23,38

t = 15,58

t = 10

| 200 | 150 | 100 | 50 | 0 |

←—— number of elements

Figure 18. Computer output of a transmission line demonstrating the propagation of pulses by which information is transferred. For 200 elements of the line the concentration of X is plotted at some times t. The number of the elements is plotted from right to left. The curves are generated by a reaction scheme to which the reaction $3Y \rightarrow Z$ has been added. The data for the oscillatory circuit are those of Fig. 15. The rate constant for the additional reaction is choosen 8.992×10^{-4}. The coupling constant of the diffusion is $q_X = 0.35$. Here the first element is triggered by a series of three pulses with different time intervals.

164

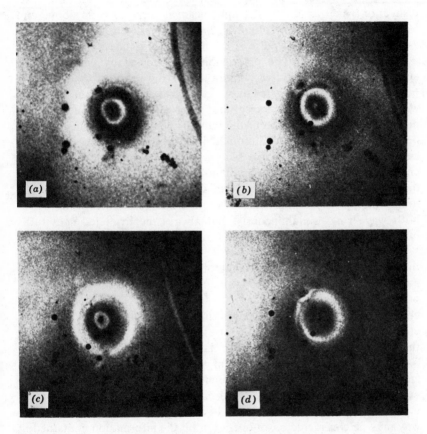

Figure 19. Propagation of a signal in an oscillatory chemical reaction system. The initial concentrations of the solutions are: $[KBrO_3] = 0.35$ M; $[H_2C(COOH)_2] = 1.2$ M; $[Ce(SO_4)]$ $= 3.9 \times 10^{-3}$ M; $[ferroin] = 0.48 \times 10^{-3}$ M in 3 N H_2SO_4. About 15 min after initiation of the reaction by addition of the catalyst, the ultraviolet light pulse is given, and shortly afterwards pictures are taken at intervals of 0.5 sec. The sequence of the pictures shown is from (a) to (d), in a clockwise direction. The white ring represents the blue outward-moving wave; the spot in its center was irradiated. For further explanation see text.

In order to demonstrate an experimental signal transmission, the oscillatory oxidation of malonic acid by bromate was perturbed by a pulse of ultraviolet light. A typical experiment is carried out in a Petri dish onto which an ultraviolet light beam is directed from above and focused to a diameter of 0.4 cm on the solution layer of approximately 2-mm thickness. A light pulse is given to the blue phase of the oscillation for 2 to 30 sec. The propagation is followed by pictures taken approximately 15 min after preparation of the solution. As shown in Fig. 19, the wave propagates outward from the center of the initiation with a nearly constant velocity until after about 1 to 2 cm the wave disappears in the oscillation of the bulk of the solution. During the next period again, at the spot illuminated before, the solution turns blue before the bulk solution turns blue, and again a wave starts from this position, disappearing in the oscillation of the bulk solution. In Fig. 19 this following wave is already forming, whereas the first one has not totally disappeared in the bulk solution. Other conditions do not show both wave rings simultaneously, but always the irridiated spot turns blue ahead of the bulk solution to form a new wave center.[37]

The common qualitative property in the reaction scheme treated in the theoretical approach and that of the experimental system is the motion on a quasistable limit cycle. This cycle must have at least one portion that can be perturbed by superimposed diffusional flow of chemical substances. Moreover, a linear one-dimensional motion is computed, whereas the experiments show moving rings. Nevertheless, the common feature required to initialize a propagating wave remains.

This experiment and computation may be considered only as a first step toward a realization of a chemical transmission line. Of course, other types of triggering in the solution could be used as an information input. Furthermore, a combination of two or more signals in a specified way may be considered as an operation on the information contained in the signals, and such a combination may lead to a network made of "chemical wires."

VI. OUTLOOK

Phenomena of spatiotemporal organization can clearly be observed in soluble, membrane-bound, and cellular systems occurring widely in living nature. The pure chemical system of malonic acid and bromate might even be an example of phenomena taking part in prebiotic evolution. None of these structures generated in space and time develop under near-equilibrium conditions. They are bound to occur far off equilibrium in open systems. Also, they have the potential to serve for signal transmission, as observed in the case of the malonic acid–bromate system or in the case of intercellular or neural communication. All these states are the result of energy dissipation in time. The elucidation of their molecular structure and kinetic properties is a challenge for future research.

References

1. P. Glansdorff and I. Prigogine, *Thermodynamic Theory of Structure, Stability and Fluctuations*, Wiley-Interscience, New York, 1971.
2. I. Prigogine and G. Nicolis, *Quart. Rev. Biophys.*, **4**, 107 (1971).
3. B. Hess and A. Boiteux, *Annual Rev. Biochem.*, **40**, 237 (1971).
4. B. Hess and A. Boiteux, Control of Glycolysis, in *Regulatory Functions of Biological Membranes*, Johan Järnefelt, Ed., *Biochim. Biophys. Acta* Library, Vol. 11, 1968, p. 148.
5. A. Boiteux and B. Hess, Control Mechanism of Glycolytic Oscillations, in *Biological and Biochemical Oscillators*, B. Chance, K. Pye, A. Gosh, and B. Hess Eds., Academic Press, New York and London, 1973, p. 243.
6. B. Hess, A. Boiteux, and J. Krüger, Cooperation of glycolytic enzymes, in *Enzyme Regul.* 7, 149 (1969) Pergamon Press, London.
7. B. Hess, H. Kleinhans, and D. Kuschmitz, Component structure of glycolytic oscillations, in *Biological and Biochemical Oscillators*, B. Chance, K. Pye, A. Gosh, and B. Hess, Eds., Academic Press, New York and London 1973, p. 253.
8. A. Goldbeter and R. Lefever, *Biophys. J.*, **12**, 1302 (1972); A. Goldbeter, *Proc. Nat. Acad. Sci.* (USA) **70** (1973).
9. B. Hess and A. Boiteux, Substrate control of glycolytic oscillations, in *Biological and Biochemical Oscillators*, B. Chance, K. Pye, A. Gosh, and B. Hess, Eds., Academic Press, New York and London, 1973 p. 229.
10. K. J. Johannes and B. Hess, *J. Mol. Biol.*, **76**, 181 (1973).
11. E. E. Selkov, *Eur. J. Biochem.*, **4**, 79 (1968).
12. B. Hess, Th. Plesser, and V. Schwarzmann, private communication.
13. A. Boiteux, H. Degn, and B. Chance, Oscillating Respiration in Mitochondria and Control, in preparation.
14. A. Boiteux and H. Degn, Hoppe Seyler's *Z. Physiol. Chem.* **353**, 696 (1972).
15. A. Boiteux, *Ergeb. d. exp. Med.* **9**, 347 (1973).
16. G. Gerisch, in *Current Topics in Developmental Biology*, Vol. 3, Hsg. A. Monroy and A. A. Moscona, Eds., 1968, p. 157, Academic Press, New York and London.
17. G. Gerisch, *Naturwiss.*, **58**, 430 (1971).
18. T. M. Konijn, J. G. C. van de Meene, Y. Y. Chang, D. S. Barkley, and J. T. Bonner, *J. Bacteriol.*, **99**, 510 (1969).
19. A. Arndt, *Roux Archiv Entw.-Mech. Org.*, **136**, 681 (1937).
20. B. M. Shaffer, in *Advances in Morphogenesis*, Vol. 2, 1962, p. 109, Academic Press, New York and London.
21. M. H. Cohen and A. Robertson, *J. Theor. Biol.*, **31**, 101 (1971).
22. G. Gerisch and B. Hess, *Proc. Nat. Acad. Sci.* (USA), **71**, 2118 (1974).
23. L. Wolpert, J. Hicklin, and A. Hornbruch, *Symp. Soc. exp. Biol.*, **25**, 391 (1971).
24. A. Gierer and H. Meinhardt, *Kybernetik*, **12**, 30 (1972).
25. B. C. Goodwin and H. M. Cohen, *J. Theor. Biol.*, **25**, 49 (1969).
26. B. P. Belousov, *Sb. Reg. Radiats. Med. za 1958*, Medgiz, Moscow, 1959.

27. H. Busse, *J. Phys. Chem.*, **73**, 750 (1969).
28. A. M. Zhabotinsky and A. N. Zaikin, in *Oscillatory Processes in Biological and Chemical Systems*, Vol. II, E. E. Selkov, A. M. Zhabotinsky and S. E. Snoll, Eds., Puschino on Oka, 1971, p. 279.
29. A. M. Zhabotinsky, in *Symposium on Oscillatory Processes in Biology and Chemistry*, Academic, New York, Moscow, 1967, p. 252
30. A. T. Winfree, *Science*, **175**, 634 (1972).
31. L. Bornmann, H. Busse, and B. Hess, *Z. f. Naturforschung*, **28b**, 93 (1973).
32. L. Bornmann, H. Busse, B. Hess, R. Riepe, and C. Hesse, *Z. f. Naturforschung*, **28b**, 824 (1973).
33. L. Bornmann, H. Busse, and B. Hess, *Naturforschung*, **28c**, 514 (1973).
34. R. M. Noyes, R. J. Field, and E. Körös, *J. Am. Chem. Soc.*, **95**, 1394 (1972).
35. R. J. Field and R. M. Noyes, private communication.
36. B. Hess, E. M. Chance, H. Busse, and B. Wurster, Simulation of Enzymes, in *Analysis and Simulation of Biochemical Systems, FEBS 8th Meeting*, Vol. 25, H. C. Hemker and B. Hess, Eds., North-Holland/American Elsevier, Amsterdam, 1972, p. 119.
37. H. Busse and B. Hess, *Nature*, **244**, 203 (1973).
38. G. F. Oster, A. S. Perelson, and A. Katchalsky, *Quart. Rev. Biophys.* **6**, 1 (1973).

THEORETICAL MODELS FOR
BACTERIAL MOTION AND
CHEMOTAXIS

JEAN-PIERRE BOON*

Faculté des Sciences, Université Libre de Bruxelles,
1050 Bruxelles, Belgium

CONTENTS

Abstract

The motion of motile living cells is a complex phenomenon that arises as the consequence of a cascade mechanism initiated by the presence of an attractant in the medium. Chemotaxis leads to motility processes at the microscopic level of individual cells and to migration phenomena at the macroscopic level of large numbers of microorganisms. The present review is essentially devoted to the latter phenomena, which have attracted the most attention from the theoretical viewpoint. Those experiments for which theoretical analyses have been proposed are briefly reviewed. The corresponding models are reviewed and discussed. A succinct literature survey is presented in which the bibliographical references are classified according to the different aspects of the subject.

*Chercheur qualifié au Fonds National de la Recherche Scientifique (F.N.R.S), Belgium.

I. INTRODUCTION

The movements of living microorganisms (e.g., bacteria or sperma-tozoa) are quite different—qualitatively and quantitatively—from the motion of dead cells, which exhibit merely the dynamical characteristics of a system of Brownian particles. We are therefore led to distinguish between *mobility* and *motility*: whereas Brownian particles are mobile, living cells are said to be motile, when they are able of autonomous displacements. Motility arises as the result of a mechanism that can be summarized as depicted in Fig. 1. The existence of a sequence of such processes in living systems as simple as a bacterium renders the latter particularly appropriate for the study of chemotaxis and of chemorecep-tion. Several authors have pointed out the importance of investigating these phenomena not only per se, but also in view of the analogy between the chemotactic response of the bacteria and the mechanism of sensorial chemoreception in higher organisms (cf. the Weber-Fechner law). As viewed according to the scheme given in Fig. 1, the sequence *stimulus→response* is initiated by the presence of some *attractant* in the medium and gives rise to specific displacements of the microorganism. The successive processes leading to motility seem to take place at the membrane level, but one of the steps in the sequence, the "recognition" of the attractant by the binding protein, plays a key role in both *chemoreception* and *transport*. However, it is not the purpose of the present chapter to analyze the biochemical aspects of these different steps (most of which are still unexplained to date), but rather to review the theoretical studies in which an attempt has been made to describe the resulting effects of the chemotactic mechanism. Here a distinction should be made between the phenomena studied at the microscopic level and those observed macroscopically: (*1*) *Motility* is meant to designate the microscopic dynamics of the living cell responding to an external stimulus. (*2*) Because of the large number of cells in an actual sample (typically $\gtrsim 10^6/\text{cm}^3$), collective (or cooperative) effects are observed on a macroscopic scale; the bacterial population travels in bunches, which phenomenon will be called *migration*, to designate the collective motion.

We first briefly review some of the most striking experimental obser-vations of bacterial movements. Restriction is made to those experi-ments for which theoretical models have been proposed. Mobility and migration phenomena are then discussed, with emphasis on the collec-tive motion effects, which have attracted the most attention from the theoretical viewpoint. The last section is devoted to a brief literature survey: This (far from exhaustive) bibliography divides the references

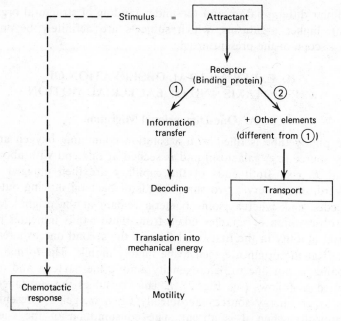

Figure 1. Schematic representation of the stimulus→response mechanism, initiated by the presence of an attractant in the medium and leading to bacterial motility. The central position of the receptor indicates the key role of the binding protein in both chemotaxis and transport.

into categories according to the different aspects of the subject (whether or not treated in this chapter).

There is no doubt that the problems of bacterial motion and chemotaxis exhibit strong analogies with similar problems encountered in other microorganisms, in particular spermatozoa on the one hand and slime-mold amoebae on the other. However, the complexity of the dynamics of the spermatozoon, as well as the scarce quantitative data on spermatozoa motility, renders theoretical analysis even more difficult and more speculative than for bacterial systems. We shall restrict ourselves to a couple of references on the dynamics of spermatozoa given in the bibliographical section.

Slime-mold amoebae have received wide attention in recent years, and a number of theoretical studies have been devoted in particular to the life cycle of the species *Dictyostelium discoideum*. Here great effort has been focused on the analysis of aggregation phenomena, which are discussed in the chapter by Hess et al. in the present volume (p. 137). These investiagtions may be considered as a first step towards the study of cellular differentiation and the understanding of structural organization in higher organisms, which subjects are definitely beyond the limited scope of the present article.

II. EXPERIMENTAL OBSERVATION OF CHEMOTAXIS AND OF BACTERIAL MOTION

A. One-Dimensional Migration

A capillary tube is filled with a solution containing oxygen and an energy source (e.g. galactose) and is seeded at one end with about 10^6 cells of *E. coli*. Both ends of the capillary are then plugged. Soon afterward, one observes two sharp bands of bacteria moving out from the seeded end, whereas some bacteria remain at the origin.[1] Microscope observation of samples taken from these bands indicates higher bacterial motility in the first band than in the second one, whereas the cells left at the origin are seen to be merely mobile. The formation of two bands is not due to heterogeneity among the bacteria and can be explained as follows (see Fig. 2). *E. coli* consumes the substance that serves as an energy source—aerobically when oxygen is present, and anaerobically when it is absent. The consumption of the chemical creates a concentration gradient, and the bacteria move preferentially in the direction of higher concentrations. A first band appears, which consumes all the oxygen to oxidize the energy source and is followed by a second band, which uses the residual sugar anaerobically, when the latter is in excess over the oxygen. Steric hindrance could be held responsible for the remaining at the origin of a certain number of bacteria. The studies reported by Adler and co-workers have demonstrated that *E. coli* is chemotactic toward oxygen and energy sources such as sugars (galactose, glucose, ribose, fructose, maltose) and amino acids (serine, aspartate, etc.).[4,5] It is important to mention that it has been proved experimentally that functional metabolism is not a necessary factor for chemotaxis.[3]

B. Radial Migration

Bacteria are deposited at the center of an agar plate containing an energy source. Consider in practice the case where the experiment is performed on a low-agar-concentration plate (so that the bacteria can swim easily), containing the same medium as used in the capillary-tube experiment described in Section II.A. One observes that the bacteria migrate in the form of a ring centered at the point of deposit.[1] The ring diameter grows in time, and the radially migrating band goes

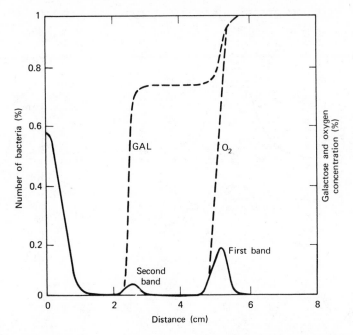

Figure 2. Migration of *E. coli* bands shown as the population fraction (solid curves) versus the distance traveled after 6 hr of incubation. The corresponding consumption of oxygen and of the energy source (GAL) is indicated by the dashed curves. (After Adler[1]).

throughout the depth of the agar. In the presence of one energy source alone (e.g., galactose), only one ring forms where the bacteria consume all the attractant. It is observed that increasing the concentration of the energy source induces a decrease in the migration velocity. Since in the present experiment oxygen always remains in excess over the energy source, the experimental results can be represented essentially as in Fig. 2, with the horizontal axis now denoting the ring diameter, and omitting the second band peak and the oxygen concentration curve.

C. One-Dimensional Migration in the Presence of an Attractant Concentration Gradient

A bacterial population is suspended in a vertical tube containing a physiological solution where the cells swim easily. A vertical gradient of attractant (e.g. serine) is superimposed on the initially uniform bacterial distribution. After a short time (a few minutes) one observes a redistribution of the bacteria (*Salmonella typhimurium*).[6] The bacteria tend to

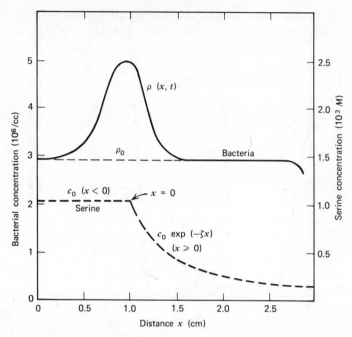

Figure 3.   Representation of the response of Samonella typhimurium to an exponential gradient of serine, as obtained after 15 min, starting initially with a uniform bacterial distribution (After Dahlquist, Lovely, and Koshland[6]).

swim from the region of lower to the region of higher attractant concentration. An example is shown in Fig. 3 for the case of an exponential gradient of serine: A peak in the bacterial distribution appears, which is centered about the point of steepest concentration change. The peak area and the peak intensity grow steadily in time at the expense of the low-attractant-concentration region.

D. Motility Measurement by Path Tracing

Two methods are in use for observing the motility of microorganisms by direct tracing of their displacements. The first technique consists in microscope dark-field observation by stroboscopic photomicrography such that with a suitable choice of flashing rate successive pictures are generated to yield the tracks of the bacterial displacements. This method has been used in particular by McNab and Koshland[7] to investigate the gradient-sensing mechanism in bacterial chemotaxis. In these experiments a bacterial population is subjected to a sudden temporal change in the attractant gradient, ∇c. It is found that when $\nabla c = 0$ (uniform attractant concentration), bacteria often show "wobble" as they travel. When $\nabla c > 0$, longer runs are observed, that is, the bacteria exhibit smooth linear trajectories (supercoordinated swimming); whereas for $\nabla c < 0$, MacNab and Koshland find a significant increase of tumbling (poor coordination manifesting itself as frequent erratic changes in direction).

Another method has recently been developed which analyzes three-dimensional tracking by means of a microscope with a device that follows individual cells. Cartesian projections are obtained in this fashion that allow one to reconstruct the three-dimensional motion of the microorganism (see Berg and Brown[8]). The prime object of such studies has also been to investigate the chemotactic response as a function of the changes in the chemical environment. Some conclusions could be drawn as to the modifications of the bacterial motion in response to changes in the attractant concentration. According to Berg and Brown, in the absence of a chemical gradient, typical motility tracks exhibit an alternating sequence in the motion of the bacteria which consists of a running phase (essentially translational motion) followed by a twiddling phase (nontranslational displacements) in a temporal proportion of 15 to 85% for the twiddle to the run. In the presence of an attractant gradient this ratio tends to decrease, as that component of the trajectory that is directed along $\nabla c > 0$ becomes elongated as compared to its mean value in the absence of a gradient, whereas the components that have no positive contribution in the ∇c direction remain essentially unchanged.

Despite the lack of accurate quantitative results, one conclusion can be drawn from the above investigations: The twiddling in the bacterial motion should appear as a crucial phase of the sequence. It now remains an open question to elucidate the following alternative: Either the twiddling appears to be a purely random process, in which case the dynamics of motion is determined during the runs, or the twiddling reflects the possible existence of a search phase during which information is collected and used by the bacterium to determine its subsequent dynamical behavior for the next temporal sequence.

E. Motility Measurement by Light Scattering

The development of laser light scattering techniques during the past decade has proved to be of great interest for the investigation of dynamical properties of physical, chemical, and biological systems. In particular, when the scatterers consist of particles with dimensions of the order of the wavelength of the light (a few micrometers), laser light scattering provides direct information about single-particle motion, and appears therefore as an appropriate tool to study the movement of microorganisms like viruses, bacteria, spermatozoa, and so on.[9, 10, 13] The method consists essentially in shining a laser beam on a suspension of microorganisms, which scatter light in all directions. The analysis of the intensity of the light scattered and detected at a given (variable) angle provides the velocity autocorrelation function of the scatterers, which function constitutes essentially the solution of the equation of motion of these scatterers.

One of the virtues of the method lies in the fact that the information obtained from the scattered intensity yields a significant statistical measure of the dynamical phenomenon investigated. Because an attractant in the medium will play the role of an external force, the presence of a concentration gradient will modify the equations of motion and hence the autocorrelation function of the directed velocity of the bacteria. As a consequence the spectrum of the scattered light will depend on the factors determining the changes in the chemical environment and will be modified accordingly. The spectral analysis thus provides qualitative and quantitative information about such effects as the influence of an attractant gradient, as well as the "fine structure" of the bacterial motion (e.g., the nontranslational components of the velocity or the twiddling, which are very difficult to analyze by means of the other techniques). Although the use of light scattering has not yet revealed dramatic new features in the study of bacterial motion, this method appears to be quite promising and efficient for the reasons mentioned above. This will be briefly discussed in the last section.

III. THEORETICAL MODELS

A. Preliminary Remarks

The theoretical analyses of bacterial motion start essentially from a diffusionlike equation where the current includes a term describing the directed motion arising from chemotactic response. In the light of the available experimental observations, it appears that the explicit expression for the chemotactic current depends crucially on the type of situation discussed. This point needs to be stressed, as there seems to be some confusion in the literature regarding this essential fact: It should be realized that (1) there is no general analytical form of the chemotactic current; (2) the analytical form is guessed from what appears to date as experimental evidence, if any. We shall therefore first describe the general—but nonexplicit—formulation of the transport equation

that serves as the starting point, and then consider the different cases
that correspond to the experimental situations described in the previous
section.

B. Transport Equation for Chemotactic Motion

The temporal evolution of the spatial distribution of the bacteria can
be described by the continuity equation

$$\partial_t \rho(\mathbf{r}, t) + \nabla \cdot \mathbf{J}(\mathbf{r}, t) = G(\mathbf{r}, t) \tag{3.1}$$

with $\rho(\mathbf{r}, t)$ the bacterial density and $\mathbf{J}(\mathbf{r}, t)$ the current, and where $G(\mathbf{r}, t)$
accounts for the growth of the bacterial population. Equation (3.1)
merely describes the conservation of matter, and $\mathbf{J}(\mathbf{r}, t)$ consists of two
terms,

$$\mathbf{J}(\mathbf{r}, t) = \mathbf{J}_D(\mathbf{r}, t) + \mathbf{J}_C(\mathbf{r}, t) \tag{3.2}$$

Here $\mathbf{J}_D(\mathbf{r}, t)$ is a diffusionlike term pertaining to isotropic mobility,

$$\mathbf{J}_D(\mathbf{r}, t) = -\mu \nabla \rho(\mathbf{r}, t) \tag{3.3}$$

with μ the diffusion (or mobility) coefficient of the cells, which depends
on the physical properties of the solution; and $\mathbf{J}_C(\mathbf{r}, t)$ describes the
directed motion due to chemotaxis,

$$\mathbf{J}_C(\mathbf{r}, t) = \mathbf{v}_C(\mathbf{r}, t)\rho(\mathbf{r}, t) \tag{3.4}$$

where the chemotactic velocity $\mathbf{v}_C(\mathbf{r}, t)$ depends on the chemoattractant
concentration $c(\mathbf{r}, t)$. It can be shown, for instance, that a modified
Langevin equation for the directed (or chemotactic) velocity yields an
expression for $\mathbf{v}_C(\mathbf{r}, t)$ in terms of the *chemotactic force* $\mathbf{F}_c(\mathbf{r}, t)$, which
depends on the attractant concentration.[14] To a good approximation,
one has

$$\mathbf{v}_C(\mathbf{r}, t) = \eta^{-1} \mathbf{F}_C(\mathbf{r}, t) \tag{3.5}$$

where η is the friction coefficient related to viscous drag experienced by
the bacteria as they move through the medium. Equation (3.1) can now
be rewritten as

$$\partial_t \rho(\mathbf{r}, t) = \nabla \cdot \{ \mu \nabla \rho(\mathbf{r}, t) - \eta^{-1}[\rho(\mathbf{r}, t)\mathbf{F}_c(\mathbf{r}, t)] \} + G(\mathbf{r}, t) \tag{3.6}$$

where the explicit dependence of the chemotactic force on the attractant
concentration still remains unspecified. Once $\mathbf{F}_C(\mathbf{r}, t)$ is expressed in
terms of $c(\mathbf{r}, t)$, Equation (3.6) has to be supplemented by an equation

governing the degradation of the attractant. It is clear that an explicit expression for $G(\mathbf{r}, t)$ is also required to solve Equation (3.6).

C. One-Dimensional Step Model (Keller and Segel)

The chemotactic response of the cell can be viewed as analogous to Brownian motion, but here the motion of the cell is influenced by the presence of the attractant through chemical interaction in addition to the direct-impact characteristic of Brownian collisions.[16] The basic idea of the step model consists in a "pseudostochastic walk": the microorganisms can take steps of length Δ to the left or to the right, and the average frequency of steps, $f(c)$, depends on the average attractant concentration at the propelling edge of the bacterium, which is represented by a one-dimensional rod of length d oriented along the x-axis. The current $J(x, t)$ is then approximated in the same fashion as in theoretical studies of Brownian motion to yield

$$J(x,t) = \Delta^2[-f(c)\partial_x\rho(x,t) + (\alpha - 1)\rho(x,t)\partial_c f(c)\partial_x c(x,t)] \quad (3.7)$$

where $c(x, t)$ indicates the space-time dependence of the attractant concentration and $\alpha = d/\Delta$. Equation (3.7) can be compared with the general expression as given in Section III.B when reduced to one-dimensional space. One obtains

$$J(x,t) = -\mu\partial_x\rho(x,t) + \chi\rho(x,t)\partial_x c(x,t) \quad (3.8)$$

where the mobility coefficient is given by

$$\mu = \Delta^2 f(c) \quad (3.9)$$

and the chemotactic coefficient is defined by

$$\chi = (\alpha - 1)\Delta^2\partial_c f(c) = (\alpha - 1)\partial_c \mu(c) \quad (3.10)$$

One thus sees from Equation (3.4), (3.5), and (3.8) that

$$F_C(x,t) = \gamma\partial_x c(x,t) \quad (3.11)$$

where $\gamma = \eta\chi$ can be interpreted as the linear response of the bacteria to the chemotactic force. Equation (3.11) with $\gamma = \text{const}$ seems to hold in that range of situations where there is no *externally imposed* gradient of concentration, and where such a gradient is created by the consumption of the attractant by the bacteria as they move. Actually it seems logical to interpret the phenomenon as follows: It is because the bacteria consume the attractant that an attractant gradient is created, which induces bacterial motion toward a region of higher concentration. (In particular, this is quite obvious when the medium is depleted in the attractant after the "passage" of the bacteria.)

In the absence of growth Equation (3.6) now becomes

$$\partial_t \rho(x,t) = \partial_x \left\{ \mu \partial_x \rho(x,t) - \chi[\rho(x,t)\partial_x c(x,t)] \right\} \qquad (3.12)$$

It is now assumed that the attractant concentration is governed by the equation[17]

$$\partial_t c(x,t) = -k\rho(x,t) + D\partial_x^2 c(x,t) \qquad (3.13)$$

with k the consumption rate of the attractant by the bacteria, and D the diffusion coefficient of the attractant in the medium. The second term on the right-hand side of Equation (3.13) is the usual diffusion term, which can be ignored in general, as the limiting factor is the rate of depletion and not the availability of attractant through diffusion in the case of traveling bands or rings. (Note that the situation is quite different in the case of capillary assays as described by Adler.[3])

The following assumptions are now introduced:

1. The bacterial mobility is independent of the attractant concentration, that is, $\mu = $ const.

2. Since μ is usually large compared to the diffusion rate of a substance in the medium, D is set equal to zero for the reasons explained above.

3. Similarly, because bacteria are capable of consuming "up to the last molecule" of attractant at their disposal, k is taken to be constant in Equation (3.13).

4. With the hypothesis of the existence of a minimum value of c at which chemotaxis can take place, and on the basis of the argument that $\chi(c)$ must be sufficiently singular for Eqs. (3.12) and (3.13) to yield traveling bands as a solution,[17] χ is tentatively taken to be of the "least singular" form

$$\chi(c) = \frac{\delta}{c - c^*}, \qquad \delta = \text{const}$$

or

$$\chi(c) = \frac{\delta}{c} \qquad (3.14)$$

if c is reinterpreted as the difference between the observed concentration and the (presumably small) threshold value c^*.

Notice that the present assumption introduces an explicit change in the form of the chemotactic term in the expression for the current. Indeed, substitution of Equation (3.14) into Equation (3.8) yields

$$J_C(x,t) = \delta\rho(x,t)\partial_x \ln c(x,t) \qquad (3.15)$$

If this expression for J_C seems to hold for situations where the bacteria are subject to an exponential gradient (see Sections II.C and III.E), it remains questionable in the absence of an external concentration gradient, and the validity of assumption 4 should therefore be subject to caution.

With the above assumptions and with the appropriate initial-boundary conditions, namely,

$$c(x,0) = c_0 = \text{const}, \qquad \rho(x,0) = \rho_0 \qquad (3.16)$$

$$\partial_x c = \partial_x \rho = 0 \qquad \text{at} \quad x = 0 \quad \text{and} \quad x = L \to \infty \qquad (3.17)$$

(where L is the length of the capillary tube, which is long compared to the width of the band), Equation (3.12) and (3.13) can be solved, provided one considers solutions of the form[17]

$$\rho(x,t) = \rho(\xi), \qquad c(x,t) = c(\xi), \qquad \text{with} \quad \xi = x - v_B t \qquad (3.18)$$

Here v_B, the speed of the traveling band, is assumed to be constant, which represents quite a restriction, in that in doing so one "injects" a priori into the solution a given type of dynamical behavior. It will become clear in Section III.C that this assumption is consistent with the hypothesis of no growth, but represents only a particular case (not very commonly encountered) in actual experimental situations. One then obtains

$$c(\xi) = c_0 \left[1 + \exp\left(-\frac{v_B \xi}{\mu} \right) \right]^{-\mu/(\delta-\mu)} \qquad (3.19)$$

and

$$\rho(\xi) = \rho_0 \left(\frac{c(\xi)}{c_0} \right)^{\delta/\mu} \exp\left(-\frac{v_B \xi}{\mu} \right) \qquad (3.20)$$

with

$$\rho_0 = \frac{c_0 v_B^2}{k(\delta - \mu)} \qquad (3.21)$$

and

$$v_B = \frac{kN}{c_0} \qquad (3.22)$$

where N is the total number of bacteria in the band.

The primary purpose of the Keller-Segel model consists essentially in setting up a theoretical framework for the description of chemotactic bands. Furthermore, some numerical computation is possible from the final results, Equations (3.19) to (3.22). For instance, evaluation of v_B from Equation (3.22) shows semiquantitative agreement with experimental data.[2] Similarly, the width of the traveling band can be evaluated from Equations (3.19) and (3.20), as well as the change in the "wings" of the bacterial distribution as a function of the ratio δ/μ. One finds that the smaller the ratio, the sharper the band, and that $\rho(\xi)$ decreases more rapidly toward the front of the band than toward the rear when $\delta/\mu > 2$, whereas the opposite occurs for $\delta/\mu < 2$. The final results, Equations (3.19) to (3.22), of course depend strongly on the basic assumptions 1 to 4 discussed above. Among the latter, assumption 4 is certainly the most important, being largely responsible for the analytical form of the final results. Such an assumption, in accordance with the Weber-Fechner law, implies that the chemotactic flux is proportional to the relative concentration gradient, that is, $v_c \propto \nabla c/c$. It needs further justification. Attention should also be drawn to the hypothesis that no growth is taking place during the displacement of the traveling band. Finally, the analyses show that, given the assumptions, steadily traveling bands appear only when δ exceeds μ, that is, when the ordering of motion caused by the attractant gradient is sufficient to outweigh the disordering diffusion effect due to random motion. In other words, the organizational tendency of a collection of cells arises as a consequence of symmetry breaking of the random distribution under the influence of an external agent. This observation suggests the possibility of an interesting connection with the concept of dissipative structures discussed elsewhere in the present volume.

D. Boundary Motion of Bacterial Populations (Revised Nossal Model)

Consider the transport equation (3.6) where the chemotactic force is given by

$$\mathbf{F}_C(\mathbf{r}, t) = \gamma \nabla c(\mathbf{r}, t) \qquad (3.23)$$

with $\gamma = \text{const}$, a logical assumption to make in the absence of any external concentration gradient. Furthermore, the degradation of the chemotactic agent can reasonably be assumed to be directly proportional to $\rho(\mathbf{r}, t)$ alone, as discussed in Section III.C. Then Equation (3.13) has the solution[18] (with $D = 0$)

$$c(\mathbf{r}, t) = c(\mathbf{r}, 0) - k \int_0^t \rho(\mathbf{r}, t') \, dt' \qquad (3.24)$$

which is substituted with (3.23) into Equation (3.6) to yield

$$\partial_t \rho(\mathbf{r}, t) = \chi k \nabla \cdot \left[\rho(\mathbf{r}, t) \int_0^t \nabla \rho(\mathbf{r}, t') \, dt' \right] + G(\mathbf{r}, t) \qquad (3.25)$$

To obtain Equation (3.25), the assumption has been made, according to experimental observation, that the speed of the chemotactic band is much larger than that due to isotropic mobility, that is, $\mu \ll \chi$, where χ is a constant independent of $c(\mathbf{r}, t)$. Consequently, the diffusion term [namely, the first term on the right-hand side of Equation (3.6)] can safely be ignored. Notice also that the term proportional to $\nabla c(\mathbf{r}, 0)$ vanishes here, as the concentration is uniform at the initial time.

From Equation (3.25) two cases should be considered. For one-dimensional migration—that is, pertaining to migrating bands in a capillary tube, as described in Section II.A—Equation (3.25) becomes

$$\partial_t \rho(x, t) = \chi k \rho(x, t) \left([\partial_x \ln \rho(x, t)] \int_0^t \partial_x \rho(x, t') \, dt' \right.$$

$$\left. + \int_0^t \partial_x^2 \rho(x, t') \, dt' \right) + G(x, t) \qquad (3.26)$$

whereas for radial migration—that is, for the experimental situation as described in Section II.B—one has

$$\partial_t \rho(r, t) = \chi k \rho(r, t) \left[[\partial_r \ln \rho(r, t)] \int_0^t \partial_r \rho(r, t') \, dt' \right.$$

$$\left. + \int_0^t \left(\frac{1}{r} \partial_r \rho(r, t') + \partial_r^2 \rho(r, t') \right) dt' \right] + G(r, t)$$

$$(3.27)$$

with r the radial distance from the point of inoculation.

Here it is the motion of the advancing boundary of the traveling band that is really of interest, and since such bands usually exhibit a sharp spatial distribution (see, e.g., Fig. 2), one assumes the following model for the spatial dependence of the bacterial population in the neighborhood of the sharp boundary x_B. The traveling band has a sharp exponential shape,[1] and in the linear geometry $\rho(x,t)$ is characterized as

$$\rho(x,t) = \phi(t)\exp\{-\beta[x - x_B(t)]\} \tag{3.28}$$

with β large (i.e., $\beta x \gg 1$), and where $\phi(t)$ represents the growth factor, related to $G(x,t)$ by

$$\phi(t) = \int_0^t G(x_B, t')\,dt' \tag{3.29}$$

Substitution of Equations (3.28) and (3.29) into Equation (3.26) yields an amazingly simple expression for the equation describing the evolution of the boundary $x_B(t)$. One obtains[18]

$$x_B(t) = (2\chi k)^{1/2}\int_0^t \phi^{1/2}(t')\,dt' \tag{3.30}$$

Despite its apparent simplicity, this result indicates that the band speed

$$v_B(t) = \dot{x}_B(t) = (2\chi k)^{1/2}\phi^{1/2}(t) \tag{3.31}$$

is not a trivial function of the bacterial density at the boundary, even for the simple case of an exponential distribution [Equation (3.28)]. Furthermore, it is quite clear from Equation (3.30) that the growth rate is a crucial factor in the migration. Different cases may be considered. When there is no growth, $\phi(t) = \rho_0 = $ const, and one obtains

$$x_B(t) = (2\chi k\rho_0)^{1/2}t \tag{3.32}$$

In this case, v_B is a constant, and one retrieves the ansatz used in the Keller-Segel model [see Equation (3.18)]. It should be mentioned, however, that although such a case represents a rather pathological situation, chemicals essential for growth are omitted in the experiments that Keller and Segel are modelling.

For exponential growth, which seems to be the most commonly

encountered experimental case, $\phi(t) = \rho_0 e^{\lambda t}$ (with λ the growth parameter), and one finds

$$x_B(t) = \frac{2}{\lambda}(2\chi k\rho_0)^{1/2}(e^{\lambda t} - 1) \qquad (3.33)$$

Another case of interest is the situation where the bacteria are approaching a stationary state of growth during times over which measurements are performed. Then $\phi(t) = \rho_0[1 + a(1 - e^{-\lambda' t})]$, where a determines the value of the growth plateau reached at large times ($\lambda' t \gg 1$). From Equation (3.30), one finds in the present case

$$x_B(t) = (2\chi k\rho_0)^{1/2}\int_0^t[1 + a(1 - e^{-\lambda' t'})]^{1/2}dt' \qquad (3.34)$$

Linear growth can be viewed as a special case of the above result when approximated for short times. Indeed, for $\lambda' t \ll 1$, one has $\phi(t) = (1 + a\lambda' t)\rho_0$, and one obtains

$$x_B(t) = (2\chi k\rho_0)^{1/2}\frac{2(1 + a\lambda' t)^{3/2}}{3a\lambda'} \qquad (3.35)$$

A similar analysis can be carried out from Equation (3.27) for radial migration, in which case $\phi(t)$ takes the form

$$\phi(t) = r_B(t)\int_0^t G(r_B, t')dt' \qquad (3.36)$$

where r_B denotes the radius of the migrating ring. A trial solution of exponential form for $\rho(r, t)$, similar to Equation (3.28), may be attempted, provided one includes a factor $r_B(t)$ in the denominator to account for the geometrical increase in unit area as the ring moves outward. The crucial consequence of this additional factor in the expression, as well as in that one for $\phi(t)$, is that growth is here indispensable for radial migration. Indeed, it is easy to understand that in the absence of growth, the bacterial density at the edge of the ring would decrease, so that the circular structure would soon disappear. Even more striking is the obvious consideration that the inoculation of a small number of cells of a nongrowing strain at the center of a Petri dish will never induce ring formation. On the contrary, when growth takes place, however small the number of bacteria inoculated, ring migration will always occur.

Subsequent experimental measurements have been performed—in particular, to test the square-root dependence of the ring speed on the growth factor, that is[19]

$$\dot{r}_B(t) \propto \left[\chi k \phi(t) \right]^{1/2} \tag{3.37}$$

From these experiments, it is also possible to analyze the time dependence of the ring diameter and of the bacterial density. One observes that for a fairly large temporal range, linear growth holds to a good approximation (see Fig. 4). Consequently, according to Equation (3.35), the ring diameter should grow in time as $t^{3/2}$. This is also illustrated in Fig. 4. Combination of the data displayed in Fig. 4 yields the relationship between $\dot{r}_B(t)$ and $\phi(t)$, which is found to be in good agreement with the theoretical prediction, Equation (3.37), as shown in Fig. 5.

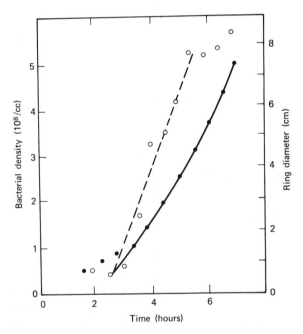

Figure 4.   The migration of an *E. coli* chemotactic ring as a function of time. The dashed line (with open circles) represents the bacterial density at the ring boundary, $\rho(r_B, t)$; the continuous curve (with dots) represents the increase of the ring radius, $r_B(t)$ [Equation (3.35)]. (After Nossal.[19])

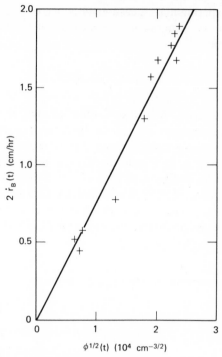

Figure 5. The radial velocity $\dot{r}_B(t)$ as a function of the square root of the bacterial density $\phi(t)$. The crosses are obtained by combination of the data given in Fig. 4. The solid line corresponds to Equation (3.37). (After Nossal.[19])

E. Bacterial Chemotaxis in the Presence of an External Concentration Gradient (Nossal-Weiss Model)

Consider the experimental situation where bacteria are subjected to an externally imposed concentration gradient of an attractant, as described in Section II.C. In the light of their experimental observations, the authors came to the conclusion that bacteria in such situations cannot be responding solely to the absolute gradient of attractant. Indeed, the complicated chemotactic response of the bacteria to a linear gradient suggests that they actually respond to proportional changes in concentration, $\nabla c/c$. Then using an exponential gradient, with $\nabla c/c$ =const, should elicit a constant bacterial response.[6] It seems thus logical to assume in the present case that the chemotactic force has the form

$$F(x) = \sigma \frac{\partial_x c(x)}{c(x)} \qquad (3.38)$$

where σ is a weakly varying function of c and may thus be taken as constant. The transport equation now becomes[20]

$$\partial_t \rho(x,t) = \partial_x [\mu \partial_x \rho(x,t) - \kappa \rho(x,t) \partial_x \ln c(x)] \qquad (3.39)$$

where μ and $\kappa = \sigma/\eta$ are constants, and where the growth term $G(x,t)$ has been ignored. In the case of a constant exponential concentration gradient (see Fig. 3),

$$c(x) = \begin{cases} c_0 & \text{for } x < 0 \\ c_0 e^{-\zeta x} & \text{for } x \geqslant 0 \end{cases} \qquad (3.40)$$

the bacterial current consists of a chemotactic term and a diffusion term for $x \geqslant 0$, whereas for $x < 0$, only the latter is present. Equation (3.39) can therefore be separated into two equations,

$$\partial_t \rho(x,t) = \mu \partial_x^2 \rho(x,t) = -\partial_x J_-(x,t) \qquad \text{for } x < 0 \qquad (3.41a)$$

$$\partial_t \rho(x,t) = \mu \partial_x^2 \rho(x,t) - v \partial_x \rho(x,t) = -\partial_x J_+(x,t) \qquad \text{for } x \geqslant 0 \qquad (3.41b)$$

Here $v = -\kappa \zeta$ is the chemotactic velocity and is a negative quantity as it is oriented in the direction opposite to $Ox > 0$. This is quite logical, since it is indeed expected that the bacteria will swim up gradient, and $c(x)$ decays exponentially with x [Equation (3.40)]. The solutions to Equations (3.41a,b) are most easily given in term of $\rho^*(y,\tau)$ defined as

$$\rho^*(y,\tau) = \frac{\rho(y,\tau) - \rho_0}{\rho_0} \qquad (3.42)$$

where ρ_0 is the initially uniform bacterial distribution, and y and τ are the dimensionless variables

$$y = \left(\frac{\omega}{\mu}\right)^{1/2} x, \qquad \tau = \omega t, \qquad \text{with} \quad \omega = \frac{(\kappa \zeta)^2}{4\mu} \qquad (3.43)$$

One obtains

$$\rho^*(y,\tau) = \frac{y e^y}{2\sqrt{\pi}} \int_0^\tau \rho^*(0, \tau - \tau') \frac{e^{-y^2/4\tau' - \tau'}}{\tau'^{3/2}} d\tau' \qquad (y \geqslant 0) \quad (3.44a)$$

$$\rho^*(y,\tau) = \frac{-y}{2\sqrt{\pi}} \int_0^\tau \rho^*(0, \tau - \tau') \frac{e^{-y^2/4\tau'}}{\tau'^{3/2}} d\tau' \qquad (y < 0) \quad (3.44b)$$

$$\rho^*(0,\tau) = \left(\frac{4\tau}{\pi}\right)^{1/2} - 1 + e^{-\tau/2} \left[(1+\tau)I_0\left(\frac{\tau}{2}\right) + \tau I_1\left(\frac{\tau}{2}\right)\right] \qquad (3.44c)$$

where $I_i(\tau/2)$ are Bessel functions of imaginary argument. Numerical computation based on these results shows good qualitative agreement with the experimental results of Dahlquist et al. The experimental bacterial profile as shown in Fig. 3 increases in time and is qualitatively well reproduced by Equations (3.44a, b). In particular, the increase of the bacterial concentration at the origin as a function of time is quite well represented. The available experimental points fall in the short-time range, so that they can be described by using a short-time expansion (τ small) of Equation (3.44c), yielding

$$\rho(0,t) = \rho_0 \left[1 + 2\left(\frac{\omega t}{\pi}\right)^{1/2} + \tfrac{1}{2}\omega t + O\big((\omega t)^2\big) \right] \qquad (3.45)$$

from which excellent agreement is obtained with the experimental data.

IV. CONCLUSIONS

Considering the present status of the study of the related problems of bacterial motility and chemotaxis, one faces the following—slightly paradoxical—situation. A fairly large number of experimental results are available, among which a nonnegligible number of facts remain unexplained to date. The amount of experimental information certainly exceeds the number of theoretical interpretations presently available, but, on the other hand, remains amazingly insufficient to cover the wide range of questions raised by these experimental studies. Theoretical analyses have been performed on the basis of simple models, which nevertheless remain subject to revision in the light of realistic experimental facts. Reciprocally, more experimental information is needed to attempt more accurate theoretical treatments.

As discussed in the introduction, two classes of problems should be considered. Those problems related to bacterial migration (that is, essentially the macroscopic aspect of chemotaxis) have been investigated quite extensively from the experimental point of view, and the few corresponding theoretical analyses have been reviewed in the present chapter. These theoretical models certainly appear as valuable attempts, which have shed some light to this aspect of the problem. However, it quite clearly appears that without further information on the microscopic mechanism of chemotaxis and motility, several basic questions will remain indefinitely open. This remark calls for extensive biochemical investigations of the chemotactic mechanism as schematically represented in Fig. 1. On the other hand, the biophysical aspect of the problem, which is more directly related to the dynamical study of the motion (i.e., the motility itself), is of equally great importance. This is the reason why Sections II.D and E were devoted to the experimental possibilities offered by currently available techniques to attack the problem of the physical mechanism involved in the chemotactic response. The fact that Section III lacks the corresponding theoretical analyses is essentially due to the very recent development of such methods.

It is, however, very likely that in the near future these microscopic techniques, described in Sections II.D and E, should, in combination with new biochemical information, lead to significant progress towards the understanding of the coupled mechanisms of chemotactic response and motility.

Acknowledgments

We thank R. Lavallé and R. Nossal for stimulating and enlighting discussions, and L. A. Segel and G. Nicolis for valuable comments on the manuscript.

References

Chemotaxis in Bacteria (Including Experiments on One-Dimensional Migration and on Radial Migration)

1. J. Adler, *Science*, **153**, 708 (1966).
2. J. Adler and M. M. Dahl, *J. Gen. Microbiol.*, **46**, 161 (1967).
3. J. Adler, *Science*, **166**, 1588 (1969).
4. G. L. Hazelbauer and J. Adler, *Nature N. Biol.*, **230**, 101 (1971).
5. R. Mesibov and J. Adler, *J. Bacteriol.*, **112**, 315 (1972).

Bacterial Motility and Chemotaxis in the Presence of an Attractant Gradient

6. F. Q. Dahlquist, P. Lovely, and D. E. Koshland, *Nature N. Biol.*, **236**, 120 (1972).
7. R. M. MacNab and D. E. Koshland, *Proc. Nat. Acad. Sci. (U.S.A.)*, **69**, 2509 (1972).
8. H. C. Berg and D. A. Brown, *Nature (Lond.)*, **239**, 500 (1972).

Motility Measurements by Light Scattering

Viruses

9. H. Z. Cummins, F. D. Carlson, T. J. Herbert and G. Woods, *Biophys. J.*, **9**, 518 (1969).

Spermatozoa

10. P. Bergé, B. Volochine, R. Billard and A. Hamelin, *C. R. Acad. Sci. Paris*, **265**, D-889 (1967).
11. P. Bergé and M. Dubois, *Rev. Phys. Appl.*, **8**, 89 (1973).
12. R. Combescot, *J. Phys.*, **31**, 767 (1970).

Bacteria

13. R. Nossal, *Biophys. J.*, **11**, 341 (1971).
14. R. Nossal and S. H. Chen, *J. Phys.*, **33**, C-1-171 (1972).
15. J. P. Boon, R. Nossal and S. H. Chen, *Biophys. J.*, **14**, 847 (1974).

Theoretical Models

16. E. F. Keller and L. A. Segel, *J. Theor. Biol.*, **30**, 225 (1971).
17. E. F. Keller and L. A. Segel, *J. Theor. Biol.*, **30**, 235 (1971).
18. R. Nossal, *Math. Biosci.*, **13**, 397 (1972).
19. R. Nossal, *Exptl. Cell. Res.*, **75**, 138 (1972).
20. R. Nossal and G. H. Weiss, *J. Theor. Biol.*, **41**, 143 (1973).
21. L. A. Segel and J. L. Jackson, *J. Mechanochem. Cell Motility*, **2**, 25 (1973).

THE MOLECULAR VARIATIONS OF CYTOCHROME *c* AS A FUNCTION OF THE EVOLUTION OF SPECIES*

E. MARGOLIASH

Department of Biological Sciences, Northwestern University, Evanston, Illinois

CONTENTS

I. PROSPECTS OF MOLECULAR STUDIES OF EVOLUTION

When over a decade ago it was deduced that the amino acid sequences of protein polypeptide chains were nothing but images of segments of the genome and reflected the fine structure of the corre-

---

* Reproduced by permission from the Harvey Lectures, *Series 66*: 177 (1972) (Academic Press, Inc., New York, N.Y.)

sponding stretches of DNA according to simple and inflexible rules,[21, 155] it became clear that the structures of proteins were the current molecular end products of the totality of the evolutionary variations undergone by their genes. As eloquently stated by Francis Crick,[20] "Biologists should realize that before long we shall have a subject which might be called 'protein taxonomy', the study of the amino acid sequences of proteins of an organism and the comparison of them between species. It can be argued that these sequences are the most delicate expression possible of the phenotype of an organism and that vast amounts of evolutionary information may be hidden away within them." This prediction, dating back some 16 years, has since been amply confirmed. However, because of the labor involved in obtaining numerous amino acid sequences and the difficulty of gathering large enough samples of pure proteins from many less common species, the data collected in the best of cases are as yet minimal. Only a few proteins have been subjected to significant comparative studies, and in many cases vast groups of species have been totally ignored. The 1972 edition of the *Atlas of Protein Sequence and Structure*,[22] though already out of date, is nevertheless fairly well representative of that situation. We are still largely at the threshold of the new biological era of "protein taxonomy." Nevertheless, since the development of techniques for the determination of the nucleotide sequences of nucleic acids,[46] one can already see the beginnings of its even more difficult extension to the other classes of informational macromolecules.

The difficulties are largely technical. Human operators are singularly ill suited to the sort of tedious repetitive functions that constitute sequence determinations, and it has become obvious that if we are to delve systematically into the vast stores of evolutionary information encoded in the structures of the informational macromolecules of extant forms of life, the determination of sequences will have to be fully automated. In addition to speed and reliability within set limits, automation lends itself to the use of smaller and smaller amounts of material as techniques and instrumentation are perfected, an advantage that could well be crucial for many evolutionarily important groups of organisms. The successes already recorded in the development of such automation for important portions of the necessary overall system of procedures[31, 29] augur an early solution to these problems and a consequent flowering of the various fields of evolutionary biology.

Taxonomy or phylogeny via the structures of informational macromolecules is from several points of view more satisfactory than organismal morphological taxonomy or than molecular taxonomy based on substances other than informational macromolecules. This is considered below. Nevertheless, such advantages are not the only attractive or even the most important features of molecular studies of evolution. Not only do proteins represent the most delicate phenotypic expression of an organism, but their chemistry is related in a direct and stable fashion to the fine structure of their genes. Their changes are largely, if not entirely, simple expressions of the fundamental unit changes of evolution, namely, the evolutionary fixation of single nucleotide substitutions in the corresponding segments of the genome. Definite constraints are, therefore, placed on protein evolutionary variations, which though necessarily present, cannot possibly be deciphered in morphological evolutionary changes representing complicated changes often encompassing numerous genes. Similarly, changes of molecules other than informational macromolecules, which are products of whole assembly lines of enzymes, are not much less complex from the genetic standpoint than morphological evolutionary events.[157] Protein evolutionary changes, on the other hand, are, in an observable fashion, restricted by both the genetic code and the phenotypic requirement of maintaining a functional structure at every step of the variation. This second constraint can conceivably be bypassed when more than one gene is used to code for the same protein. One gene can maintain function while another can undergo evolutionary changes presenting eventual advantages, but having intermediate stages that could not be tolerated if it was alone in supporting function.

In general, one can therefore expect that a thorough understanding of the structure-function relations of a protein, together with the information that can be extracted by the statistical techniques developed over the last few years[43, 41, 45, 34] from the amino acid sequences of the protein from a suitable range of species, may well lead to a view of evolution far more precise than otherwise possible. This would entail a detailed reconstruction of the complete evolutionary tree of the protein, a delineation of its evolutionary relations to other gene products, and quite possibly an understanding of the rules and regulations that govern molecular evolution. Whether these last are in all details identical to the rules that govern organismal evolution is open to question. This is exemplified by the recent and unresolved controversy[53] as to whether there are evolutionarily neutral variations in protein structure, that is, changes that present no selective advantage or disadvantage, when such changes have long been considered to be nonexistent in organismal evolution.[97] The question of possible neutral mutational changes in the evolutionary variations of cytochrome *c* is discussed below.

Just how far such developments may lead, when they are extended to numerous sets of homologous proteins and possibly to nucleic acids, is difficult to judge. The remarkable successes scored so far on the basis of the amino acid sequences of proteins from a relatively minute sample of species is most encouraging. The structures of an evolutionarily relatively slowly changing protein, such as cytochrome c, have permitted derivations of the phylogenetic relations of widely dispersed species, including in a single taxonomic scheme higher plants, fungi, and animals[43,41,92] (see below). Similar approaches with a rapidly changing structure, such as the fibrinopeptides A, have led to a satisfactory classification of a much smaller group of species, namely artiodactyls.[100,92] Protein taxonomies approximate not only species phylogenies, but also gene phylogenies, as first shown in the case of the relative evolutionary relations of the genes for the various hemoglobin chains and for myoglobin.[61, 59, 60, 156, 157, 43] Protein structures have also revealed the evolutionary relations of genes which cannot a priori be suspected of homology, as in the now classical case of lysozyme and α-lactalbumin.[10] All in all, one can readily foresee that the accumulation of protein acid sequence information will lead to definitive solutions of old unsettled phylogenetic questions. Among them one may list the precise interrelations of prokaryotes and eukaryotes, the taxonomic relations with groups of prokaryotes for which morphological or metabolic criteria are unsatisfactory and the phylogenetic relations between such groups, and in what invertebrate group the vertebrates originated. Ideally, this approach will eventually yield a drawing of the complete tree of biological evolution depicting the relations of many protein structural genes and, superimposed on this ground picture of gene flow, the topology of the descent of species in the course of biological evolution. Such developments, particularly when linked to our increasingly sophisticated understanding of protein structure-function relations, are likely to provide insights into the mechanisms of molecular evolution well beyond our present naive and largely qualitative views. Moreover, just as one can estimate the amino acid sequence of the ancestral form of a protein from a statistical phylogenetic tree of the corresponding homologous set of structures[43] (see below), a complete description of the evolutionary relations of numerous sets of proteins should enable the approximation of ancestral polypeptide structures dating back to the obscure period of biological history that saw the appearance of the early replicating structures and possibly the establishment of the mutual relations between proteins and nucleic acids. There is no reason, other than present technical difficulty, that a similar approach should not eventually be possible for nucleic acid sequences, leading to an independent identification of the nucleic acid counterparts of early protein sequences, and to tests of hypotheses concerning the development of the initial successful macromolecular machinery of biological systems.[41, 85] With the perfection of procedures for the laboratory synthesis of long polypeptide chains, one can imagine that structure-function relations of ancestral forms of present-day proteins will be amenable to direct experimentation, just as proteins prepared from ordinary biological sources are utilized today.

In summary, the approach to biological evolution through the study of the structures of informational macromolecules is likely to lead to a description of the process in a completeness of detail and with a precision far beyond what can be attained by more classical means. Since molecular changes are subject to genetic and functional restrictions that make them amenable to direct computational and experimental attack with the proteins of extant and possibly with those of ancestral extinct species, one may expect to develop a fundamental understanding of the mechanisms of molecular evolution, of the means and pathways that led to the present living systems, and perhaps even of the extent to which the more or less successful solutions that they represent are unique. A good proportion of the techniques needed to acquire the necessary information, chemical as well as statistical, have already been developed, and we can only hope that the remainder will not present insuperable difficulties.

II. CYTOCHROME c PROTEIN STRUCTURES AND FUNCTION

Cytochromes of c type have an essentially ubiquitous biological distribution. They are found in the mitochondrial respiratory chain of all eukaryotes, in the oxidation systems of many prokaryotes, and in the photosynthetic membranes of both eukaryotes and prokaryotes, including those that do not utilize oxygen as the terminal oxidant of metabolism.[67,66,88,65] So far, following the determination of the primary structure of horse cytochrome c in 1960[94] (Fig. 1), the amino acid sequences of over 50 cytochromes c of the eukaryotic or mitochondrial type from different species have been examined (see legend to Fig. 2) through the continued efforts of several groups of investigators. The taxonomic coverage achieved to date is very wide, including 14 mammals, 6 birds, 2 reptiles, 1 amphibian, 5 fish, 4 insects, 6 fungi, 11 higher

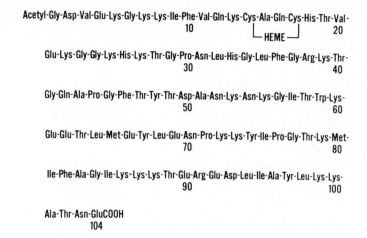

Figure 1. The amino acid sequence of horse cytochrome c, according to Margoliash et al.[94] The basic residues lysine, arginine, and histidine, and the hydrophobic residues tryptophan, tyrosine, phenylalanine, leucine, isoleucine, valine, and methionine, to a large extent occur in basic and in hydrophobic clusters along the amino acid sequence. There is no obvious overall grouping of acidic residues in the amino acid sequence.

plants, and 2 protists, far in excess of corresponding information available for any other protein. All of these proteins are small, having chain lengths slightly over 100 residues. They are strongly basic and have extensive similarities of amino acid sequence (Refs. 93, 83, 127, 82, 126, 89, 88, 81, 43, 125, 86, 103, 92, 130, 41, 85). All those tested to date react identically in cytochrome c–depleted mitochondria from mammalian species[12] and with cytochrome c oxidase and reductase preparations from mammalian species.[132, 131] They have been commonly classified as "mammalian-type" cytochromes c, though the present usage of "mitochondrial" or "eukaryotic" cytochrome c seems more appropriate. Even these terms may not be fully satisfactory, as borderline cases such as blue-green algae, which do not have distinct nuclei (are prokaryotes) or any other cell organelles, appear to carry all three varieties of c-type cytochromes: the photosynthetic green-plant type, an acidic bacterial type, and even a trace of a basic c-type cytochrome that may be of the so-called eukaryotic or mitochondrial type.[58] The structure of none of these has yet been determined, and until this is done their relationships cannot be assesssed with certainty.

The properties of bacterial c-type cytochromes are much more variable than those of the eukaryotic proteins,[65] many being acidic, while

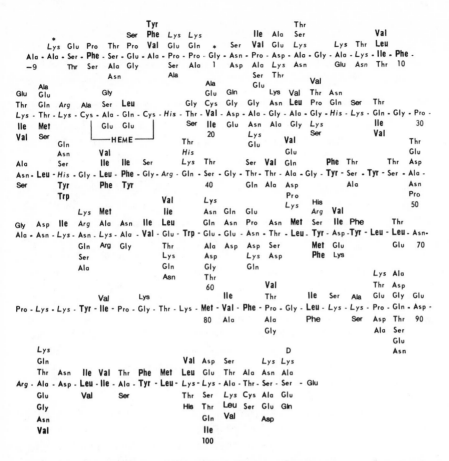

Figure 2. Composite amino acid sequence of eukaryotic cytochromes c. The continuous sequence of 112 residues from residues −8 to 104 is that of the wheat-germ protein. The alanyl residue at position −9 occurs in the iso-2 cytochrome c of baker's yeast. The glutamic acid at position 105 is found in the ginkgo protein, the only one to extend that far. The asterisk at position −8 indicates that this residue is N-acetylated in the higher plant cytochromes c, while the asterisk at position 1 similarly denotes the acetyl present in all vertebrate cytochromes c. The hydrophobic residues—valine, leucine, isoleucine, tyrosine, phenylalanine, tryptophan, and methionine—are in boldface type; the basic residues —lysine, arginine, and histidine—are in *italics*; and D marks the position at which a gap has been introduced in some proteins to maximize similarity. The cytochromes c tabulated are those of man; the chimpanzee; the rhesus monkey *Macaca mulatta*; the horse, donkey, cow, pig, sheep, dog, elephant seal, rabbit, California grey whale, camel, great grey kangaroo, chicken, turkey, pigeon, duck, emu, rattlesnake, snapping turtle, bullfrog, tuna, bonito, carp, dogfish, and Pacific lamprey; the moth *Samia cynthia*; the tobacco hornworm moth; the fruit fly *Drosophila melanogaster*; the screwworm fly; the fungi *Neurospora*, baker's yeast (iso-1 and iso-2 cytochromes c), *Debaromyces*, *Candida krusei*, and *Ustilago sphaerogena*; the higher plants wheat, mung bean, sunflower, sesame, castor, cauliflower, buckwheat, pumpkin, *Abutilon*, cotton, and ginkgo; and the protists *Physarum* and *Euglena*. References to the structures of these proteins are listed in Ref. 92 (see also Refs. 103,149,22), except for the proteins of the carp[49]; the elephant seal[5a] and emu;[5b] (Augusten and Webb, (1972 & 1973), the camel;[134] the bonito;[101] the tobacco hornworm moth;[14] *Debaromyces*;[142] *Ustilago sphaerogena*;[7] castor, sesame, mung bean, and sunflower;[8] pumpkin, buckwheat, and cauliflower;[140] *Abutilon* and cotton;[141] ginkgo;[111] and *Physarum* and *Euglena*[80].

some are strongly basic, such as the cytochromes c_3 of sulfate-reducing organisms. These have long been known to be essentially unreactive with cytochrome c oxidase preparations from mammalian sources and were, therefore, generally considered to be totally unrelated to the eukaryotic cytochromes c. More recently, as the amino acid sequences of several of these proteins were established,[65, 1] similarities between their structures and those of the eukaryotic proteins became obvious, to a degree that is clearly beyond what could be expected on a random basis.[28, 45, 38] Thus, it could turn out that the c-type cytochromes of eukaryotes, prokaryotes, and photosynthetic organelles as well are all phylogenetically related, or *homologous*, proteins—namely, that they all had at one time a common ancestral gene in the early stage of biological evolution. Even more, in view of the similarities of structure detected between cytochrome c and cytochrome b_5,[92] the same homologous groups could include cytochromes of other than the c type. Cytochromes could well be the proteins of choice, not only to examine the phylogenetic relationships of various groups of prokaryotes, but to determine unambiguously the evolutionary relations between prokaryotes and eukaryotes, and possibly to obtain an insight into the history of that most important of evolutionary transitions, the organization of intracellular organelles, events that led to the enormous diversification of life on this planet.

These are the tasks of the immediate future. At present the relative paucity of amino acid sequence data for the prokaryotic cytochromes c precludes any significant consideration of these problems. The following discussion will be limited to the evolutionary information that has accrued from a study of the structure of eukaryotic cytochromes c. This study, the work of numerous investigators, not only is of intrinsic biological interest, but has served as a model system to develop and test the conceptual framework and the statistical and other techniques that permit the extraction of evolutionary information from the structural data for a set of homologous proteins. It is likely to remain for some time to come a main basis for extensions of knowledge of the mechanism of evolution at the molecular level well beyond the present restricted horizon.

A. The Variability of Cytochrome c Primary Structures

The amino acid sequences of the first eukaryotic cytochromes c to be determined were those of the horse,[94] the pig, rabbit, and chicken,[16, 83] man,[127] the tuna,[78] and baker's yeast.[102] At that time, well before comparable evidence was obtained for other proteins, it was already clear that the primary structure of the molecule was capable of accommodating extensive variations, and that, even more remarkably, the degree of molecular variability in terms of numbers of residue differences, was demonstrably related to the phylogenic distance between the species carrying the protein.[93, 83, 82] It was noticed that the cytochromes c of mammals were roughly equally different from the cytochromes c of birds, that the proteins of mammals and birds together were equally different from those of fish, that those of all vertebrates were equally different from those of insects, and that the cytochromes c of all animals, vertebrates and invertebrates, were roughly equally different from those of fungi.[83, 89] Thus, the time elapsed since the divergence of lines of evolutionary descent seemed to be the main parameter that decided by how many residues the cytochromes c of any two species differed, implying that residue changes in the protein were fixed in the course of evolution at a roughly constant rate.[83] Therefore, using as an internal standard the known time of divergence of the line of descent that led to birds from the one that led to mammals, it was possible to define the *unit evolutionary period* for cytochrome c, namely, the evolutionary time required to effect a single change in the cytochromes c of two diverging lines of evolutionary descent.[89] In terms of residue changes this was calculated to be 26.4 million years, and in terms of nucleotide changes in the corresponding codons the value was 21.4 million years[86, 103] (Table I). One could also draw a curve relating

TABLE I

Average Numbers of Variant Residues and Minimal Replacement Distances Between Eukaryotic Cytochroces c—Comparison by Taxonomic Groupings[a]

Groups compared	Average number of variant residues	Average of the minimal replacement distance
Mammals to birds	10.6 ± 1.5	13.6 ± 1.9
Mammals and birds to reptiles	14.5 ± 4.5	18.6 ± 6.8
Mammals, birds, and reptiles to fish	17.9 ± 1.8	26.5 ± 2.4
Vertebrates to invertebrates	22.1 ± 2.0	30.4 ± 3.2
Animals to fungi	43.1 ± 2.3	59.9 ± 2.8
Unit evolutionary period (years)	26.4×10^6	21.4×10^6

[a]According to Margoliash and Fitch.[86] Each number is followed by its average deviation from the mean. The *unit evolutionary period*[89] is defined in the text.

cytochrome c residue or codon variations to the time elapsed since the divergences of major lines of evolutionary descent (Fig. 3) and correct it statistically for the chance occurrence of more than one change per residue, a situation that could not be detected by the direct comparison of amino acid sequences.[89, 86] The results were satisfactory in that they accorded in general with other estimates. For example, the divergence of the chain of descent that led to fungi on the one hand, and other taxonomic groups considered on the other, was estimated at about 1.5 billion years.

These then were the first observations that demonstrated that the expected concordance between protein amino acid sequences and the phylogenetic relations of species[20, 158] was indeed a reality. Moreover, they made clear that the problems of unraveling the molecular concomitants of evolutionary transitions and of understanding the mechanisms of these changes could be attacked on a quantitative basis. They also indicated that on the basis of the genetic code, at that time only very partially determined, it would be possible to approximate the structure of ancestral forms of the protein.[83] Thus, these studies laid a good part of the conceptual foundations for the more sophisticated approaches that followed.

Even though the concept of the molecular evolutionary clock[158, 83, 89] has recently been reworked in terms of the much more extensive amino-acid-sequence material now available and more precise paleontological standards,[23] leading to much the same conclusions, it can only represent an approximation, which becomes less and less satisfactory as we develop an interest in more and more accurate descriptions of molecular evolutionary history, and attempt to understand the underlying fundamental biological mechanisms.

A simple examination of the composite picture of the amino acid sequences of all the cytochromes c that have been examined to date (Fig. 2) shows that some 26 residue positions have remained unvaried, and that approximately 75% of the molecule has undergone residue substitutions can be subdivided, according to evolutionary behavior, into two classes of positions.[126] There are those at which the alternative residues are so similar in structure or physicochemical properties that one can readily believe they are capable of identical functions, constituting so-called *conservative substitutions*. Examples are given by residue 13, which seems to be strictly confined to either of the two available strongly basic amino acids, lysine and arginine; and residue 35, which can apparently accommodate only hydrophobic

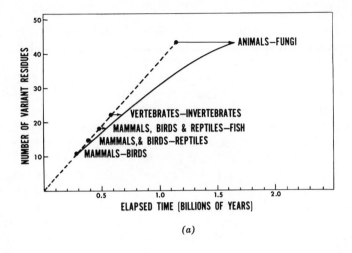

(a)

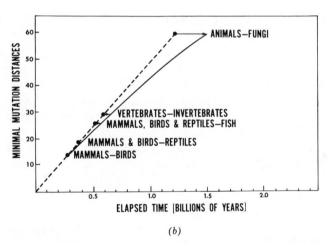

(b)

Figure 3. Relation of number of variant residues (*a*) among cytochromes *c* from different classes and phyla of organisms to the time elapsed since the divergence of the corresponding lines of evolutionary descent, and (*b*) the similar relation of the number of variant nucleotide positions in the respective structural genes. The straight dashed lines were calculated on the basis of a value of 280 million years for the time elapsed since the divergence of the mammalian and avian lines of descent, corresponding to unit evolutionary periods of 26.4 million years for residue replacements and 21.4 million years for nucleotide replacements. The solid curved lines show the relationship corrected for multiple changes in residue (*a*) and nucleotide (*b*) positions, according to the formula $r = (n\ln n)/(n-\lambda)$[89], assuming a total of 76 variable residues. According to Margoliash and Fitch[86].

201

amino acids such as valine, isoleucine, leucine, and phenylalanine and tyrosine. Other locations seem capable of being occupied by amino acids that are so dissimilar that at first glance they appear unlikely to be importantly involved in maintaining the tertiary structure or the function of the protein. These are termed *radical substitutions*, and many locations carrying such substitutions are obvious in Fig. 2, such as residues 60, 89, and 92.

It must not be overlooked, however, that this simplistic approach to the significance of evolutionary residue changes in proteins assumes that all similar residues have a similar function, while chemically different residues necessarily have different functions. Either assumption can be wrong. For example, in a position at which aromaticity is the essential attribute that needs to be conserved, a tyrosine-phenylalanine substitution is conservative, but in a position at which the tyrosyl residue provides an essential hydrogen bond, the same substitution is radical. As long as, at a minimum, we do not know the tertiary structure of the protein examined, or, ideally, until we understand the functional role of each residue, any classification of evolutionary variable residues can be thoroughly misleading. At best, negative conclusions can be drawn. The fact that a certain residue can be substituted by others provides evidence that a function attributable only to the particular residue cannot be occurring. For example, the observation that the only constant histidine in cytochrome *c* was that at position 18, that commonly at position 33 being absent in the kangaroo,[104] snapping turtle,[15] and tuna[78, 77] proteins, while that commonly found at position 26 is not present in the iso-2 cytochrome *c* of baker's yeast[135] and the protein from the fungus *Neurospora crassa*,[55, 54] provided the first definitive evidence that the cytochrome *c* hemochrome is not a diimidazole hemochrome constituted by the side chains of two histidyl residues coordinating with the heme iron from both sides of the porphyrin plane, as had long been supposed.[139, 138, 108, 107, 32, 51, 50] This conclusion has since been amply confirmed by x-ray crystallographic[27, 26, 25] and other studies,[3, 2, 143, 117, 52, 121, 150, 98, 48, 112] which demonstrate that the sulphur of methionyl residue 80 and the imidazole of histidyl residue 18 are the protein groups bound to the heme iron.

With regard to invariant positions the situation is better, in that strict invariance very likely indicates that the residues involved have a function that is compatible only with their particular structures. Moreover, when such invariance extends over a very broad taxonomic spectrum of species, as in the case of cytochrome *c*, the function subserved must be general—not one adapted only to the needs of each species—and therefore important. But, on this basis, and without independent lines of evidence, to attempt to decide what are the functions of the invariant residues is unwarranted. An excellent example is provided by the stretch of 11 residues (residues 70 to 80), which had been found to be strictly invariant until the cytochromes c of two protists were recently examined. To note that the next longest invariant segment is only two residues long (see Fig. 2) emphasizes the most unusual character of residues 70 to 80. This extreme degree of conservatism was attributed either to close contact intramolecularly with the one structure in cytochrome c that is necessarily invariant in evolution, namely the heme; or to contact with another protein whose reaction with cytochrome c is functionally essential, such as the cytochrome c oxidase or the reductase of the mitochondrial respiratory chain, or to both types of interaction. A surface of close contact with another protein is likely to be evolutionarily conservative, since amino acid substitution in one protein will require appropriately inverse substitutions in the other to maintain close contact. The x-ray crystallographic solutions for the structures of horse and bonito ferricytochrome c[25] and of tuna ferrocytochrome c[137] have indeed shown that the carboxyl end of the invariant stretch is near the heme, the methionyl residue at position 80 providing one of the two hemochrome-forming groups (Figs. 4 and 5). Also, as long as the areas of interaction with the oxidase and the reductase have not been directly identified on cytochrome c, one may consider the invariant segment a likely candidate for such functions. However, an entirely novel reason for possible structural invariance was overlooked in the evolutionary argument, and subsequently beautifully brought out by the crystallographic work. The invariant segment, located at the lower left side of the protein (see Figs. 4 and 5) is one that appears to be capable of considerable change of conformation, as evidenced by the preferential reaction of the internally located tyrosyl residue 67 with tetranitromethane.[123, 119, 120] The reagent has no access to the residue unless the peptide chain comprising the invariant segment swings away from its normal position. Moreover, the immediately following segment of residues 81 and 82 is involved in the most dramatic movement linked to oxidoreduction.[137] This entire section is the so-called "weak side" of the protein.[25] Thus, conformational mobility—which is probably crucial to function, and therefore likely to be of a very specific variety—may be among the most stringent constraints imposing evolutionary invariance on protein structure. The invariant segment of cytochrome c would well represent the first case of such a mechanism for constancy.

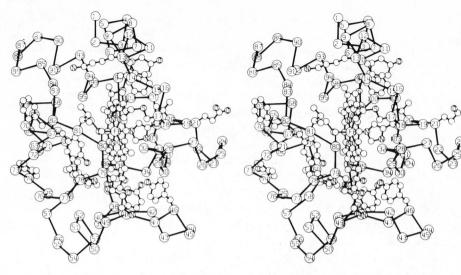

Figure 4. Stereoscopic α-carbon diagram of tuna cytochrome *c* in the ferric form, according to Dickerson et al.[25] The side chains that are indicated are those of residues 10 (phenylalanine), 13 (lysine), 14 (cysteine), 17 (cysteine), 18 (histidine carrying heme-iron-linked imidazole), 20 (valine), 21 (glutamic acid), 46 (tyrosine), 48 (tyrosine), 59 (tryptophan), 67 (tyrosine), 74 (tyrosine), 78 (threonine), 79 (lysine), 80 (methionine carrying heme-iron-coordinated sulfur), 81 (isoleucine), 82 (phenylalanine), and 97 (tyrosine). The heme is the square structure inserted in the front of the molecule. It is seen nearly edge on, the left surface being just visible.

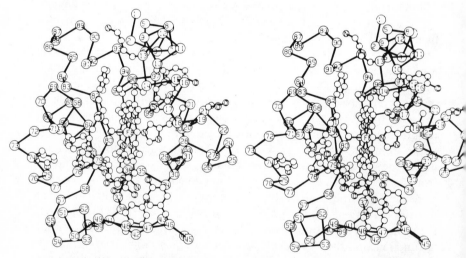

Figure 5. Stereoscopic α-carbon diagram of tuna cytochrome *c* in the ferrous form, according to Takano and Dickerson[136] (see also Takano et al.[137]. The protein has been kept in the same general relative orientation as in Fig. 4, so that changes in conformation on oxidoreduction can be observed by direct comparisons of Fig. 4 and 5. The same side chains are marked as in Fig. 4.

B. The Invariance of Cytochrome c Tertiary Structure and Function

With respect to evolutionary considerations, the amino acid sequence of a protein is operationally near the genetic end of the spectrum of biological phenomena. By itself it has no phenotypic meaning. This is acquired only when the polypeptide folds into its own spatial or tertiary structure, binds cofactors or prosthetic groups as required, and appears as a fully fledged molecular machine fit to perform its designated function in the specific cellular milieu provided, and subordinated to the various levels of organization that make up a live organism, or even a population of organisms. Since evolutionary selection operates at the phenotypic level, a most thorough understanding of the spatial structure of a protein and of its structure-function relationships in the broadest biological sense is necessary if we are ever to unravel the historical significance of molecular evolutionary changes and the mechanisms by which they occur.

For cytochrome c, perhaps the most striking result of the x-ray crystallographic determination of structure and of the examination of the functional activities of the cytochromes c from a wide variety of species is that both structure and function remain essentially the same throughout. In the case of function this statement is well documented. The first-order rate constants for the oxidation of ferrocytochrome c by solubilized beef cytochrome c oxidase remains, within experimental error, the same for the protein of the cow and four other mammals, three birds, one fish, and one insect, even though the amino acid sequences differ by as much as 27 residues out of 104.[132] The same relation is maintained over a wide range of cytochrome c concentrations, and an essentially identical result is obtained with the cytochrome c reductase system.[131] Cytochrome c function can also be measured under conditions that approximate the *in vivo* situation even far better than for most proteins. It is possible to remove as much as 80 to 85% of the cytochrome c of mitochondria, without damaging the mitochondrial inner membrane, by procedures similar to that developed by Jacobs and Sanadi.[63] The recovery of function can then be measured on adding graded amounts of cytochrome c. In such cytochrome c–repleted depleted mitochondria, the oxidation of substrates and oxidative phosphorylation are well coupled when the repletion procedure is carried out at low ionic strength, and function can then be estimated from the rate of oxidation of substrates in the presence of ADP. Again here it is observed that with rat-liver mitochondria, for example, it makes no difference whether the added cytochrome c is the horse, cow, rabbit, human, chicken, pigeon, snapping-turtle, tuna, lamprey, moth, or baker's-yeast protein. Identical titration curves of functional recovery are obtained.[12, 11]

This is in strong contrast with the differences in the amino acid sequences of these cytochromes c, some of which differ by nearly half their residues. One such repletion titration curve is shown in Fig. 6, in which tuna cytochrome c is used to reactivate cytochrome c–depleted rat-liver mitochondria. When the mitochondrial inner membrane has been damaged in the osmotic manipulations of the depletion, the titration curves do not go to a saturation plateau, but the rate of oxidation continues to increase as more cytochrome c is added, in a manner similar to the behavior of the commonly used particle preparations from disrupted mitochondria. Interestingly, the amount of cytochrome c required to restore maximal function under coupled conditions is not very different from that present in native mitochondria. Nevertheless, the cytochrome c bound to the repleted depleted mitochondria is most probably in equilibrium with cytochrome c in solution, as the amount of the protein needed for maximal reactivation varies with the temperature at which rates of oxidation are measured, being about twice as high at 37°C as at 25°C.[11] Since this is not the case for native mitochondria, and in the depletion procedure the outer membrane is damaged, these observations can be interpreted to indicate that one function of the mitochondrial outer membrane is to prevent cytochrome c from diffusing away from the organelle.

Comparisons of different cytochromes c with respect to their ability to maintain the function of the mitochondrial respiratory chain had earlier been carried out by Jacobs and Sanadi[63] when they showed that beef cytochrome c could restore the oxidation of substrates and some 75% of the rate of oxidative phosphorylation of cytochrome c–depleted rat-liver mitochondria. Mattoon and Sherman[96] later utilized a genetic variant of baker's yeast (cy_{1-12}), which carries a chain-terminating codon at position 20 of the structural gene for cytochrome c (corresponding to position 15 of the horse cytochrome c amino acid sequence in Fig. 1). This strain made mitochondria devoid of cytochrome c but otherwise normal. On repletion with either yeast or horse cytochrome c, oxidative phosphorylation was recovered, the horse protein being only some 10% less effective than yeast cytochrome c. However, these cytochrome c–deficient yeast mitochondria bound many times more cytochrome c than required to saturate function, which, as noted above, carefully depleted rat-liver mitochondria do not. Whether this is due to a tissue or species difference, or results from the more drastic conditions required for the preparation of yeast mitochondria, cannot be decided.

Moreover, the cytochrome *c* preparations were not exhaustively de-ionized, and the species comparison was limited to just two preparations of the protein. For all these reasons it is difficult to judge whether these results do or do not represent an evolutionarily significant difference.

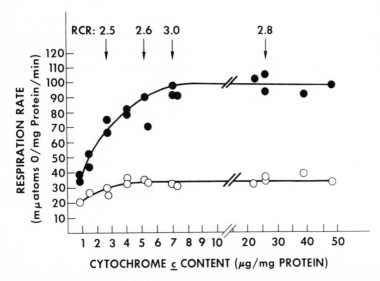

Figure 6.   Repletion of cytochrome *c*–depleted rat-liver mitochondria with tuna cytochrome *c*. Respiration in the presence of ADP (filled circles) and in the absence of ADP (open circles), using succinate as substrate, is plotted against the amount of added tuna cytochrome *c*, in micrograms per milligram of mitochondrial protein. The numbers indicated by RCR at the top are the respiratory control ratios, estimated at the corresponding arrows. The maximal rate of O_2 uptake in the presence of ADP is the same as that obtained with the native untreated mitochondria, and the amount of cytochrome *c* required to attain this rate is approximately that present in native mitochondria.

Indeed, correct functional comparisons of cytochromes *c* from different species were made possible by the finding that a large variety of anions bind to ferricytochrome *c*, while cations bind to the ferroprotein,[6,91] and that these can be removed by electrodialysis to the isoelectric point, dialysis at the isoelectric point (pH 10), or gel filtration at pH 10. Unless all cytochrome *c* preparations are so treated, their activities in the enzyme systems or in intact mitochondria are very erratic, indicating the presence in many preparations of varying amounts of a potent inhibitor of cytochrome *c* function.[132] This inhibitor has not yet been identified. Observations indicating differences in the functional activities of the cytochromes c of various eukaryotes[154, 33, 152, 151, 153] are probably vitiated by this and other types of artifact.[84, 87]

There is much less information with regard to spatial structures. The one fully documented case is the comparison of the structures of the horse and bonito ferricytochromes c by x-ray crystallography.[25] Both proteins have essentially identical polypeptide backbone conformations, even though they differ by 17 residues out of 104. The only differences are the expected ones in the location of some of the side chains. For example, when a tryptophan in the bonito protein replaces the histidyl residue at position 33 in the horse protein, the indole side chain is folded back into the hydrophobic interior of the protein, while the imidazole remains on the external, hydrophylic surface. In view of the quantitative identity of functional cross reactivity of the cytochromes c from a very wide taxonomic spread of eukaryotes noted above, it seems probable that this is the general situation. If any spatial structural differences exist at all, they are likely to be so minor that they can hardly be considered to modify the general rule of essential structural identity.

C. The Role of Evolutionary Selection

From the evolutionary point of view this situation immediately poses a fundamental problem. Evolutionary selective processes can only operate by distinguishing between the functional effects of the genes being selected for or against. When the biological results of the products of two genes are precisely identical, no selection can be effected, and the mutational event that led from one gene to the other is termed a *neutral* mutation. If the evidence so far developed is taken at face value, it would appear that for cytochrome c the vast majority of evolutionary fixations of mutations, if not all of them, belong to this category. Indeed, the proteins of all species tested are functionally indistinguishable, and though amino acid sequence variations are as extensive as about 50% of the entire chains, functional differences, if any at all, amount to less that 5%, that being roughly the estimated error of the methods of assay discussed above. The approximately 75% of residue positions that have been observed to vary as among the cytochromes c of different species could, on this basis, be considered to have been subject to functionally irrelevant changes, purely the result of random events. Quite possibly the metabolic machinery of cytochrome c, once perfected in the early ancestral form from which eukaryotes are descended, has not required any further significant functional changes, so that outside of the limitations of structure imposed by its immutable function, the rest of the changes are selectively neutral.

The difficulty is that the above contention is in essence an argument from ignorance, namely, since we have not been able to observe any functional differences between the cytochromes *c* from different species, there are none. As such it is subject to several objections. In the first place, our tests of function may not reveal evolutionarily important differences, as natural selection could well operate on properties not necessarily related to electron transport or oxidative phosphorylation that is, cytochrome *c* could have biological roles other than the classical ones. Alternatively, functional differences that are quantitatively so small as to be below the limits of sensitivity of our tests could have large cumulative evolutionary effects. This second possible objection is to some extent unsatisfactory, since the tests of cytochrome *c* function are carried out under conditions that are very close to those that obtain *in vivo* in nearly intact mitochondria, and the more artifacts of preparation or testing are eliminated[132] and tests are made rigorously quantitative, the more the cytochromes *c* of different species have appeared to be functionally identical, as discussed above. In contrast, the first of the above two objections to the concept of selective neutrality in cytochrome *c* variations is of course very strong, bearing with it the full force and extensive background of Darwinian morphological evolutionary studies. The few cases in which evolutionary changes at the morphological level were at first considered to be the result of selectively neutral variations have so far always been shown on further study to possess clear-cut positive selective value.[97]

An important test of whether the evolutionary changes that do occur in cytochrome *c* are preponderantly selectively neutral or not would be to determine if suitable randomly interbreeding populations of a single species show an appropriate degree of polymorphism in the cytochrome *c* gene. If no polymorphism exists one would have to assume that the evolutionary changes are actively selected for. The major complicating factor in any such test is that cytochrome *c* varies particularly slowly in evolution. Thus, Kimura[69] calculates that if all mutations fixed in a gene are selectively neutral, such mutations would be established in the population with a probability equal to their probability of occurring in an individual. On this basis one can estimate the expected degree of polymorphism according to Kimura and Crow.[70] For a mammal with an average generation time of 4 yr, the observable neutral mutation rate (μ) in the cytochrome *c* gene with approximately one nucleotide substitution every 21.4 million years (see above) would be about 10^{-7} per generation per gamete. With an effective population size (N_e) of 2.5×10^4, the probability that a randomly sampled individual would be homozygous in the cytochrome c gene is

$$p = \frac{1}{1 + 4N_e\mu} = 0.99$$

Conversely, only one individual in 100 is likely to carry an "abnormal" cytochrome c. Thus, to test whether selective neutrality is of common occurrence with cytochrome c variations would require the determination of the complete amino acid sequences of the cytochromes c of several hundred individuals of a properly chosen population. The relatively small numbers of individually tested human and horse preparations[92] are much too small to detect such an expected low level of polymorphism.

This argument as to the possibility that variations of primary structure fixed in the course of evolution in cytochrome c may be of no selective value is just one part of a multifaceted discussion that has been proceeding vigorously during the last few years over the question of the presence or absence of selectively neutral variations in proteins in general (Refs. 73, 64, 130, 128, 129, 105, 4, 69, 68, 71, 72, 106, 9, 18, 19, 113, 145, 45, 35).

Parts of the discussion have been summarized by Fitch and Margoliash[41] and more recently by Harris.[53] There is no need to reproduce it here. However, how the concept of selective neutrality of protein evolutionary changes has fared in the case of cytochrome c, as evolutionary and functional information was accumulated, is an important contribution of this study to our understanding of protein evolutionary transitions in general. It will be further considered below.

III. STATISTICAL PHYLOGENETIC TREES

If one is to attempt to examine the mechanisms by which structures have changed one into the other, the first requirement is to obtain at least an approximation of the topology of their relationships, in biological terms a phylogenetic tree. A simple, but comprehensive, statistical procedure for obtaining such a tree from amino acid sequences was first developed in 1967,[43] and it has since proved to be very fruitful in examining the evolutionary information content of proteins, as described below (see also Refs. 41, 85, 103). Other procedures to estimate the topology of structural relationships and similar parameters have also been used.[22, 99, 133, 76, 116, 109] Of these, the better developed present both some advantages and some disadvantages over the original statistical technique, none of which is crucial, and indeed, the final products are mostly the same for all techniques.

However, if the derivation of the similarity relations of a set of amino acid sequences is to have any biological significance, two prior conditions must be met: (*1*) it must be shown that the structures considered possess a degree of similarity that is greater than random, and (*2*) the proteins considered must be *homologous*, namely, all descended from a common ancestral form.

A procedure for systematic search for significant similarity between amino acid sequences proposed by Fitch in 1966,[40] and more recently improved further,[38] consists in comparing all possible segments of a definite length of the two chains under consideration. For each comparison the minimal mutation or replacement distance is computed. This is the minimal number of single nucleotide substitutions required to transform the gene segment coding for one protein segment into the coding for the other. The replacement distances are then plotted against the number of times each occurs in all the comparisons, to give a curve such as that in Fig. 7. This represents a comparison of human and baker's-yeast *iso*-1 cytochrome *c*. The comparisons indicating random similarity fall within the confines of the Gaussian portion of the distribution curve. The summit of that portion of the curve is near a replacement distance of 45, as expected, since 30 residue segments were compared, and on the average it takes about 1.5 mutations to transform any codon to any other according to the genetic code.[40] The tail of the distribution curve to the left consists of those comparisons for which the replacement distances are smaller than expected for purely random similarities, and its presence therefore, indicates the existence of a significant degree of similarity between human and baker's-yeast cytochrome *c*. The same data can be plotted on a probability scale, as shown in Fig. 8. Here the Gaussian distribution curve becomes a straight line, and it is the deviation from that line to the lower left that denotes comparisons for which the replacement distances are smaller than for random similarities. The more recent developments of the method,[38] among other things, make it possible to assign a probability that the conclusion of significant similarity is in error, namely, that notwithstanding appearances, the similarity observed is a chance event. In the case of Fig. 7 that probability, as indicated, is less than 10^{-80}.

Evidence of significant similarity can be taken as a presumption of homology for the proteins examined. However, lack of such evidence by this test does not necessarily mean that the proteins did not at one time

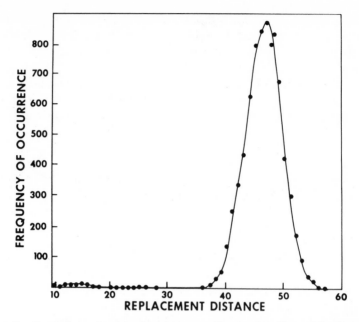

Figure 7. Comparison of minimal replacement distance for all possible 20-residue segments of human cytochrome c and baker's yeast iso-1 cytochrome c (7565 comparisons) by the procedure of Fitch[40]. The numbers of times various values of minimal mutation distances occur in the comparisons are given on the ordinate. The Gaussian portion of the curve on the right represents the random part of the distribution. The extended line on the left indicates those comparisons for which the minimal mutation distances are less than expected for a random distribution and hence suggest ancestral homology. The probability that such a distribution occurred by chance is $10^{-80.38}$

have a common ancestral form. It merely indicates that, even if they did enjoy such a common ancestry, the changes they have undergone since their divergence from the common line of descent are so large that it is impossible to detect any evidence of their homology in the present-day structures. As just noted, significant similarity is only presumptive evidence of homology. Indeed, structures, whether they are front limbs or polypeptide chains, may appear to be similar either because of divergence from a common ancestral form, as for the cases marked "recent divergence" and "parallel evolution" in Fig. 9, or because of so-called functional convergence from independent evolutionary stocks, as in the third diagram of Fig. 9. In this last case the proteins considered would tend to acquire the degree of similarity of structure necessitated by their similarity of function. For example, it is not inconceivable that

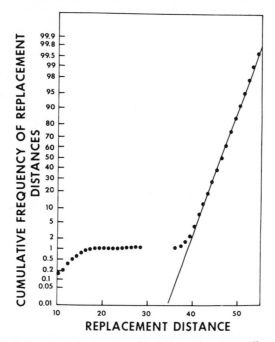

Figure 8.   Probability plot of the data from Fig. 7, according to Fitch[40]. The random part of the distribution is represented by the straight line. The comparisons for which the minimal replacement distances are less than expected for a random distribution are represented by the points that deviate from the straight line at the lower left.

fungal and mammalian cytochromes c could have had quite separate phylogenetic origins. If that were so, one would have to interpret the very considerable degree of similarity indicated by Figs. 7 and 8 as meaning that, in order to carry out its function, a relatively large proportion of the protein is specified with narrow limits. For this reason, even if deriving from independent lines of descent, the baker's-yeast and human cytochromes c could have tended toward structural similarity.

A final determination of whether two sets of proteins are similar because of common ancestry or through functional convergence can be made only if one can read their structures at two distinct times in the history of their evolutionary descent. This would enable one to decide whether during the course of evolutionary time they have become less similar, that is, are diverging from each other, or more similar, that is, are undergoing evolutionary convergence. To do so requires the reconstruction of ancestral sequences. Since this has been done from the

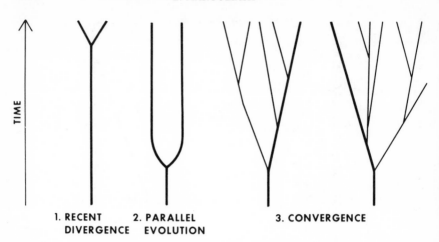

TIME

| 1. RECENT | 2. PARALLEL | 3. CONVERGENCE |
| DIVERGENCE | EVOLUTION | |

Figure 9.   Possible evolutionary reasons for the similarity of present-day polypeptide chains.

statistical phylogenetic trees yet to be described, the distinction between *homologous* (similar because of common ancestry) and *analogous* (similar though of different phylogenetic origins) protein structures is discussed below.

The topology of relations described by the phylogenetic tree in Fig. 10 is based on the amino acid sequences of the cytochromes c of the species listed. Strictly, the only other biological information employed in the computation is the genetic code. It is remarkable that on such a modest basis one obtains a phylogenetic tree that is so close to trees generated by more conventional morphological evolutionary criteria. The procedure is that of Fitch and Margoliash.[43] The minimal replacement distances are calculated for all possible comparisons of the complete amino acid sequences of the cytochromes c employed. For n sequences there are $n(n-1)/2$ replacement distances. One can start by joining together the two proteins that show the smallest replacement distance, and calculate the average replacement distances of all other proteins to the first two, now considered as a single subset. The next protein to be joined to the tree is the one with the smallest replacement distance to the first two. It now joins their subset, and the procedure is repeated until all proteins have been linked. The final tree is merely a graphical representation of the order in which proteins were joined. The number of nucleotide substitutions required between various branch points on the tree can be calculated, and one can then reconstruct the replacement distances between any two proteins by summing the appropriate segments of the tree. This yields the so-called "output" replacement distances, which will differ from the "input" replacement distances, those calculated directly from the amino acid sequences of the proteins and used in the initial step of tree construction. The occurrence of these differences stems from the procedure itself in that, once two proteins are joined, only average distances from all other proteins to the two in the first subset can be used, and such averages are necessarily employed throughout the computation. Because "input" and "output" replacement distances do not exactly match, the initial tree produced does not necessarily provide the most effective utilization of the data. Several possible procedures can be employed to choose an optimal tree. The most useful involves calculating a percent standard deviation from the replacement distances reconstructed from the tree to those determined directly from the amino acid sequences, and examining alternative trees to minimize that criterion of difference. Since the number of possible trees is very large (10^{39} for the proteins of 29 species[42, 85]), not all alternatives can be examined, and common numerical taxonomic methods are used to eliminate unlikely prospects and pick trees for examination.[43, 133, 22] It should not be overlooked, of course, that other criteria can also be employed for deciding which is the "best" tree. For example, one can pick the tree for which the total number of nucleotide replacements required to derive all present-day sequences from the presumed common ancestral form is a minimum.

In the case of cytochrome *c*, the phylogenetic tree obtained is obviously a fair representation of the phylogeny of the species carrying the proteins employed in the computation. This is because the proteins are not only homologous but also *orthologous*, namely, in the most recent common ancestor of all the species considered, at the topmost apex of the tree, cytochrome *c* was represented by a single gene. In that way a one-to-one gene-to-protein relationship is retained throughout, and if the evolutionary variations fixed in this particular gene are a statistically valid representation of the evolutionary changes of the species as a whole, one obtains an accurate species phylogeny. The tree in Fig. 10 is nevertheless not perfect, as, where the primates branch off the ancestral mammalian line before the marsupial, the turtle appears to relate more closely to the birds than to the rattlesnake, and the shark is closer to the lamprey than to the tuna. Such blemishes are likely to be eliminated as the proteins of more species are taken into account. Nevertheless, it is remarkable that as accurate a representation of phylogeny can at all be extracted from the changes of one small gene, out of the millions that constitute a total genome. Without these

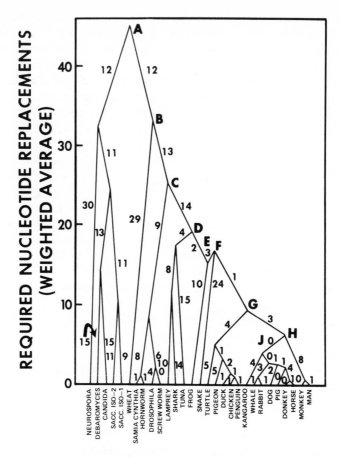

Figure 10. Statistical phylogenetic tree based on the minimal replacement distances between the cytochromes *c* of the species listed, according to Fitch and Margoliash.[43] Each number on the figure is the replacement distance along the line of descent as determined from the best fit to the data so far found. Each apex is placed at an ordinate value that is the average of the sums of all mutations in the lines of descent from that apex, weighting equally the two lines descending from any one apex. References to the amino acid sequences of the cytochromes *c* are given in the legend to Fig. 2. All the proteins in the figure are *orthologous*, with the exception of the *iso*-1 and *iso*-2 cytochromes *c* of baker's yeast, which occur in the same cells and are, therefore, paralogous. As this is the only such relationship, it does not introduce any errors in the rest of the tree.

molecular procedures, a phylogeny as precise as this cannot be obtained even from a single morphological trait, possibly comprising the influence of many genes. One must assume that this is because an accounting of the number of mutations fixed in the course of evolution, even when restricted to a single gene, yields more precise estimates of the degree of evolutionary divergence than can be obtained from single morphological traits. In as wide a sweep of taxonomic groups as represented in the cytochrome c phylogenetic tree, it is unlikely that one can find single morphological characters that are present throughout all species considered, from man to baker's yeast. Similarly, difficulties also arise at the other extreme, when small groups of organisms are all morphologically so similar that it is difficult to distinguish between them. Both problems are readily obviated by the molecular approach. To cover widely divergent phylogenetic relations one would choose a slowly varying protein, while for a small, only slightly divergent group of species one would choose a protein that tends to fix evolutionary changes at relatively short intervals. The exciting prospects presented by this non-dependence of the molecular approach to phylogeny on the complexities and inadequacies of morphology, together with the quantitation inherent in these procedures, the indefinitely increasing precision likely to accrue to them when more and more appropriate sets of proteins are examined, and the possibility of eventually introducing a valid time parameter in considerations of molecular evolution, have been noted above.

Homology does not necessarily imply orthology, as defined above. Indeed, homologous genes may have duplicated but remained side by side in the same species and proceeded to undergo independent evolutionary variations, both varieties being transmitted and continuing to change independently in the descendant species. Such genes have been termed *paralogous* (from *para*, meaning *in parallel*.[41,85]) An examination of the phylogenetic relations that may be derived from the structures of the products of such genes yields not a species phylogeny, but rather a gene phylogeny. This is the case, for example, with the relations postulated by Ingram[60] between myoglobin and the various hemoglobin chains, or later between the two segments of the light chains and between the light and heavy chains of λG-globulins.[122, 57, 56, 114, 30] In the case of hemoglobin, for example, one could, if enough information were available, generate a species phylogeny from the structures of the α chains of the proteins of mammals, and an independent phylogeny from the β chains. However, the most recent common ancestor of mammals already carried both α and β hemoglobin chains, and these chains had by that time undergone considerable independent evolutionary divergence. If one attempted to obtain a species phylogeny by considering the α chains of some mammals and the β chains of others, one would necessarily obtain an absurd result: the species would be primarily segregated depending on whether their α or β chains had been employed in the computation.

IV. ANCESTRAL AMINO ACID SEQUENCES

Once the topology of a set of presumably homologous proteins has been decided, the resultant statistical phylogenetic tree can be used to obtain a variety of derived items of evolutionary information.[41] Among the most informative of these are the amino acid sequences of ancestral forms of the protein at each of the branching points. Originally, as proposed by Fitch and Margoliash,[43] this was accomplished by following a series of rules, according to which each amino acid residue in an ancestral sequence was chosen so that during its phylogenetic descent the corresponding codon required the smallest number of mutations, the fewest segments of the phylogenetic tree containing multiple mutations, or the like. The procedure was directed toward obtaining an approximation to actual evolutionary transitions. It permitted a complete reconstruction of the cytochrome c phylogenetic tree, as well as a test of the reliability of the procedure by reconstructing satisfactorily a model case, consisting of a computer-derived random phylogeny.[42] From the phylogenetic tree for the cytochromes c of 20 species it was possible to show, for example, that the ancestral primate cytochrome c differed from the ancestral mammal protein at residues 17, 18, 21, 56, and 89; while the line of descent of human cytochrome c differed from the *Macaca mulatta* protein line of descent by a single mutation which had occurred in the human line, not the monkey line; and so on.[43]

This was nevertheless a statistically cumbersome process, and Fitch[39] has more recently developed a simplified version according to which, given the topology of relations defined by the statistical phylogenetic tree, the descent of each nucleotide for each codon representing the amino acid sequences is reconstructed separately, as follows. The nucleotide, or nucleotides, assigned to each coding position at an apex is one that is common to both the immediate descendants (the intersection of the descendant nucleotide sets); or when intersection leads to an empty set (namely, there is no common nucleotide among the descendants), the apex is given to the totality of all nucleotides in the two immediate descendants (the union of the descendant sets). At a second stage, a nucleotide replacement is noted whenever an intersection would have yielded an empty set, in such a way that the minimal number of such replacements or "mutations" are required to account for the total phylogeny at the particular coding position under consideration, from the topmost apex to all the descendants. A diagrammatic example of this process is given in Fig. 11. It can be extended to all coding positions to yield a complete reconstruction of the statistical phylogenetic tree with a minimal total number of nucleotide replacements, each assigned to a specific codon and a particular branch of the tree.

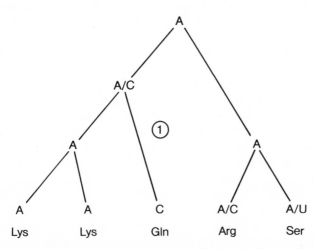

Figure 11. Determination of the ancestral nucleotide for the first nucleotide of the codons for the amino acids marked in the bottom row, given the indicated topology of protein relationships. The procedure is described in the text. The circled 1 denotes that one mutation in the marked branch is the minimum required to derive the ancestral nucleotide.

Whatever the technique employed, a reconstruction of the complete pathway of the evolutionary fixation of mutations will yield a codon and hence an amino acid sequence for every ancestral form of the protein located at each branch point or apex in the phylogenetic tree. Such a reconstruction of the amino acid sequence of the ancestral form at the topmost apex of the tree is given in Fig. 12. Presumably, this represents an approximation to the structure of the cytochrome *c* of the most recent common ancestor of all eukaryotes, which probably already utilized a terminal respiratory chain similar to that so generally distrib-

uted today. More than one residue marked for an amino acid position
in Fig. 12 indicates that the total number of "mutations" in the tree
would not change whichever alternative residue is assigned to that
position, making it impossible to distinguish between them according to
the criterion employed for the reconstruction of ancestral forms. It is, of
course, likely that when enough cytochrome c amino acid sequences
become available for computation, such ambiguities will disappear.
Interestingly, the ancestral sequence has a somewhat smaller preponderance of basic residues than the present-day proteins. This is a
character one would expect for an ancestral protein, since the extreme
basicity of extant cytochromes c is most probably an adaptation to
function from an originally less specialized precursor more like the
majority of proteins with near neutral isoionic points. When the remaining ambiguities in the sequence of the ancestral form have been resolved, a particularly important investigation will be the laboratory
synthesis of that protein with a view to examining its functional activities in present-day enzyme systems. Possibly, this would gain an experimental basis for the study of the functional concomitants of
evolutionary molecular transitions and might serve to define and
quantitate the role of selective as contrasted with possible nonselective
events. These not too distant possibilities of extending the transformation of evolutionary molecular studies into a fully experimental science
were mentioned in the introductory section above.

Applying the newer technique of reconstructing ancestral sequences,
Fitch[39] was able to develop a method for distinguishing analogous from
homologous sets of proteins. Phylogenetic trees are independently computed for two sets of proteins. The codon sequences of the corresponding ancestral forms are obtained and examined to determine whether
these are more similar or less similar to each other than the present-day
sequences at each coding position. If at a certain position the ancestral
forms are the same but the descendants are different, then the latter
have undergone evolutionary divergence, by definition, and can be
taken to be homologous. If, on the other hand, the ancestral forms are
different, any similarities developed among the descendants are due to
evolutionary convergence, and they can be considered to be only
analogously related. Extending this to the entire ancestral nucleotide
sequence gives how many positions in the two sets of proteins are
divergently and how many are convergently related. An appropriate
calculation performed with two sets of totally unrelated sequences yields
an estimate of how many divergent and how many convergent relationships would be expected on a purely random basis. When the two

```
        Ala   Glu         Thr                    Pro
Ala -   Lys - Ser - Ser - Ala - Gly - Val - Ser  Ala - Gly - Asn - Ala - Lys - Lys - Gly - Ala - Lys  Leu - Phe - Lys
-9    [Glu][Ala]  Phe  Ser              Phe [Thr] Pro                      Glu                   Asn   Ile   10
      [Thr]                                 [Ala]  1
```

```
        Lys                                     Glu                    Lys
Thr -   Cys - Ala - Gln - Cys - His - Thr - Val - Glu - Gly - Gly - Gly - Thr - His - Lys - Val - Gly - Pro - Asn
      Arg                                       Ala                  [Arg]
         └── HEME ──┘                      20                                                        30
```

```
                                  Ser
                                  Lys       Gln
Leu - His - Gly - Leu - Phe - Gly - Arg - Thr - Gly - Ala - Glu - Gly - Tyr - Ser - Tyr - Thr Asp - Ala
                                  Gln        [Pro]      Ala
                                  40                                                    50
```

```
        Lys              Val   Lys         Glu
Asn - Lys - Lys - Lys - Gly - Ile - Thr - Trp - Asp - Glu - Asn - Thr - Leu - Phe - Glu - Tyr - Leu - Glu - Asn - Pro
      Asn                                                                                            70
                                        60
```

```
                                                                              Glu
                                         Ala       Gly                        Pro
Lys - Lys - Tyr - Ile - Pro - Gly - Thr - Lys - Met - Phe - Gly - Leu - Lys - Lys - Ala - Lys - Asp - Arg
                                         Val       Ala                        Ala
                                  80                                   [Gln]  90
```

```
Glu
Ala                                                          Ala   Glu
Thr - Asp - Leu - Ile - Ala - Tyr - Leu - Lys - Lys - Ala - Thr - Ser - Ala
Lys                Thr         Met           100               [Thr] Ser
Asn                                                                  104
```

Figure 12. Amino acid sequence of the ancestral form of cytochrome *c* at the topmost apex of the phylogenetic tree. Any of the alternative amino acids shown would permit the evolution of the 29 descendent cytochromes *c* in the minimum number of 366 nucleotide replacements, assuming the topology shown in Fig. 10. Amino acids in brackets have not yet been observed in any present-day cytochrome *c*.

sets of proteins examined have a significantly greater number of divergently related coding positions over that resulting from chance, then one can be certain that the two sets are homologous. An excess of convergently related positions demonstrates that they are analogous.

The results of a comparison of fungal and nonfungal cytochromes *c* by this procedure are graphed in Fig. 13. It is clear that the two sets of cytochromes have such a large preponderance of divergences that the probability that they are not homologously related is only 1 in 10^{10}, when a total of 24 different cytochromes *c* is utilized for the two phylogenetic trees in the comparison. The extensive similarities of amino acid sequence between the cytochromes *c* of a wide taxonomic range of species were noted very early in these studies and used to suggest a common evolutionary origin for them.[83, 126, 89] However, these comparisons could not, in fact, distinguish structural similarity due to common ancestry from similarity resulting from evolutionary convergence. As just described, this has now been accomplished by Fitch.[39]

His procedure provides for the first time a solid statistical demonstration that all eukaryotic cytochromes c are indeed the descendants of a common ancestral form, thereby giving a definitive proof that all living forms on the earth, as we know them, are the result of a single occurrence, the basic tenet of the so-called unitary theory of the origin of life.

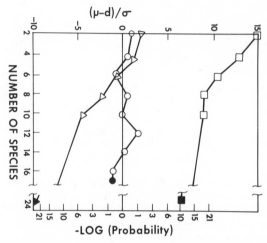

Figure 13. Convergence and divergence as a function of the number of species. The abcissa gives the total number of sequences examined. Open symbols indicate that equal numbers of sequences were present in the two groups compared; closed symbols, that they were divided unequally between the two groups. On the left, the ordinate gives the deviation δ from expectation μ-d, in units of standard deviation σ; on the right, it gives the equivalent probability of a result being due to chance, as negative powers of 10. Points above the zero line represent an excess of divergent comparisons; below the line, an excess of convergent comparison. Random sequences of amino acids are shown by circles (O—O), convergent sequences by triangles ($\triangle$—$\triangle$), and divergent sequences of squares ($\square$—$\square$). The convergent sequences were obtained by computer simulation. The divergent sequences compare fungal with nonfungal cytochromes c. According to Fitch.[39]

Contrary to appearances, the technique for resolving homology from analogy in protein structure does not contradict the principle that knowledge of any system at one point in time by itself cannot define the direction in which the system is changing as a function of time. Indeed, it cannot distinguish between pure divergence and pure convergence. In the case of the descent of the cytochromes c plotted in Fig. 13, for example, on this basis one cannot tell the difference between divergence from a single common ancestor and convergence of the 24 proteins toward an identical descendant from 24 independent evolutionary origins. What is ruled out absolutely is any mixture of divergence and convergence. If one accepts that the proteins of each of two small taxonomic groups of species are homologous within each group, one can proceed to demonstrate whether or not homology is also the case for the two groups together. Biologically, the evidence is overwhelming that at least within a well-defined taxonomic group, such as mammals, a highly conservative structure such as cytochrome c does not have multiple phyletic origins. Moreover, from the molecular point of view, to postulate an independant ancestral origin for each cytochrome c would make it impossible to understand how the structures of such a set of proteins could possibly yield a statistical phylogenetic tree (Fig. 10) so close to that deduced from classical zoological criteria. In essence, the argument is like that between the concepts of the evolutionary origin of species and of separate creation for each living form. The latter becomes less and less tenable as more species are considered, just as pure convergence becomes less and less likely as similar results are obtained with further sets of proteins.

V. RATES OF EVOLUTIONARY CHANGE IN CYTOCHROME c CODONS

A cursory examination of the composite amino acid sequence of eukaryotic cytochromes c (Fig. 2) and its comparison with the sequences of the individual proteins makes it obvious that the tendency to vary in the course of evolution changes over a very wide range for different positions. As already noted, some positions are invaried, others accept few alternatives and remain unvaried over fairly extensive taxonomic groupings of species, and still others are most commonly changed whenever any amino acid substitutions whatsoever occur. Calculations of overall rates of evolutionary residue substitutions or of their corresponding nucleotide substitutions in the structural gene for the protein[89,86,23] can only yield averages of many rates, which differ over a wide range. As described above, an approximately linear relation of cytochrome c variations to elapsed evolutionary time is manifested and serves to define the concept of *unit evolutionary period*.[89] Nevertheless, such calculations obviously cannot help in specifying why some codons change while others do not, and what are the mechanisms operating to restrict evolutionary variability, as to rate and as to acceptable alternatives, at the level of individual positions in the protein. To these ends it is first necessary to attempt to classify the codons for cytochrome c into classes with respect to the ease with which they undergo evolutionarily effective nucleotide substitutions, and to attempt to identify the codons in each class. One would then be in a position to examine the interdependence of structure-function relations as between the corresponding residues of each class, which will hopefully lead to a complete recognition of the series of evolutionary events transforming one protein into another along the branches of the phylogenetic tree. This is the sort of historical description of molecular evolutionary phenomena likely to provide the background required for the unraveling of mechanisms operating at that level, as referred to in the first section of this paper. Only a fraction of this program has so far been achieved.

A. Hypervariable, Normally Variable and Invariant Codons

The first approach to the classification of cytochrome c codons according to their apparent rates of evolutionary change[44] depended on a statistical phylogenetic tree for 20 different cytochromes c[43] and the following assumptions:

1. There are particular amino acids in the protein so vital to structure-function that the probability that a strain would survive evolutionarily a mutation in the codons for these residues is essentially nil. The probability of such a variant occurring in a tree depicting the descent of the amino acid sequence of the protein is therefore also near zero. These are commonly termed "invariant codons," and any mutation in them is defined as *malefic*, since any line carrying such a mutation will because of it eventually fail to survive.

2. The vast majority of variable amino acid positions correspond to codons that undergo mutations that are retained in the course of evolution at rates sufficiently similar that each codon can be considered to have as much chance as any other to fix the next mutation. Thus, the frequency of occurrence of such mutations in these codons will approximate a Poisson distribution. These are defined as "normally variable codons."

3. The relatively few remaining codons each carry so high a probability of undergoing an evolutionarily effective mutation that the number of changes observed in the corresponding residues is large enough to permit their ready identification and exclusion from the set of normally variable codons. These are the so-called "hypervariable codons."

The reconstruction of the ancestral sequences from the first phylogenetic tree,[43] the 20 cytochromes c of which showed 35 unvaried positions (compare with Fig. 2), identified the number of mutations that had occurred in every codon along each branch of the tree, for a total of 231. There were five codons showing nine or more mutations each. These were considered to be hypervariable. Eliminating them left 182 mutations, which could distribute themselves over a number of codons between 105 (if the invariant class defined in paragraph 1 above did not exist at all) and 70 (if the invariant class was as large as the 35 unvaried positions observed in the 20 cytochromes c). On this basis, one could identify the Poisson distribution that most closely fitted the distribution of mutations observed on the reconstructed phylogenetic tree, namely, the number of codons with 1, 2, 3,... mutations. The best fit was found for 27 to 29 invariant codons, with the remaining 76 to 78 positions representing so-called normally variable codons. This result clearly demonstrated that a very large proportion of the unvaried residues so far observed in the cytochromes c of eukaryotes were in fact invariant, namely, would never be observed to change, however many cytochromes c of different eukaryotes were examined. This is in sharp contrast with the situation for myoglobins, for example, in which even with a much narrower taxonomic coverage, no more than three residues are found to be unvaried.[103]

More recently, Fitch and Markowitz,[45] and Markowitz[95] employing a statistical phylogenetic tree for 29 cytochromes c, have further developed this type of approach to estimates of codon variability. The reconstruction of ancestral sequences from that phylogenetic tree (see Fig. 10) gave 366 mutations to be distributed over 113 codons. The improved method does not arbitrarily exclude any codons and can deal simultaneously with two Poisson-distributed classes. It was found that the distribution of codons having undergone 1, 2, 3,... evolutionary fixations was adequately fitted by a model that assumes 32 invariant residues, 65 "normally variable" residues, and 16 "hypervariable" residues. This last class fixes mutations in the course of evolution about 3.2 times more often than the normally variable set. On the basis of the statistical phylogenetic tree employed, more than two classes of variable residues could not be distinguished. This, of course, does not mean that with many more amino acid sequences and a correspondingly larger tree, more classes of codon variability will not be uncovered. The descriptive limitations are merely imposed by limitations in the sequence data and in the sensitivity of the method, as well as by the degree to which the property examined is statistically well behaved.

The above estimates of numbers of invariant residues were carried out before the more recent additions to our catalog of cytochrome c amino acid sequences, at a time when 34 residues had not been observed to vary. However, with the amino acid sequences of the proteins from the two protists listed in the legend to Fig. 2 and from the ginkgo tree, variations were observed in eight previously unvaried positions. Five are from the sequence of *Euglena* cytochrome *c*, one from that of the *Physarum* protein, one from both these cytochromes *c*, and one from the ginkgo protein. Perhaps the least expected are the alanine in position 14 replacing the usual thioether heme-bonded cysteine, leaving the heme covalently attached only through one cysteinyl residue, and the variants at positions 75 and 77 in the center of the heretofore inviolate invariant segment extending from residue 70 to residue 80. It is particularly interesting that these changes have occurred in the cytochromes *c* of species outside the range of taxonomic groups previously explored, and it is obviously important to determine whether these cytochromes *c* do or do not have the functional characteristics common to all the other cytochromes *c* examined to date. If they do, then clearly the previously observed lack of variation at the positions now showing variants cannot be ascribed to requirements constant throughout the eukaryotic range. If they do not, then these cytochromes *c* would become prime objects of studies aimed at understanding the fundamental mechanisms of all the eukaryotic proteins. As always, the exploitation of manipulations performed by nature, when they can be recognized as such, is likely to be particularly fruitful.

B. Concomitantly Variable Codons—Covarions

The estimate of the number of evolutionarily invariant codons in the gene for the cytochromes *c* of eukaryotes was obtained by utilizing all of the phylogenetic tree under the topmost apex in Fig. 10, apex *A*. The extent of invariance corresponds to nearly 30% of the cytochrome *c* gene. When the calculation is done again, excluding the cytochromes *c* of fungi, namely, utilizing only that portion of the tree under apex *B*, the degree of predicted invariance increases. It increases further and further as the taxonomic span of species is contracted by gradually excluding the proteins of groups remote from mammals. The values obtained for the set of species under apices A, B, C, and so on of Fig. 10 are plotted in Fig. 14 as a function of the weighted averages of the replacement distances for all species taken into account for each re-calculation, that is, the ordinate values at which the apices are located in Fig. 10. A roughly linear regression is obtained, which on extrapolation to a replacement distance of zero exhibits an evolutionary invariance of over 90% of the gene. This particularly important application of their technique for estimating the extent of invariance by Fitch and Marko-witz[45] thus leads to the quite unexpected demonstration that for any one mammalian cytochrome c at the present time (corresponding to a replacement distance of zero), only about 10 residues are available to evolutionary amino acid substitutions. Any mutation leading to a change of amino acid in any other than these approximately 10 allowed positions is by this evidence necessarily malefic, resulting sooner or later in the extinction of the line that carries it. The calculations were carried out with a view to contracting the taxonomic coverage in the direction of mammals, only because cytochromes c from various groups of mammals were much more generously represented than any other group in the phylogenetic tree, affording many points, and thus greater preci-sion, near the ordinate toward which extrapolation was required. At

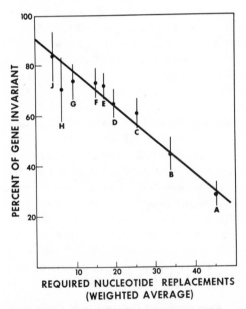

Figure 14. Estimation of the number of concomitantly variable codons (*covarions*) for cytochrome c. The percent of the gene found to be invariant is plotted as a function of the weighted average of required nucleotide replacements (heights of apices in Fig. 10). Letters A to J represent the groups of cytochromes c indicated in Fig. 10. The line at each point is an estimate of the standard deviation of the ordinate value of the point. A weighted least-squares fit to the results for cytochrome c is extrapolated to the abcissa to estimate the fraction of the cytochrome c gene for which all mutations are lethal or malefic[44] in any one mammalian species. The codons that do not fall in this category are the covarions. According to Fitch and Markowitz.[45]

present there are nearly enough cytochromes c of known amino acid sequence from other groups, such as plants, fungi, and insects, to make it possible to direct calculations toward them. It will, therefore, soon be possible to determine directly whether the number of residues capable of accepting evolutionary variations is the same or different for the proteins from different taxonomic groups of species. The codons corresponding to those amino acid positions, which in any one species and at any one time in the course of evolution are free to fix mutations, are termed *concomitantly variable codons*, a phrase contracted in current usage to *covarions*.[45]

More recently, employing a quite different statistical approach, Fitch[34] has estimated that the number of covarions for cytochrome c is 4.5, confirming the essential previous result that the evolutionary variability of cytochrome c is very severely restricted. The method depends on the number of double mutations, defined as the fixation of two mutations within a single codon between two successive branch points (apices or nodes) in the statistical phylogenetic tree. The number of double mutations will depend on the number of covarions and the persistence of a codon among the set of covarions after it has fixed a first mutation. The reconstruction of the phylogenetic tree yields the total number of mutations and the number of double mutations in each internodal segment, and hence the distribution of double mutations as a function of the number of mutations observed in all such segments. The number of covarions and their persistence of variability that yields a distribution most closely approximating that observed on the statistical phylogenetic tree is taken to represent the actual situation. The best fit was found for 4.5 covarions with a persistence of variability of 0.04. This indicates that, on the average, the likelihood of a covarion losing its place among the set of covarions, namely, becoming invariable as a result of the fixations of mutations among other covarions, is as high as or higher than 0.75. Thus, not only is the number of covarions small, but also their rate of turnover is high as compared to the numbers of evolutionary variations fixed in cytochrome c.

C. Evolutionary Significance of Covarions

Notwithstanding the remarkably small number of residues that appear to be amenable to evolutionary substitutions at any one time in the history of the descent of any one cytochrome c, the comparison of the amino acid sequences of the over 50 cytochromes c of known primary structure shows that more than 75% of the residue positions have varied during evolution (see Fig. 2). Clearly then, one must assume that when a mutation is fixed in a particular covarion this will in many cases also change some of the residues in the current set of covarions. Some residues previously invariable become variable, while others previously variable now lose their status as covarions and become invariable. Over prolonged periods of evolutionary history over two-thirds of the molecule can be made to change in this fashion. As noted above, the statistical calculation[34] does indeed show that the turnover of covarions is remarkably rapid in cytochrome c, indicating that the variable positions in the protein are to a large extent interdependent. A change in one tends to affect the other currently variable positions in a large proportion of cases.

Cytochrome c has, therefore, a very low tolerance to evolutionary change. Only some 4 to 10 residues, on the average, are available for substitution at any one time, a picture that is very different from the common simplistic view that since a residue has changed at some time in the evolution of one particular cytochrome c, it is available for change in any cytochrome c at all times. This is possibly the most clarifying contribution that the concept of covarions has brought to our general understanding of the evolutionary changes of protein structure. The number of covarions can be considered an inverse expression of the stringency and extent of structure-function requirements, and as such, this number is a fundamental parameter expressing the quantitative effect of function on the evolutionary behavior of a protein.[85]

The rate of evolutionary change of a protein is indeed related to the number of its covarions. Thus, the calculations,[34-36,45] which yielded 10 as the number of covarions for cytochrome c, also showed that in fibrinopeptides A (the segment of fibrinogen removed by the action of thrombin at the initiation of clotting), 18 of the 19 residues are covarions, and that in the α chain of hemoglobin there are 50 covarions out of 141 residues, while in the β chain of hemoglobin there are 39 covarions out of 146 residues. Since in all these cases the proteins for the horse and pig are of known amino acid sequence and had been included in the statistical phylogenetic trees which served to estimate the number of covarions, one could calculate the rates of evolutionary fixation of mutations per codon and per covarion in these two lines of descent since their most recent common horse-pig ancestor.[35] For cytochrome c these rates turned out to be 0.048 per codon and 0.50 per covarion, for α hemoglobin 0.156 per codon and 0.44 per covarion, for β hemoglobin 0.212 per codon and 0.82 per covarion, and for fibrinopeptide A 0.684 per codon and 0.72 per covarion. The rates of evolutionary fixations per codon are highly variable, reflecting the slow changes of

cytochrome c structure, the intermediate rates of change of the hemoglobin chains, and the very rapid variation of fibrinopeptides. The
corresponding unit evolutionary periods (see above) are about 20 million years for cytochrome c, 5.8 million years for hemoglobin, and 1.1
million years for fibrinopeptide A.[23] In sharp contrast with this spread,
the values for residue fixations per covarion are so near each other that,
within the error of the method, they may well be considered to be the
same. In other words, as a function of the number of positions capable
of accepting mutations without leading to the disappearance of the line
of evolutionary descent, four different proteins show the same rate of
evolutionary change, even though their unit evolutionary periods vary
over a range of about 20-fold.

Here one comes up once more against the argument as to the
presence or absence of neutral mutations in proteins in general, and in
cytochrome c in particular. If a large proportion of mutations fixed in
the course of evolution in proteins are selectively neutral, as maintained
by Kimura,[69] then covarions are the only sites at which mutations can
manifest themselves. All other codons are prevented by selective forces
from undergoing any evolutionarily effective changes whatsoever.
Moreover, since neutral mutations are the result of random occurrences,
all proteins should have the same rate of neutral mutation per covarion.
This is sufficiently similar to what is observed for cytochrome c, the α
and β chains of hemoglobin, and fibrinopeptides A,[35] as outlined above,
that these results can be taken to suggest that, unless a radically
different explanation for the similarity of rates of protein change per
covarion is forthcoming, neutral mutations may well account for a
significant proportion of all protein evolutionary changes.

In summary, in the case of cytochrome c, one is left with the concept
of a protein whose structure is very severely restricted by structural or
functional requirements, so that in any one of its many evolutionary
forms, only a small number of positions (4 to 10, approximately, out of
the over 100 residues) are available for at least some possible
evolutionary variations. These positions are, moreover, strongly interre-
lated, so that when one is changed, the others are likely to be affected in
such a way that some become invariable, while others that were previously invariable are now open for at least some variation. All other
positions are required to carry precisely the amino acid residue found in
them at the time. Selective influences are so effective that the occurrence of any other residues would inexorably lead to the extinction of
the line carrying such forbidden variants. Natural selection is thus
conceived only as a force that constrains the variability of the protein in
evolution, eliminating the deleterious, but it is not seen providing any

advantage by selecting for cytochromes c better adapted to their function or internal milieu. In support of this view, the excellent functional evidence available appears to indicate that no cytochrome c is any better adapted to its own mitochondria than to those of any other species. Moreover, the calculations of the rates of evolutionary residue substitution per covarion, described above, yield similar rates for four different proteins, a result compatible with the idea that the majority of evolutionary changes in these proteins are selectively neutral.

With this fairly sophisticated picture of the possible course of evolution in cytochrome c, it is interesting to examine the molecular models of the ferric and ferrous forms of the protein, recently achieved by x-ray crystallography,[25, 137] to determine whether one can observe in them the structural expressions of the expected widespread evolutionary constraints.

VI. THE SPATIAL STRUCTURE OF CYTOCHROME c AND CONSTRAINTS ON EVOLUTIONARY VARIATIONS

An understanding of protein evolutionary variations necessarily includes a rather thorough knowledge of protein function, since selection must operate via functional characteristics. In the case of cytochrome c we are fortunate that the extent of knowledge of the amino acid sequences of the protein from many species, leading to the effective statistical approaches to molecular evolutionary phenomena already outlined, is matched by satisfactory solutions for its spatial structure in both functional states, the ferric and the ferrous.[25, 137] That all this information has not yet led to a much completer and generally accepted description of the evolutionary transitions of the protein and of their mechanisms is due in part to the fact that knowledge of spatial structure is merely a first basis upon which one may plan the experiments required to understand functional activities. Furthermore, it is also possible that identified physiological functions, in the case of cytochrome c electron transfer in the terminal oxidase segment of the respiratory chain, may not be the only or even the most important property with respect to evolutionary selection. Other properties conceivably related to the integration and maintenance of the protein in its functional environment, to its biosynthesis, or even to unknown physiological functions, and so on, may play crucial evolutionary roles. Nevertheless, it is clear that present knowledge of its spatial structure has already given some insight into a few of the evolutionary aspects of cytochrome c variations and provided possibly useful limits in regard to others. These are described below.

A. The Structures of Ferric and Ferrous Cytochrome c

The structures of horse and bonito cytochromes c in the ferric form have been obtained to a resolution of 2.8 Å.[25] The molecules are prolate spheroids of about $30 \times 34 \times 34$ Å if one includes the side chains, and represent the most perfect illustration yet of the "oil drop" model of protein structure, with the hydrophobic side chains packed in the inside, and the polar residues at the surface. The heme is held in a deep crevice perpendicular to the surface termed the "front" of the molecule, and in Figs. 4 and 5 is seen mostly along the front edge of the heme plate (that of pyrrole ring II above and III below). In Fig. 5, which represents an α-carbon diagram of ferrous tuna cytochrome c,[137] the protein has been kept in the same relative position as in Fig. 4, which represents ferric tuna cytochrome c. However, in the conformation range attendant to oxidoreduction, the heme has slightly shifted, and one can now see a little less of its left surface. Cysteines 14 and 17, which are bound to the vinyl side chains of rings I and II, respectively, are at the upper right edge of the heme. The hemochrome-forming groups—the imidazole of histidyl residue 18 and the sulfur of methionyl residue 80—extend from the right and the left, respectively, to the central heme iron atom to coordinate at positions 5 and 6. The propionyl side chains of the heme point downwards to the bottom of the crevice. The interior or posterior propionic acid is buried in the hydrophobic interior of the molecule and held there by hydrogen bonds to the phenolic hydroxyl of tyrosine 48 and the indole of tryptophan 59. The front propionyl side chain is in the polar external surface. Interestingly, the imidazole of histidine 18, which is coordinated to the heme iron by its ε-nitrogen, is also hydrogen bonded by its δ-nitrogen to the main-chain carbonyl of residue 30 at the lower front right of the molecule. This probably serves both to hold the imidazole ring in a rigid conformation for reaction with the heme iron and also to transmit any movements imparted by heme changes to the coordinated imidazole, such as in oxidoreduction, to the right half of the molecule. Indeed, the imidazole-to-residue-30 hydrogen bond is one of several that appear to make the right lower part of cytochrome c the most rigid portion of the whole, relatively plastic molecule. In contrast (see Section II.B), the left lower part of the molecule, on the side of the hemochrome-forming methionyl residue 80, appears to be more readily deformable and capable of considerable movement.

The molecule seems to be made in roughly two halves: Residues 1 to 47, including the segment of α-helix of residues 1 to 11, comprise the right side of the molecule; residues 48 to 86, the left; while most of the remainder, residues 87 to 102, form the second stretch of α-helix across the back of the molecule from right to left, like the strap of a suitcase. There are two areas at which the hydrophobic interior seems to meet the surface. These have been called "channels," although they present no access of the solvent to the interior. The left "channel" is bordered by residues 52 to 74, while the right "channel" is surrounded by residues 6 to 20 and the last portion of the C-terminal α-helix.

Other than the helical regions, the molecule is essentially a shell of extended chain wrapped around the heme. Hydrophobic side chains are nearly all kept in the interior of the molecule, in contact with each other or with the heme. These contacts appear to provide a good proportion of the forces stabilizing the native conformation, as the overall number of hydrogen bonds is smaller than for several proteins of the same size. Such a structural arrangement also explains the early observation of evolutionary stability of the hydrophobic segments of cytochrome *c*.[83,89] Internal residues, which have to fit in a limited space, and whose close packing with other side chains or with the heme is structurally important, are likely to be preserved as long as a change in one residue—by leaving more free internal space, for example—has not made it possible for another to vary. This is very much the situation described by the existence of few covarions in cytochrome *c* (see above). Those positions that show a wide spectrum of acceptable substituents are external, as expected from the point of view of evolutionary selection. However, some external residues show conservative substitutions, and some are invariant. Thus, the external versus internal localization is far from sufficient to explain the evolutionary variability of residues, and one must seek other parameters to explain the behavior of external residues.

Polar residues are in general external. The exceptions are few, and all seem to be of functional significance. Among them is the interior propionyl side chain of the heme, which is held down in the bottom of the hydrophobic crevice by hydrogen bonds. This is an energetically expensive arrangement. It is very different from that in the hemoglobins and myoglobins, in which the propionyl side chains are in the external solvent, and is likely to represent an important, if still unknown, functional requirement of cytochrome *c*. The imidazole of histidyl residue 18 is internal, as it is coordinated to the heme iron. An interesting case is that of the evolutionarily invariant tyrosyl residue 67. It is entirely internal. Its hydroxyl is hydrogen bonded to threonyl residue 78, and the aromatic ring is, as it were, poised between contact with the sulfur of the heme-coordinated methionyl residue 80 in front of it and with the ring of tyrosyl residue 74 at the back left surface of the protein, behind it. Such a remarkable architecture, coupled with invariance, invites a functional or structural explanation. The hypothesis that this residue is involved in electron transfer inside the protein molecule,[25] according to the mechanism proposed by Winfield[148] and by King, Looney, and Winfield,[74] does not, at this time, appear very likely. It requires the transient presence of a free-radical form of tyrosyl residue 67 in the process of cytochrome c reduction. Since acetylation of this tyrosine does not inhibit the electron-transfer activity of the protein,[62] the participation of any such free radical would appear to be ruled out. The recent observation of a phenylalanine at position 67 in a protist cytochrome c yields the same conclusion.

The strongly basic and acidic amino acid side chains are all on the exterior of the molecule and distributed in a remarkable fashion. The lysines are largely in the vicinity of the right and left "channels" described above, making the two sides of the molecule strongly cationic. In contrast, a large proportion of all acidic residues (in horse cytochrome c, for example, 9 out of 12) are located in a rather circumscribed patch at the upper half of the back of the molecule, about halfway between the positively charged regions. What is conserved in evolution is the acidic character of this area, not the acidic character of individual residues in it, all of which, except for residue 90, can also accomodate nonacidic amino acids. Basic residues never invade the negative patch, which, at a minimum, contains six acidic side chains. This type of evolutionary conservatism is similar to that observed for the clusters of hydrophobic residues in the amino acid sequences of the cytochromes c of different species,[83,89] except that the conservation of the acidic patch occurs in regard to the spatially folded native molecule, not the primary linear structure. Here again, an explanation of this architecture is needed. Whether the ideas that the basic sides are involved in reaction with the oxidase and reductase enzyme systems, and that the acidic patch binds with a mitochondrial membrane cytochrome c site,[25] have any validity remains to be determined.

As the chain folds to cover the molecular surface, there are several sharp bends and abrupt reversals of chain direction (see Figs. 4 and 5). Some of these involve the proline residues, while others occur at bends resembling single turns of the tightly hydrogen-bonded 3_{10} helix. There are six such bends in ferricytochrome c, three of which are in one and three in the other of the two conformations such bends may take.[147] Invariant glycines occur in tight corners in which there is no place whatsoever for side chains, and in horse cytochrome c glycines occur in the positions necessary to make possible the three type-II 3_{10} bends. Here again, this is insufficient to explain evolutionary invariance, as residue 45 is a constant glycine and it is on the surface of the molecule.

The distribution of aromatic residues is also remarkable. In the ferric protein, tyrosine 74 and tryptophan 59 have their rings parallel. The indole side chain points straight into the molecule from the lower third of the middle of the back and hydrogen-bonds to the internal propionyl side chain of the heme, while the ring of the tyrosine is flat on the surface at the left back corner of the protein. The ring of the completely internal tyrosine 67 is at an angle to these two aromatic rings and, as described above, seems to be bridging the gap between the two parallel rings and the heme-iron-bound sulfur of methionine 80. Two other pairs of aromatic rings are also parallel and more or less in close proximity in ferricytochrome *c*. One of these, consisting of tyrosine 48 and tyrosine or phenylalanine 46, is at the bottom of the molecule directly below the heme, with their aromatic rings perpendicular to the heme plane. Tyrosine 48 is also hydrogen bonded to the posterior heme propionyl side chain. The other pair is not strictly parallel and slightly further apart than those already mentioned. It consists of the invariant phenylalanine 10 and the tyrosine or phenylalanine at position 97. They are located on the top right side of the molecule and form the top boundary or project into the top of the right channel. The only other invariant phenylalanine, residue 82, is held in the external solvent in the middle of the left side, in the ferric protein. The immediately preceding residue 81 —commonly isoleucine, but also sometimes valine or alanine—is abutting the surface of the protein, but not projecting into the external medium. This leaves a sizable solvent-accessible pocket on the left side of the upper front part of the heme plane, in the vicinity of pyrrole ring II, but also possibly extending farther in and higher.

This is the area in which the most dramatic conformation change occurs on reduction (see Fig. 5). Indeed, other than the aromatic pair at the upper right of the molecule (residues 10 and 97), the other aromatic rings mentioned above are all in some way involved in the ferric⇌ ferrous functional transition of the protein. On reduction, there are large rotations in the region of residues 80 to 82, resulting in the aromatic ring of phenylalanine 82 being moved away from the external medium and inserted into the solvent pocket near the front left top of the heme, filling it up and leaving the phenyl and the heme rings more or less parallel. At the same time residue 81 is rotated off the surface, ending up hanging completely in the external medium, like a projecting tail. The solvent exclusion and organization effects involved in these movements and those of other hydrophobic groups are likely to be important components of the energetics of oxidoreduction in cytochrome *c*. The hitherto invariant segment, residues 70 to 80, partly surrounds this area from below, and as already noted (see Section II.B) shows evidence of having considerable mobility. This property could well be important for function, and if in addition, as previously suggested,[92,25] this surface is one of interaction with an enzyme system, the relative evolutionary constancy of residues 70 to 80 would appear to be explained.

The only other obvious conformation change in the ferric-to-ferrous transition is performed by the loop of residues 21 to 23 near the middle of the right side of the protein. This segment moves up, so that in the reduced protein the right "channel," which is wide open in the ferric protein, is partly blocked. It may be noted that this movement is the one that is likely to explain the greater resistance of the reduced protein to proteolysis by an enzyme such as chymotrypsin, whose first point of attack is at phenylalanyl residue 10, and the second at tyrosyl residue 97, both near the top of the right channel.[88]

In summary, present knowledge of the spatial structures of cytochrome c in both its oxidation states clearly points to the protein being a highly specific piece of molecular machinery, undergoing remarkable changes in protein conformation as it transfers electrons. Satisfactory explanations of protein evolutionary variations are, therefore, certain to be closely linked to our understanding of the workings of this machine.

B. Molecular Distribution of Evolutionarily Variable Positions

From the evolutionary point of view, one of the most useful products of the solution of the structure of cytochrome c by x-ray crystallography is the possibility of examining the distribution of variable and invariant residues in the folded native molecule. In the preceding section, the structural reasons why some residues appear to be invariant were pointed out, and at the same time it was made clear that in no case were satisfactory explanations available for the invariance of every single residue of any one type or category.

With regard to variable residues, a most interesting observation can be made, namely, that the positions at which evolutionary residue substitutions are fixed are different for different taxonomic groups of species.[90] Moreover, and quite strikingly, these locations are clustered into rather well-defined tertiary structure regions for each such group. An example of this sort of distribution is given in Fig. 15(a), which shows the positions at which the cytochromes c from mammals vary. For this taxonomic group, residue substitutions are limited to a semicircular band in front, above, and behind the heme, leaving both lateral surfaces of the molecule largely unvaried. As noted in Fig. 15(b), (c), and (d), the distribution of variable positions for insect cytochromes c is quite different, and that for plant cytochromes c, though partially overlapping with that for insects, has yet another distinct clustered distribution, as has the distribution of variant residues in fungal cytochromes c.

That such differences are statistically significant has been demonstrated by Fitch[37] for an example of four fungal and four metazoan species. A previously developed technique is employed. It permits the estimation of the numbers of the evolutionarily invariant positions, of those that are variable but have not yet been observed to vary, and of those that have varied in a protein, from the reconstruction of the corresponding statistical phylogenetic tree[45,95] (see above). The orthologous amino acid positions of the proteins of two groups of species are of four possible types with respect to the evolutionary fixation of amino acid substitutions: those that have not fixed any substitutions in either group of proteins, those that have fixed mutations in one group, those that have fixed mutations in the other, and those that have fixed mutations in both. The numbers of these four types of positions can be computed if one assumes a number for the positions that are invariant in common for both groups of proteins. Computing for every admissible number of these invariant positions gives a series of results, which are compared with those observed directly from the reconstructed phylogenetic tree for the cytochromes c of the two groups of species. The best-fitting set of values is taken to give the best estimate for the number of common invariant positions. For four fungal and four metazoan cytochromes c this was found to be 41, corresponding to about 29 for the invariable metazoan codons that are variable in the fungal proteins, about 17 for the codons variable in the metazoan proteins but invariable for the fungi, and about 23 for the codons variable in both groups. Clearly, there is a partial overlap of variable and invariant positions, but also considerable numbers of positions are variable in one group but not in the other, and vice versa.

These observations make it clear that not only are selective forces operating against changes of cytochrome c structure to prevent any but a very few residues from varying at any one time of evolutionary history in any particular species, but that these forces in fact operate, as they must, at different sets of molecular sites for different taxonomic groupings of species. Indeed, without such a situation one could not explain how a proportion of the protein of over 75% has been found to differ in the cytochromes c of all species examined to date, when statistical calculations of variability demonstrate that only 4 to 10 residue positions are covarions, that is, can fix evolutionary changes in any one protein at any one time (see above). In contrast, the quite unexpected

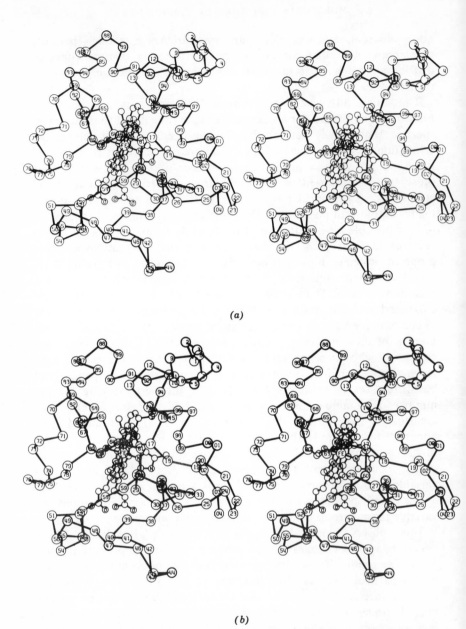

(a)

(b)

Figure 15. Stereoscopic α-carbon diagrams of horse cytochrome *c* in the ferric form, according to Dickerson et al.[25] The residues marked by heavy circles are those that vary among the cytochromes *c* of mammals (14 species) in (*a*), of insects (4 species) in (*b*), of higher plants (11 species) (*c*), and of fungi (6 species) in (*d*).

238

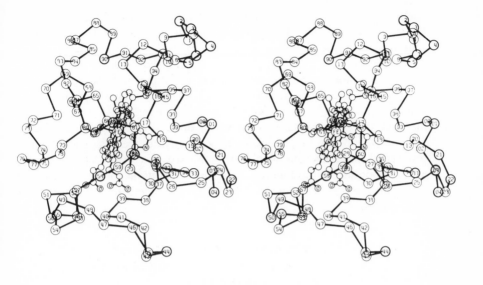

(c)

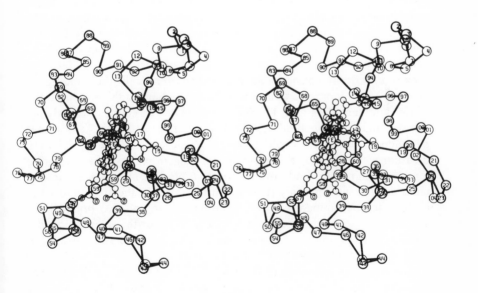

(d)

Figure 15. (*cont.*)

result deriving from the examination of the locations of variability on the molecular models of the protein is that the variable positions are spatially clustered, and have distinct distributions in the cytochromes c of different taxonomic groups of species. This is difficult to understand if one is to maintain that all evolutionary variations in cytochrome c are the result of selectively neutral events. For example, why should insect cytochromes c not be permitted to change in the regions of the molecule in which the cytochromes c of mammals do, and vice versa? If all these variations had no functional biological significance whatsoever and proceeded randomly, one would expect to see, at a first approximation, a correspondingly random distribution of positions of variability in different taxonomic groups, or at least a complete interdigitation of such positions.

VII. ION-BINDING PROPERTIES OF CYTOCHROME c AS A BASIS FOR SPECIES VARIABILITY

As the classical electron-transport properties of cytochrome c give no indication of serving to provide functional distinctions between the proteins of different species, even when tested in nearly intact mitochondrial membranes,[11, 12, 132, 131] it is imperative to attempt to discover whether other functional attributes of cytochrome c exist that could serve as a basis for evolutionary selection. Apart from the amino acid sequences themselves, the one set of properties that has so far been shown to exhibit considerable species differences is the binding of anions.[6] Many inorganic anions were found to bind to the protein in the ferric state, including Cl^-, I^-, SO_4^{2-}, and PO_4^{3-}. In contrast, cacodylate showed no detectable binding, nor did any of the cations tested, so that *tris*-cacodylate buffers could be used whenever a strictly nonbinding medium was required. By free-boundary electrophoresis, two chloride ions were shown to bind, and the relative binding affinities at the two chloride sites were roughly reflected in the ratios of the areas under the electrophoretic boundaries corresponding to one and two bound ions. This ratio was found to be 0.8, 1.8, 6, 9, and 32 for the screw worm-fly, human, hog, pigeon, and moth (*Samia cynthia*) proteins, respectively, implying large differences in binding affinity.

A more recent extension of this study[91] has led to a rather remarkable correlation between the ion-binding properties of cytochrome c and the ion-transport properties of mitochondria.[110, 75] It was found that ions that are normally excluded from the mitochondrial matrix, typified by chloride, bind to ferricytochrome c but not to the ferrous protein; that those that are forcibly carried into the matrix by the use of chemical energy, such as calcium ions, bind to ferrocytochrome c but not to the ferric protein; and that those that flow in or out of the mitochondrial matrix in response to the concentration gradient, but whose movement is carrier mediated, such as phosphate or ADP, bind to cytochrome c in both oxidation states. Many examples of each class were tested and observed to follow this general rule. Some are given in Table II. On this basis it was proposed that cytochrome c may act as a carrier for these ions in the inner membrane of mitochondria.[91] A similar conclusion was reached by Schejter and Margalit[118] from the effect of a few of these ions on the oxidation-reduction parameters of the protein. Since that time, attempts to demonstrate directly an ion-carrier function for cytochrome c in mitochondria have failed. This may be because it has not proved possible[11] to deplete the organelles of their cytochrome c to an extent of more than about 80% without disrupting the inner membrane. The remaining cytochrome c appears to be in a compartment different from the majority of the protein, as it does not reequilibrate readily with it. If the nonextractable fraction were to prove to be in the lipid phase, it could well represent the totality of the cytochrome c involved in some way in ion transport—hence the failure to observe any effect of ordinary cytochrome c depletion of mitochondria on such phenomena.

TABLE II

Electrophoretic Mobility of Deionized Horse Cytochrome c Towards the Cathode in the Presence of Various Anions[a]

	Electrophoretic mobility at 1°C $(10^5 \text{ cm}^2/(\text{sec})(\text{V}))$	
Added ions	Ferricytochrome c	Ferrocytochrome c
None	6.1	4.6
Cl^-	3.7; 4.2	4.6
K^+	6.1	4.6; 5.1
Ca^{2+}	6.1	5.1; 5.4
PO_4^{3-}	4.3	2.6
Citrate	1.5	1.3
α-Ketoglutarate	1.9	2.9

[a]Solutions contained 0.407 to 1.017 mM cytochrome c, 0.10M $tris$-cacodylate buffer, pH 7.2, and 0.05 M of added ions, brought to pH 7.2 with $tris$ or cacodylic acid as required. When two electrophoretic boundaries occur, both mobilities are listed according to Margoliash, Barlow, and Byers.[91]

Notwithstanding these difficulties and negative results, the possible functional involvement of cytochrome c with ions remains an attractive possibility. If this were in regard to the transport of ions, it would very likely suffice to provide the selective forces necessary to explain how the cytochromes c of different species acquired their particular amino acid sequences. Indeed, the ion-transport and ion-binding properties of the mitochondria of different species have been known to vary quite extensively, even though no thorough systematic studies have been carried out so far.[110,79,13,17,144,146] The transport of ions across the inner mitochondrial membrane affects the intra- and extraorganelle concentrations of such metabolites as citrate, ADP, and ATP, all known to have crucial metabolic regulatory effects.[5,47]

Furthermore, if the cytochrome c in the mitochondrial membrane were divided into pools of several interacting molecules, rather than individually strung on separate respiratory chain assemblies, ion movement via such cytochrome c pools could proceed by successive reequilibrations as the ion passes from one protein molecule to another on its way through the pool.[91] As a result, large differences in the ability to transport various ions would arise from small changes in affinity, since the ratio of the rates of transport of two ions would be proportional to the ratio of their binding affinities to the carrier, to the power of the number of carrier molecules encountered in one passage. Such a situation could similarly elucidate the very large temperature coefficients for the mitochondrial transport of ATP and ADP.[75]

It should also not be overlooked that the relation of cytochrome c to ions need not necessarily be as a transport carrier to account for the evolutionary phenomena. Conceivably, the protein—a macromolecular component of the inner mitochondrial membrane that is available on its outer surface and can readily move in and out of it[88,146]—could act to regulate the flow of ions to their transmembrane carriers, if the cytochrome c sites or pools were in some way interposed between ion carriers and the external medium. Moreover, the ion-binding properties of cytochrome c may be important in connection with its reaction with the oxidase and reductase enzyme systems, as there are large differences in the ion-binding properties of the two oxidation states of the protein (see Table II). However, on such a basis it would be difficult to see how structural differences could have developed, since cytochromes c of widely differing amino acid sequences react identically with these systems, as discussed above.

Almost any physiological role that can be ascribed to the binding of ions to cytochrome c will also probably resolve the problem of the clustering of positions of variability in the proteins of different taxonomic groups, if such a role is shown to differ for the different groups of species. The mitochondria of any such group are likely to be physiologically similar, and relatively different from those of another taxonomic group. If different locations on the surface of cytochrome c bind different sets of ions, the mitochondria of the species in one taxon, because of their metabolic similarity, are likely to tolerate changes in the binding of certain ions but not in those of others, while in another taxon, required and facultative bindings will relate to different ions and hence to different surface clusters on the molecule. In any case, it is to be hoped that, whatever the fate of the hypothesis of the connection of cytochrome c with these phenomena, extensive comparative studies of the mitochondrial activities of different species will be forthcoming, to serve as a basis for the understanding of the functional variables affecting the evolutionary transformations of the molecular components of the organelle.

VIII. CONCLUDING SUMMARY: CHARACTER OF THE EVOLUTIONARY TRANSFORMATIONS OF CYTOCHROME c

Notwithstanding the controversy that still rages about whether evolutionary residue substitutions in proteins are selected for or are the result of so-called neutral mutational events, there is no question that the examination of the amino acid sequences of a protein in a wide taxonomic range of species has yielded considerable and often novel insight into both molecular evolutionary processes and structure-function relations. Cytochrome c remains to date the prime example of the successful application of this approach. It has served and continues to serve as a model case for the development of the statistical techniques for extracting evolutionary information from amino acid sequences.

As soon as a few cytochrome c sequences were available,[83] it became obvious that the numbers of variant residues tended to be constant in comparisons of the protein from large taxonomic groupings, all cytochromes c from mammals being, for example, roughly equally different from all bird cytochromes c, all vertebrate cytochromes c equally different from insect cytochromes c, and so forth. The protein thus appeared to vary at a constant rate in the course of evolution, permitting the definition of the *unit evolutionary period* as the time required to allow, on the average, one residue substitution to occur in the cytochromes c of two diverging lines of phylogenetic descent. By comparison with the known times of major phylogenetic divergences, one could then show that cytochrome *c* was a slowly varying protein, carrying a characteristic unit evolutionary period of about 20 million years.[89] However, such calculations could only represent rough approximations, and any further advance was predicated on the computation of precise topologies of structural relationships, which, in the case of an orthologous set of proteins, should correspond to a species phylogeny. Following the development of a statistical technique for distinguishing nonrandom from random degrees of sequence similarity,[40] this was accomplished by comparing amino acid sequences in terms of minimal replacement distances, the minimal number of single nucleotide changes required to transform the gene coding for one protein into that for another. A simple procedure yielded statistical phylogenetic trees in remarkably good accord with expectation, even though the only biological information they contained was the amino acid sequences of a numer of cytochromes *c* and the genetic code.[43] These trees provided the solid basis upon which the further conceptual and technical developments have rested. Among these may be mentioned the derivation of the amino acid sequences of ancestral forms of cytochrome *c*,[43,39] which has brought within reach the synthesis of the protein from extinct and very primitive forms of life and the possible attendant clarification of the mechanisms of molecular evolutionary developments. It also became possible to demonstrate quite unequivocally that all cytochromes *c* are descended from one ancestral form.[39] This is the first time a unique gene has been followed to very early periods of evolutionary history, providing direct proof of the concept of the unitary origin of life on this planet.

The reconstruction of a statistical phylogenetic tree gives the distribution of codons having undergone 1, 2, 3, ... mutations, and on this basis it is possible to estimate how many codons are invariant and how many are included in Poisson-distributed variable classes. The data from the phylogenetic tree fits an invariant class of about one-third of the protein, a normally variable class, and a "hypervariable" class.[44,45] As the taxonomic span of the species considered in the computation is gradually decreased, the invariant class increases till it comprises over 90% of the cytochrome *c* of any one species. The remaining variable codons, termed *covarions*, represent only 4 to 10 residues,[45,34] and correspond to different amino acid positions in the proteins of different groups of species.[37] Their rate of turnover is high,[34] so that when a covarion is changed, other covarions tend to lose their status and become invariable, while previously invariable residues tend to join the groups of covarions. This contagiousness of variability explains how more than 75% of the protein has varied over the whole span of its evolutionary history, while only 4 to 10 residues are variable at one time in any one species.

A related observation is that the distribution of variable residues on the x-ray crystallographic model of the protein[25] takes the form of distinct and only partially overlapping clusters for the proteins of separate taxonomic groups.[90] The surprising element is that these variable residues are clustered, rather than randomly distributed as one would expect if precisely the same set of structure-function selective constraints were to determine the spreading of covarions during the evolutionary history of all groups of species. The number of covarions is indeed an inverse quantitative expression of the extent and stringency of selective constraints on the evolutionary variability of proteins, so that the rate of residue substitutions per covarion appears to be more or less constant for four proteins whose rates of change as a function of the total number of residues varies from fastest to slowest over a range of about 20-fold. Respectively, these are fibrinopeptides A, the hemoglobin α and β chains, and cytochromes c.[35]

The question of the importance of selectively neutral residue changes arose from the observation of the apparent constancy of the rate of evolutionary change of the protein[83,89] and the great contrast between the extensive structural variability and ostensible functional identity among the cytochromes c of various species.[92] A possible test of the existence of selectively neutral variants—namely, the expected consequent polymorphism of the protein in a suitable population—has not been carried to the point of giving a dependable result (see above). As already discussed, the newer approaches to the study of evolutionary structural variations have also given equivocal answers. The constancy of the rate of evolutionary fixation of mutations per covarion, for four different proteins, argues for the neutralist point of view. In contrast, the observed clustering of variable positions seems to point to the proteins of different taxonomic groups of species having different functional requirements, thereby making selection possible. The hypothesis that the ion-binding properties of the protein are in some way related to the distribution or the movement of ions in mitochondria[91] has not been proved, though it still remains an attractive possibility as a basis for such selective effects.

Moreover, it should be noted that the apparent randomness of evolutionary changes in a group of variable codons, as evidenced from the fitting of the data by Poisson distributions, does not imply either the

absence or the presence of natural selection.[37] Appropriate fixations in a
group of codons could eventually lead to a selective advantage, but the
order in which these fixations occur may well be partially determined by
which codon first mutated, and that could be a purely random event.
On the other hand, had no fit to Poisson distributions been observed,
one could have eliminated the possibility of neutral mutations. The
precise order of a series of evolutionary changes in protein structure is
at least in part determined by previous changes, so that every mutation
fixed restricts the further evolutionary pathways available. This
phenomenon is rather similar to developmental processes, for which
subsequent events depend upon prior ones having occurred. Such a
situation is implicit in the existence of limited numbers of covarions
having a high rate of turnover. Ordered mutation series may have
simple structural backgrounds, as when particular charges have to
remain neutralized, or internal space is limited so that a bulky group
must be eliminated before a new one can be introduced.[45] The same
types of quasidevelopmental series could conceivably even explain the
clustered distributions of variable positions among the cytochromes c of
separate taxonomic groups, shown in Fig. 15. If the contagiousness of
variability described above were largely local with respect to the spatial
structure of the protein, one would end with spatially more or less
segregated clusters within which the proteins of one taxonomic group
would have changed. Moreover, the older the taxon, the more widespread one would expect the locations of variability to be. This last
point is in agreement with the large number and relatively scattered
distribution of variable positions in the six fungal cytochromes c examined to date [Fig. 15(d)], as compared with the few and strictly
clustered locations of change among 15 mammalian cytochromes c [Fig.
15(a)].

Whatever the final compromise between the pure neutralist and pure
selectionist positions that will be found to express reality in the
evolutionary variations of cytochrome c, it is already quite obvious that
the molecule is under extremely stringent selective pressures that prevent all but a very few residue substitutions from being acceptable for
any species at any one time in its evolutionary history. Structural
reasons for some of these selective pressures were apparent from the
spatial structure of the protein, as noted above, but it was also clear that
such explanations encompass only a fraction of the residue positions.
The reasons for the constancy of some residues and for the variability of
others could not be deciphered on this basis. A major difficulty may be
that in the case of cytochrome c all effects of natural selection have

been conceived only as eliminating undesirable variants. Positive adaptive features have so far not been detected, and their existence remains a matter of faith or disbelief.

Cytochrome c is an antique structure. It is recognizably present in all eukaryotes. The recent detection of amino acid sequence and possible spatial structure similarities between the cytochromes c of eukaryotes and those of prokaryotes[45,41,24] Fitch and Margoliash, 1970; Dickerson, makes it likely that the family tree of the protein extends much further back than previously suspected. Even more remarkably, a direct crystallographic study has shown that exactly the same folding is present in *Rhodospirillum* cytochrome c_2,[115] locating the structure at the very earliest of evolutionary times, even before respiratory chains of the modern type had been assembled in their full complexity. It would seem that once a successful solution to a protein structural problem is attained in evolution, the same well-tempered basic machinery is adapted to as many functions as it can satisfactorily encompass. This is certainly the case for the cytochrome c spatial structure. It can already be seen that a fully developed statistical phylogenetic tree for this structure will contain many different cytochromes, depict paralogous as well as orthologous relationships, and cover the entire living world. The fundamental three-dimensional pattern exemplified by cytochrome c is best termed the *cytochrome fold*. Such is the layout of the expected major avenue of advance of molecular evolutionary studies discussed in the first section of this presentation. The overall number of successful basic protein structures is certain to be much smaller than the number of known proteins, which have classically been identified by their functions or physicochemical characteristics without any knowledge of how these relate to their spatial patterns. A biological classification of proteins in terms of their fundamental types of folding will soon be necessary. This would be an essentially evolutionary classification and have all the advantages that phylogeny gives to other biological systems of classification. What is mainly conserved in the course of molecular evolution may well be the basic amino acid chain folding patterns. These are preciously maintained, in that only a relatively small number are available to undertake the myriad functions of life and to serve them against the many hazards of the biological milieu at all its developmental stages. Thus, our difficulties in understanding the mechanisms of evolutionary transitions in cytochrome c may well stem from our lack of knowledge of the relation between the amino acid sequence of a protein and its spatial folding into a functional structure. Not until this much-studied problem is effectively solved will it be possible to unravel the laws governing molecular evolution, a process whose main thrust seems to be the preservation of biologically successful spatial structures.

Acknowledgments

The author would like to take this occasion to thank his many colleagues, who, over the years, have borne the brunt of the amino acid sequence determination work. He is also very grateful to Dr. W. M. Fitch of the University of Wisconsin for a most pleasant and fruitful collaboration on the examination of the evolutionary information content of amino acid sequences, and to Dr. R. E. Dickerson at the California Institute of Technology for a similar collaboration on the determination of the spatial structure of cytochrome c. The present work was supported by grant GM 19121 from the National Institutes of Health.

References

1. R. P. Ambler, M. Bruschi, and J. Le Gall, in *Recent Advances in Microbiology*, A. Perez-Miravete and Dionisio Pelaez, Eds., Asociacion Mexicana de Microbiologia, Mexico City, 1971, p. 25.
2. K. Ando, H. Matsubara, and K. Okunuki, *Biochim. Biophys. Acta*, **118**, 240, 256 (1966).
3. K. Ando, H. Matsubara, and K. Okunuki, *Proc. Japan Acad.*, **41**, 79 (1965).
4. N. Arnheim and C. E. Taylor, *Nature*, **223**, 900 (1969).
5a. D. E. Atkinson, *Ann. Rev. Biochem.*, **35**, 85 (1966).
5b. R. C. Augusteyn, M. A. McDowell, E. C. Webb and B. Zerner, *Biochim. Biophys. Acta*, **257**, 264 (1972).
5c. R. C. Augusteyn, *Biochim. Biophys. Acta*, **303**, 1 (1973).
6. G. H. Barlow and E. Margoliash, *J. Biol. Chem.*, **241**, 1473 (1966).
7. K. Bitar, S. N. Vinogradov, C. Nolan, L. Weiss, and E. Margoliash, unpublished experiments (1972).
8. D. Boulter, E. W. Thompson, J. A. M. Ramshaw, and M. Richardson, *Nature*, **124**, 789 (1970).
9. S. H. Boyer, E. F. Crosby, T. F. Thurmon, A. N. Noyes, G. F. Fuller, S. E. Leslie, M. K. Shepard, and C. N. Herndon, *Science*, **166**, 1428 (1969).
10. K. Brew, T. C. Vanaman, and R. L. Hill, *J. Biol. Chem.*, **242**, 3747 (1967).
11. V. Byers, C. H. Kang, and E. Margoliash, unpublished results (1972).
12. V. Byers, D. Lambeth, H. A. Lardy, and E. Margoliash, *Fed. Proc.*, **30**, 1286 (1971).
13. E. Carafoli, W. X. Balcavage, A. L. Lehninger, and J. R. Mattoon, *Biochim. Biophys. Acta*, **205**, 18 (1970).
14. S. K. Chan, *Biochim. Biophys. Acta*, **221**, 497 (1970).
15. S. K. Chan, I. Tulloss, and E. Margoliash, *Biochemistry*, **5**, 2586 (1966).
16. S. K. Chan, S. B. Needleman, J. W. Stewart, O. F. Walasek, and E. Margoliash, *Fed. Proc.*, **22**, 658 (1963).
17. J. B. Chappell, and K. N. Haarhoff, in *Biochemistry of Mitochondria*, E. C. Slater, Z. Kaninga, and L. Wojtczak, Eds., Academic, New York, 1967, p. 75.
18. B. Clarke, *Science*, **168**, 1009 (1970).
19. B. Clarke, *Nature*, **228**, 159 (1970).
20. F. H. C. Crick, in *The Biological Replication of Macromolecules*, *Symp. Soc. Exp. Biol.*, **12**, 138 (1958).

21. F. H. C. Crick, L. Barnett, S. Brenner, and R. J. Watts-Tobin, *Nature*, **192**, 1227 (1961).

22. M. O. Dayhoff, *Atlas of Protein Sequence and Structure*, Natl. Biomedical Res. Foundation, Silver Spring, Md., 1972.

23. R. E. Dickerson, *J. Molec. Evolution*, **1**, 26 (1971).

24. R. E. Dickerson, *J. Molec. Biol.*, **57**, 1 (1971).

25. R. E. Dickerson, T. Takano, D. Eisenberg, O. B. Kallai, L. Samson, A. Cooper, and E. Margoliash, *J. Biol. Chem.*, **246**, 1511 (1971).

26. R. E. Dickerson, T. Takano, O. B. Kallai, and L. Samson in *Structure and Function of Oxidation Reduction Enzymes*, Å. Åkeson and Å. Ehrenberg, Eds., Pergamon Press, Oxford, 1972, p. 69.

27. R. E. Dickerson, M. L. Kopka, J. E. Weinzierl, J. C. Varnum, D. Eisenberg, and E. Margoliash, *J. Biol. Chem.*, **242**, 3015 (1967).

28. K. Dus, K. Sletten, and M. D. Kamen, *J. Biol. Chem.*, **243**, 5507 (1969).

29. G. M. Edelman, and W. E. Gall, *Proc. Natl. Acad. Sci., U.S.A.*, **68**, 1444 (1971).

30. G. M. Edelman, B. A. Cunningham, W. E. Gall, P. D. Gottlieb, V. Rutishauser, and M. J. Waxdal, *Proc. Natl. Acad. Sci., U.S.A.*, **63**, 78 (1969).

31. P. Edman, and G. Begg, *Europ. J. Biochem.*, **1**, 80 (1967).

32. A. Ehrenberg, and H. Theorell, *Acta Chem. Scand.*, **9**, 1193 (1955).

33. R. W. Estabrook, in *The Chemistry of Hemes and Hemoproteins*, B. Chance, R. W. Estabrook, and T. Yonetani, Eds., Academic, New York, 1966, p. 405.

34. W. M. Fitch, *J. Molec. Evolution*, **1**, 84 (1971).

35. W. M. Fitch, *Brookhaven Symp. Biol.*, **23**, 186 (1972).

36. W. M. Fitch, in *Haematologie und Bluttransfusion*, Lehmann, München, 1971, p. 199.

37. W. M. Fitch, *Biochem. Genet.*, **5**, 231 (1971).

38. W. M. Fitch, *J. Molec. Biol.*, **49**, 1 (1970).

39. W. M. Fitch, *Systematie Zoology*, **19**, 99 (1970).

40. W. M. Fitch, *J. Molec. Biol.*, **16**, 9 (1966).

41. W. M. Fitch and E. Margoliash, in *Evolutionary Biology*, Th. Dobzhansky, M. K. Hecht, and W. C. Steere, Eds., Vol. 4, Meredith, New York, 1970, p. 67.

42. W. M. Fitch and E. Margoliash, *Brookhaven Symp. Biol.*, **21**, 217 (1968).

43. W. M. Fitch and E. Margoliash, *Science*, **155**, 279 (1967).

44. W. M. Fitch and E. Margoliash, *Biochem. Genet.*, **1**, 65 (1967).

45. W. M. Fitch, and E. Markowitz, *Biochem. Genet.*, **4**, 579 (1970).

46. P. T. Gilham, *Ann. Rev. Biochem.*, **39**, 227 (1970).

47. T. W. Goodwin, *The Metabolic Roles of Citrate*, Academic, New York, 1968.

48. R. K. Gupta and A. G. Redfield, *Science*, **169**, 1204 (1970).

49. L. Gurtler, and H. J. Horstmann, *Europ. J. Biochem.*, **12**, 48 (1970).

50. H. A. Harbury and P. A. Loach, *J. Biol. Chem.*, **235**, 3640 (1960).

51. H. A. Harbury and P. A. Loach, *Proc. Natl. Acad. Sci., U.S.A.*, **45**, 1344 (1959).

52. H. A. Harbury, J. R. Cronin, M. W. Fanger, T. P. Hettinger, A. J. Murphy, Y. P. Myer, and S. Vinogradov, *Proc. Natl. Acad. Sci., U.S.A.*, **54**, 1658 (1965).

53. H. Harris, *J. Med. Genet.*, **8**, 444 (1971).

54. J. Heller and E. L. Smith, *J. Biol. Chem.*, **241**, 3165 (1966).

55. J. Heller and E. L. Smith, *Proc. Natl. Acad. Sci., U.S.A.*, **54**, 1621 (1965).

56. R. L. Hill, H. E. Lebowitz, R. E. Fellows, and R. Delaney, in *Gamma-Globulins*, J. Killander, Ed., Wiley-Interscience, New York, 1967, p. 109.

57. R. L. Hill, R. Delaney, R. E. Fellows, and H. E. Lebowitz, *Proc. Natl. Acad. Sci., U.S.A.*, **56**, 1762 (1966).
58. R. W. Holton and J. Myers, *Biochim. Biophys. Acta*, **131**, 362, 375 (1967).
59. V. M. Ingram, *Nature*, **189**, 704 (1961).
60. V. M. Ingram, *The Hemoglobins in Genetics and Evolution*, Columbia U. P., New York, 1963.
61. H. A. Itano, *Advan. Protein Chem.*, **12**, 215 (1957).
62. K. M. Ivanetich, J. R. Cronin, J. R. Maynard, and H. A. Harbury, *Fed. Proc.*, **30**, 1143 (1971).
63. E. E. Jacobs and D. R. Sanadi, *J. Biol. Chem.*, **235**, 531 (1960).
64. T. H. Jukes and J. L. King, *Nature*, **231**, 114 (1971).
65. M. D. Kamen and T. Horio, *Ann. Rev. Biochem.*, **39**, 673 (1970).
66. D. Keilin, *The History of Cell Respiration and Cytochrome*, Cambridge U. P., London, 1966.
67. D. Keilin, *Proc. Royal. Soc. (London), Ser. B*, **98**, 312 (1925).
68. M. Kimura, *Proc. Natl. Acad. Sci., U.S.A.*, **63**, 1181 (1969).
69. M. Kimura, *Nature*, **217**, 624 (1968).
70. M. Kimura and J. F. Crow, *Genetics*, **49**, 725 (1964).
71. M. Kimura and T. Ohta, *Nature*, **229**, 467 (1971).
72. M. Kimura and T. Ohta, *J. Molec. Evolution*, **1**, 1 (1971).
73. J. L. King and T. H. Jukes, *Science*, **164**, 788 (1969).
74. N. K. King, F. D. Looney, and M. E. Winfield, *Biochim. Biophys. Acta*, **133**, 65 (1967).
75. M. Klingenberg, *FEBS Letters*, **6**, 145 (1970).
76. D. E. Kohne, *Quart. Rev. Rev. Biophys.*, **3**, 327 (1970).
77. G. Kreil, *Z. Physiol. Chem.*, **340**, 86 (1965).
78. G. Kreil, *Z. Physiol. Chem.*, **334**, 154 (1963).
79. A. L. Lehninger, E. Carafoli, and C. S. Rossi, *Advan. Enzymol.*, **29**, 259 (1967).
80. D. K. Lin, Ph.D. Thesis, University of Wisconsin, 1971.
81. E. Margoliash, in *The Chemistry of Hemes and Hemoproteins*, B. Chance, R. W. Estabrook, and T. Yonetani, Eds., Academic, New York, 1966, p. 271.
82. E. Margoliash, *Canad. J. Biochem.*, **42**, 745 (1964).
83. E. Margoliash, *Proc. Natl. Acad. Sci., U.S.A.*, **50**, 672 (1963).
84. E. Margoliash, *Brookhaven Symp. Biol.*, **15**, 266 (1962).
85. E. Margoliash and W. M. Fitch, in *Homologies in Enzymes and Metabolic Pathways*, W. J. Whelan, Ed., North-Holland, Amsterdam, 1970, p. 33.
86. E. Margoliash and W. M. Fitch, *Proc. N. Y. Acad. Sci.*, **151**, 349 (1968).
87. E. Margoliash and J. Lustgarten, *J. Biol. Chem.*, **237**, 3397 (1962).
88. E. Margoliash and A. Schejter, *Adv. Protein Chem.*, **21**, 113 (1966).
89. E. Margoliash and E. L. Smith, in *Evolving Genes and Proteins*, V. Bryson and H. J. Vogel, Eds., Academic, New York, 1965, p. 221.
90. E. Margoliash, W. M. Fitch, E. Markowitz, and R. E. Dickerson, in *Structure and Function of Oxidation Reduction Enzymes*, Å. Åkeson and Å. Ehrenberg, Eds., Pergamon Press, Oxford, 1972, p. 5.
91. E. Margoliash, G. H. Barlow, and V. Byers, *Nature*, **228**, 723 (1970).
92. E. Margoliash, W. M. Fitch, and R. W. Dickerson, *Brookhaven Symp. Biol.*, **21**, 259 (1968).
93. E. Margoliash, S. B. Needleman, and J. W. Stewart, *Acta Chem. Scand.*, **17**, S250 (1963).
94. E. Margoliash, E. L. Smith, G. Kreil, and H. Tuppy, *Nature*, **192**, 1125 (1961).

95.   E. Markowitz, *Biochem. Genet.*, **4**, 594 (1970).

95a. H. Matsubara and E. L. Smith, *J. Biol. Chem.*, **238**, 2732 (1963).

96.   J. R. Mattoon and F. Sherman, *J. Biol. Chem.*, **241**, 4330 (1966).

97.   E. Mayr, *Animal Species and Evolution*, Harvard U. P., Cambridge, Mass., 1966.

98.   C. C. McDonald, W. D. Phillips, and S. N. Vinogradov, *Biochem. Biophys. Res. Commun.*, **36**, 442 (1969).

99.   G. W. Moore and M. Goodman, *Bull. Math. Biophys.*, **30**, 279 (1968).

100. G. A. Mross and R. F. Doolittle, *Arch. Biochem. Biophys.*, **122**, 674 (1967).

101. T. Nakayama, K. Titani, and K. Narita, *J. Biochem.*, **70**, 311 (1971).

102. K. Narita, K. Titani, Y. Yaoi, H. Murakami, M. Kimura, and J. Vanecek, *Biochim. Biophys. Acta*, **73**, 670 (1963).

102b. K. Narita, K. Titani, Y. Yaoi and H. Murakami, *Biochim. Biophys. Acta*, **77**, 688 (1963b).

103. C. Nolan and E. Margoliash, *Ann. Rev. Biochem.*, **37**, 727 (1968).

104. C. Nolan and E. Margoliash, *J. Biol. Chem.*, **241**, 1049 (1966).

105. P. O'Donald, *Nature*, **221**, 815 (1969).

106. T. Ohta and M. Kimura, *J. Molec. Evolution*, **1**, 18 (1971).

107. S. Paléus, A. Ehrenberg, and H. Tuppy, *Acta Chem. Scand.*, **9**, 365 (1955).

108. K.-G. Paul, *Acta Chem. Scand.*, **5**, 379 (1951).

109. L. Pauling, and E. Zuckerkandl, *Acta Chem. Scand.*, **17**, S9 (1963).

110. B. C. Pressman, in *Membranes of Mitochondria and Chloroplasts*, E. Racker, Ed., Van Nostrand–Reinhold, New York, 1970, p. 213.

111. J. A. M. Ramshaw, M. Richardson, and D. Boulter, *Europ. J. Biochem.*, **23**, 475 (1971).

112. A. G. Redfield and R. K. Gupta, *Cold Spr. Harb. Symp. Quant. Biol.*, **36**, 405 (1971).

113. R. C. Richmond, *Nature*, **225**, 1025 (1970).

114. V. Rutishauser, B. A. Cunningham, C. Bennett, W. H. Konigsberg, and G. M. Edelman, *Proc. Natl. Acad. Sci., U.S.A.*, **61**, 1414 (1968).

115. R. Salemme, S. T. Freer, and J. Kraut, personal communication, 1972.

116. V. M. Sarich and A. C. Wilson, *Science*, **154**, 1563 (1966).

117. A. Schejter and P. George, *Nature*, **206**, 1150 (1965).

118. A. Schejter and R. Margalit, *FEBS Letters*, **10**, 179 (1970).

119. A. Schejter and M. Sokolovsky, *Fed. Europ. Biochem. Soc. Ltrs.*, **4**, 269 (1969).

120. A. Schejter, I. Aviram, and M. Sokolovsky, *Biochemistry*, **9**, 5113, 5118 (1970).

121. E. Shechter and P. Saludjian, *Biopolymers*, **5**, 788 (1967).

122. S. J. Singer and R. F. Doolittle, *Science*, **153**, 13 (1966).

123. K. Skov, T. Hofmann, and G. E. Williams, *Canad. J. Biochem.*, **47**, 750 (1969).

124. E. L. Smith, in *Structure and Function of Cytochromes*, K. Okunuki, M. D. Kamen, and I. Sekuzu, Eds., University of Tokyo Press, Tokyo, 1968, p. 282.

125. E. L. Smith, *Harvey Lectures, Ser. 62*, 1968, p. 231, Acad. Press, N. Y.

126. E. L. Smith and E. Margoliash, *Fed. Proc.*, **23**, 1243 (1964).

127. E. L. Smith, H. Matsubara, M. A. McDowell, and J. A. Rothfus, *Science*, **140**, 385 (1963).

128. J. M. Smith, *Symp. Zool. Soc. Lond.*, **26**, 371 (1970).

129. J. M. Smith, *Amer. Naturalist*, **104**, 231 (1970).

130. J. M. Smith, *Nature*, **219**, 1114 (1968).

131. L. Smith, Unpublished results, 1972.

132. L. Smith, M. E. Nava, and E. Margoliash, in *Oxidases and Related Redox Systems (Proc. 2nd Int. Symp.)*, T. E. King, H. S. Mason, and M. Morrison, Eds., University Park Press, Baltimore, 1973, p. 629.

133. R. R. Sokol and P. H. A. Sneath, *Principles of Numerical Taxonomy*, Freeman, San Francisco, 1963.
134. M. Sokolovsky and M. Moldovan, *Biochemistry*, **11**, 145 (1972).
135. J. W. Stewart, E. Margoliash, and F. Sherman, *Fed. Proc.*, **25**, 647 (1966).
136. T. Takano and R. E. Dickerson, personal communication, 1972.
137. T. Takano, R. Swanson, O. B. Kallai, and R. E. Dickerson, *Cold Spring Harbor Symp. Quant. Biol.*, **36**, 397 (1971).
138. H. Theorell, *J. Amer. Chem. Soc.*, **63**, 1820 (1941).
139. H. Theorell and Å. Åkeson, *J. Amer. Chem. Soc.*, **63**, 1804, 1812, 1818 (1941).
140. E. W. Thompson, M. Richardson, and D. Boulter, *Biochem. J.*, **124**, 779 (1971).
141. E. W. Thompson, B. A. Notton, M. Richardson, and D. Boulter, *Biochem. J.*, **124**, 787 (1971).
142. K. Titani, personal communication, 1970.
143. H. J. Tsai and G. R. Williams, *Canad. J. Biochem.*, **43**, 1409, 1995 (1965).
144. A. Tulp, *Biochem. J.*, **116**, 39 (1970).
145. T. Uzzell and K. W. Corbin, *Science*, **172**, 1089 (1971).
146. K. van Dam and A. J. Meyer, *Ann. Rev. Biochem.*, **40**, 115 (1971).
147. C. M. Venkatachalam, *Biopolymers*, **6**, 1425 (1968).
148. M. E. Winfield, *J. Molec. Biol.*, **12**, 600 (1965).
149. R. Wojciech and E. Margoliash, in *Handbook of Biochemistry*, 2nd ed., H. A. Sober, Ed., Chemical Rubber Co., Cleveland, Ohio, p. C228.
150. K. Wuthrich, *Proc. Natl. Acad. Sci., U.S.A.*, **63**, 1071 (1969).
151. T. Yamanaka, *Nature*, **213**, 1183 (1967).
152. T. Yamanaka, *Ann. Rep. Biol. Works Fac. Sci. Osaka Univ.*, **14**, 1 (1966).
153. T. Yamanaka and K. Okunuki, in *Structure and Function of Cytochromes*, K. Okunuki, M. D. Kamen, and I. Sekuzu, Eds., University of Tokyo Press, Tokyo, 1968, p. 390.
154. T. Yamanaka and K. Okunuki, *J. Biol. Chem.*, **239**, 1813 (1964).
155. C. Yanofsky, B. C. Carlton, J. R. Guest, D. R. Helinski, and V. Henning, *Proc. Natl. Acad. Sci., U.S.A.*, **51**, 266 (1964).
156. E. Zuckerkandl and L. Pauling, in *Evolving Genes and Proteins*, V. Bryson and H. J. Vogel, Eds., Academic, New York, 1965, p. 97.
157. E. Zuckerkandl and L. Pauling, *J. Theoret. Biol.*, **8**, 357 (1965).
158. E. Zuckerkandl and L. Pauling, in *Horizons in Biochemistry*, M. Kasha and B. Pullman, Eds., Academic, New York, 1962, p. 198.

THE DEVELOPMENT OF PATTERN: MECHANISMS BASED ON POSITIONAL INFORMATION

L. WOLPERT

Department of Biology as Applied to Medicine,
The Middlesex Hospital Medical School, London,
England

CONTENTS

The central problem is how genetic information can be expressed in terms of patterns and forms as manifested by all the living things we see around us. If this problem is approached at the cellular level, then one might view it as the spatial organization of a variety of cellular types. The vertebrate body probably does not contain more than about 120 discrete cellular types that can be characterized by their function and form. This number comes largely from classical histology, but could probably be defined as the number of cells making unique luxury molecules as distinct from household ones, to use Holtzer's terminology. The molecular basis of how different cells make different luxury molecules—molecular differentiation—is essentially the control of how the genetic information is transcribed and translated and will not be considered here (but see Refs. 1, 2).[1,2] The 100 or so cellular types are probably not very different if we look in different vertebrates—certainly very much less so than the variety of shapes and forms that vertebrates manifest. This variety is the problem of pattern and form.

Pattern formation, or spatial differentiation, is the process whereby cells in a population are specified to undergo a particular molecular differentiation, which results in a characteristic spatial pattern. For example, the differences between a forelimb and a hindlimb of a vertebrate do not lie in the processes of molecular differentiation of, say, the muscle and cartilage cells, but rather in those processes that specify which cells will become cartilage and which muscle. It is also convenient to distinguish between the development of spatial patterns and the development of form.[3] The development of form may be regarded as involving those processes bringing about changes in shape, and its genesis requires an understanding of the forces involved. The difference between pattern and form is of course not absolute.

The link between genetic information and molecular differentiation is relatively easy to see, since the problem is posed in terms of the control of production of molecular species. For pattern and form the link is slightly less obvious, since the processes are several steps removed from gene action. It is thus necessary to know what processes are involved in pattern and form. We have argued that in order to establish the link, it is necessary to first understand the processes at the cellular level. Only then is it possible to pose the problem at the molecular level. The validity of this approach is already evident from studies on the development of form. From studies at the cellular level there is now quite substantial evidence that one of the major means of generating forces that bring about changes in shape during development is localized contraction of the cell. This occurs, for example, in cell pseudopods in the gastrulation of the sea urchin[4] and the folding of the neural tube in vertebrates.[5] There is also good reason for supposing that such contractions are associated with actin-like microfilaments.[6] This latter possibility makes it clear that is is possible and meaningful to discuss such morphogenetic processes as gastrulation and neural-tube formation at the molecular level and thus bring them very close to the processes operating at the level of gene action.

One of the interesting features that has emerged from studies on changes in form is that the number of different cellular processes involved is small. By far the most common are activities involving contraction and cell contact. In the case of sea-urchin development it is possible to account for all the early changes in form in terms of such processes, which are used again and again in different situations. This has another very important implication, which cannot be repeated too often and should be engraved on the soul of all those who work on pattern and form: one cannot obtain an indication of the mechanisms that bring about changes in form or the specification of pattern by examining the final forms. One cannot argue by analogy—this is the naturalistic fallacy. To understand development, one must look at development, not the end result. For example, examination of a gastrulated sea-urchin embryo provides no clue as to how gastrulation occurred. Worse still, it gives the impression that the development of complex patterns and forms requires complex mechanisms, and as we have repeatedly emphasized, this is not necessarily the case.[4,7] The very variability of patterns almost demands simple mechanisms. It also seems reasonable to try to find universal mechanisms for pattern formation, since to me it is inconceivable that nature should have a universal genetic system without having universal mechanisms for expressing this genetic information in terms of pattern and form.

In order to avoid being overwhelmed by the apparent complexity of the pattern problem, one might formulate the problem of pattern formation along the following lines. Starting with a single cell that grows and divides to form a line of cells (it is convenient to restrict ourselves to one dimension in the first instance), and supposing that each cell can become either blue, white, or red, then, what organization is required so that the line of cells forms a French flag, and can, under certain circumstances, regulate: that is, if part of the flag is removed, the basic pattern is restored? The types of mechanism that can be envisaged might be classified in different ways, such as whether intercellular interactions were involved, or whether cell division played a crucial role, or whether cytoplasmic localization were important. My own preference is for a classification based upon the concept of positional information, an idea first proposed by Driesch in 1895. The basic idea is that the mechanism would effectively provide a means whereby each cell in the line would have its position in the line specified by ascribing to each-cell a positional value. The cells would then interpret this positional value by becoming blue, white, or red. One would then distinguish, in the first instance, between those mechanisms that relied on positional information and those that did not, and then between the different types of mechanism providing positional information. Any mechanism based on positional information essentially involves setting up a coordinate system, and this requires reference or boundary regions; a vector to specify the direction of measurement, which also defines the polarity of the system; and a scalar in terms of which position is measured. Such a system is provided by a gradient in which the boundary regions are fixed, the scalar quantity being the value of the gradient, and polarity being given by the sign of the slope.

It is not proposed here to consider all the various mechanisms whereby positional information could be specified, but rather to give some biological examples to illustrate the problems and the mechanisms that have been proposed. Before doing so, it may be helpful to mention models that do not rely upon positional information.

The type of mechanism that is most commonly invoked is one in which a prepattern is set up that reflects the overt pattern.[8] Thus for the French flag, a steplike prepattern would be required, and it is necessary to consider how this could be established. The most obvious way would be to have cytoplasmic localization in the initial cell that reflects the prepattern, and this type of mechanism has been suggested for the so-called mosaic eggs. The other solution is related to setting up prepatterns essentially along the lines initially proposed by Turing[9]: that is, to find a set of chemical reactions that will give the required prepattern. The crucial feature of a prepattern is that it involves setting up singularities—in the case of the French flag these occur at the boundaries between the three colors. Models of this type have been proposed by Maynard Smith and Sondhi[10] and by Gmitro and Scriven.[11] Some more recent results are discussed in the chapters by Nicolis (p. 29) and by Ortoleva and Ross (p. 49). These mechanisms are to be contrasted with those based on positional information. An essential feature of mechanisms based on positional information is that the global properties of the system bear no relation whatsoever to the final pattern: the distinguishing feature in the development of different patterns is the difference in cellular response to the same global framework. It is the interpretation of positional information that is the process that gives rise to the pattern. This should be compared with those mechanism in which an attempt is made to provide a global basis for the pattern, as in the form of a prepattern or with mechanisms involving interactions between parts of the pattern themselves.

One of the great virtues of models based upon positional information is that the same mechanism can be used to give rise to an infinite variety of patterns, and since the differences depend mainly on the process of interpretation, they will be closely tied to the genomic composition of the cell. While, at this stage, virtually nothing is known about interpretation, it seems reasonable to think that a considerable portion of the genetic information is associated with this process. Also, from the point of view of evolution, it obviously provides a simple means for changing patterns.

On the present evidence it seems plausible to think of the mechanisms for setting up positional information as involving three main parameters: a positional value P, which can be thought of as some parameter defining the state of the cell; an intercellular signal, possibly a diffusible substance, S; and cell division and mitosis M. It is not unreasonable to think that these are used in different ways in different systems, and two such ways will now be briefly described.

I. POSITIONAL SIGNALING AND HYDRA REGENERATION

The regeneration of hydra[12] and early sea-urchin development[13] are systems with striking powers of regulation. Removal of the head of a hydra leads to the formation of a new head; much of the presumptive ectoderm of the early sea-urchin embryo can be removed without much alteration of the overall proportions in the larva. In these systems cell growth and division seems to play little role; regulation is morphallactic (Fig. 1), and only S and P are involved. In terms of positional information, morphallactic regulation requires reestablishment of the boundary regions and the positional value with respect to those regions. In both the hydra and the sea urchin the boundary regions have special proper-

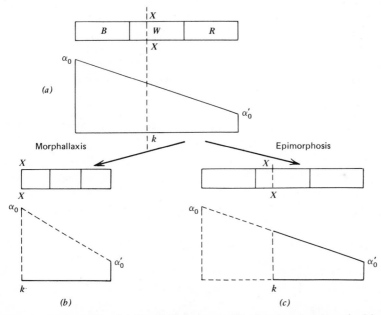

Figure 1. Regeneration of a French-flag pattern. The intact system is as in (a) and comprises three equal regions of blue (B), white (W), and red (R). The positional information is represented by the linear gradient with boundary values α_0 and α_0'. If the flag is cut at level X–X corresponding to the kth cell, and the left-hand piece removed, then morphallactic regeneration occurs as in (b). The value of the positional information at the cut now becomes α_0, and a new gradient is established (dashed line). In epimorphic regeneration (c), the positional information at the cut remains the same, and restoration of the pattern is achieved by growth.

ties. We have investigated this in some detail with respect to hydra and the regeneration of the head end. It is a basic assumption that the observed form of the hydra results from the cells' interpreting their positional value, but unfortunately the cellular activities are unknown. However, tentacle formation, for example, may be accounted for in terms of the type of cellular activities responsible for sea-urchin morphogenesis as described above. Our analysis takes no account of the distribution of cell types.[14]

Our analysis is based on grafting and determining where new boundary regions, head, and feet form. For a large and varied number of grafts we have constructed a synthetic model. We have a convenient notation for the regions of the axis in the hydra (Fig. 2). The head region, consisting of hypostome and tentacles, is represented by H, and the digestive zone by four equal-sized regions, $1,2,3,4$; B designates the budding region; 5 and 6 the distal and proximal halves of the peduncle; and F the foot or basal disk. In terms of this notation a graft of head and distal gastric region onto the subhypostome of another animal is

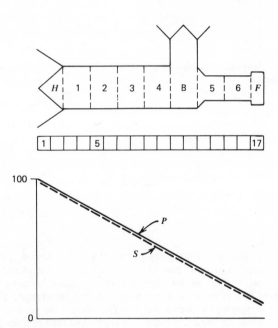

Figure 2.   Diagram to illustrate the main regions of hydra and the 17 cells used for a computer simulation. The hydra is taken to be 3 mm long. The gradients in S and P are assumed to be linear.

given as $H\,12/12\ldots F$. Before presenting the model and the experiments, it may be convenient to summarize briefly some of the results[12] with as little reference to a particular model as possible: (1) The time required to become a head increases with distance from the head end; (2) the head can prevent other regions forming a head end, and this inhibition falls off with distance; (3) the "level" of inhibitor required to inhibit a piece from forming a head end—the threshold for inhibition—decreases with distance from the head end; (4) the changes occurring during regeneration are localized very close to the cut surfaces; (5) new ends form more easily at cut surfaces.

We have investigated the spatial and temporal characteristics of these processes in some detail, and the current model postulates, with respect to head-end regeneration, two gradients (Fig. 2). The first gradient is considered to be of a diffusible substance S, which, in the steady state, is produced by a source at the head end that maintains S at a constant concentration there. S is of course broken down, and we have assumed that this only occurs at the foot end, which acts as a sink for S and maintains it as a constant concentration there. This results in a linear gradient of S along the animal. The second gradient is of P, the positional value, which is considered to be some cellular parameter, and which is not diffusible or diffuses much more slowly. Changes in P are considered to occur mainly by synthesis and breakdown. The form of the gradient in P is taken for simplicity to be identical in the equilibrium state to that of S. The rule for initiating head-end formation is that S must fall a threshold amount below P. P then increases to its boundary value and acts as a source of S again.

It may be helpful to indicate how this model was arrived at. One series of experiments that showed that some sort of threshold phenomenon was involved was that region 1 grafted in between regions 2 and 3 of an intact animal was absorbed, whereas if placed in the 56 region, it formed a head at the site of graft. In terms of the model, the high level of S in the grafted region 1 is reduced when the region is placed at a position of low concentration of S. If the concentration is sufficiently low then head formation is induced. This implies that there is a more stable gradient, not changed by grafting, with which S can be compared. One can think of P as a parameter that tells the cell what it currently is, and S as a parameter that can respond to the state of neighboring cells. This model has important similarities to that of Lawrence, Crick, and Munro[15] for the insect epidermis.

Our current model envisages the following sequence. Following the removal of the head end, the concentration of S falls, by diffusion, until it is a critical amount below the original value as "remembered" by P. We have used, arbitrarily, a gradient going from 100 to 10. The hydra is 3 mm long and is divided for computer simulation into 17 cells (Fig. 2). We have used a diffusion constant of 2×10^{-7}, as this is a plausible estimate for a small molecule[16] and was the value obtained from our own experimental studies on the time for a "signal" to be transmitted from the head end.[17] The formation of a new head-end boundary is triggered when S falls to a critical value, say 10% below P. At this value synthesis of P starts and continues until P is 100 again, but synthesis of P can be inhibited before P reaches 100 if S increases again, perhaps by diffusion from some other region, to within the critical value. When P reaches 100, it again acts as a source for S. We have carried out a simple computer simulation of this process of which Fig. 3 shows one example. Experimentally a new head end is determined from a 1 region in 4 to 5 hr with *H. littoralis* at 26°C. We have assumed that about one-third of this time is for S to fall and the remainder for P to rise; this will be discussed in detail elsewhere.[18] When S falls the critical amount (10% below P), P is synthesized. In order to ensure that only the end cell becomes determined as a head, it is necessary for this cell also to make S at a slow rate and so prevent further cells being "switched on" and becoming heads. S must not be made so fast, however, that it inhibits the synthesis of P, which will occur if S rises to within 10% of the value of P before P reaches 100. Only when P reaches 100 does rapid synthesis of S occur to reestablish the gradient. It is important to note that the changes in P are localized at the cut end in this model. A more sophisticated model[18] allows some leakage of S into the external medium at the cut surface and considers the threshold and other factors in more detail. The regeneration of a hydra in terms of this model can be represented in terms of cell states and their behavior with respect to S and P (Table I). Away from the boundary, changes in P are much slower and may involve a followup servomechanism[7] or slow diffusion of P to reestablish the steady-state linear gradient, but this is not important in boundary formation. In more general terms the model may be viewed as one involving long-range inhibition and local activation, and should be compared with the important model, based on molecular mechanisms of auto- and cross-catalysis, put forward by Gierer and Meinhardt.[19]

The long-range inhibitory feature will now be considered in more detail. In the model S is effectively an inhibitor of head-end formation, and we have investigated the propagation of this inhibitory effect by exploiting Wilby and Webster's[20] observations that one head end can

TABLE I.

The behavior of P and S for different cell states
and the transition rules.

	P	S
State A	Fixed	Diffusing
State B	$\dfrac{dP}{dt} = p_1$	$\dfrac{dS}{dt} = s_1$ (slow)
State C	100	$\dfrac{dS}{dt} = ks_1$ (fast)
State D	100	100

Rules for transitions

State $A \rightarrow$ State B	if	$1 - S/P \geqslant 0.1$
State $B \rightarrow$ State A	if	$1 - S/P < 0.1$
State $B \rightarrow$ State C	if	$P \geqslant 100$
State $C \rightarrow$ State D	if	$S \geqslant 100$

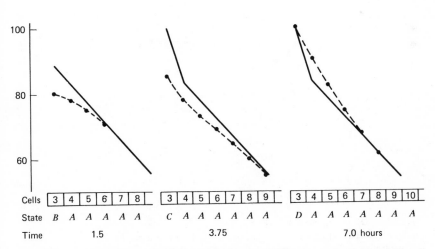

Cells	3	4	5	6	7	8		3	4	5	6	7	8	9		3	4	5	6	7	8	9	10
State	B	A	A	A	A	A		C	A	A	A	A	A	A		D	A	A	A	A	A	A	A
Time			1.5								3.75								7.0 hours			

Figure 3. Simulation of regeneration at the head end following removal of H. The cell
states and the transition rules are shown in Table I.

261

inhibit the formation of another head end over quite long distances and that this inhibition can be propagated in a proximodistal direction.[17]

The plan of the experiment using *H. attenuatta* is shown in Fig. 4. After various times the host head is removed to give $123/H$ and the animal observed to see if a new head regenerates from the 1 region. The absence of a head is taken to mean that the grafted head has inhibited head formation at the 1 region. The grafting times for 50% inhibition for $123/H$, $12/H$, and $1/H$ are, respectively, -8, -2, and $+6$ hr at $18°C$. These results show clearly that the time required for a grafted head to be able to inhibit the development of another head region at some distance from it is very significantly altered by this distance. It must be grafted on 8 hr before host-head removal for a 123, but can be grafted on 6 hr after host-head removal for a 1 (that is, 14 hr later than for the

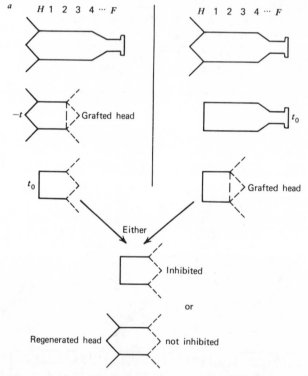

Figure 4. Plan of experimental scheme to test for propagation of inhibitory signal from the head end. The host head is removed at time 0. On the left-hand side the grafted head is placed in position before removal of the host head, while on the right-hand side it is placed in position after removal of the host head.

previous case). This dependence on distance, with time scale of hours, would seem to exclude a comparatively fast periodic signaling mechanism of the type observed in slime molds. We have extended this analysis to include the time for head-end determination, which at 18°C is 12 hr, and have carried out a simple computer-based simulation, which gave results consistent with diffusion.

It should be noted that in hydra polarity is manifested by heads' always forming at a distal cut surface. The model we have presented suggests that polarity is a global property and that it is determined by the interaction between the two gradients. If a piece of hydra gastric region is isolated, at the moment of isolation the differences in the S and P values at the distal and proximal surface may be quite small, but diffusion of S will result in S decreasing at the distal end and increasing at the proximal end. This obviously amplifies the differences and provides a simple mechanism for preventing, for example, head formation at both ends. Reversal in polarity of hydra is very slow.[20] This is consistent with the idea that reversal of polarity requires a reversal of the gradient in P, and changes in P appear to be slow away from the boundary.[12]

II. THE PROGRESS ZONE AND THE CHICK LIMB

The development of pattern formation in systems in which growth is an integral feature, such as the development of the vertebrate limb and epimorphic regulation[7] (Fig. 1) need not involve the intercellular signaling of positional information. The chick wing starts as a small bulge from the flank and grows and extends as a tongue-shaped mass. It consists of mesenchymal cells encased in a thin ectodermal jacket, which is thickened at the distal end to form the apical ectodermal ridge. We have recently given detailed attention to the mechanism by which the skeletal elements along the proximodistal axis are specified—that is, humerus, radius and ulna, carpals, and digits.[21] It is known that they are laid down in a proximodistal sequence, and the primordia of proximal structures can be seen to begin differentiation first, while the primordia of more distal structures appear at successively later times and more distal positions. But the very tip—a region of about 300 μm—remains apparently undifferentiated and continues to proliferate relatively fast. We have suggested that this region comprises the progress zone, in which there is autonomous change in P with time that is possibly linked to cell division (Fig. 5). The progress zone is specified by the apical ectodermal ridge. Removal of the ridge leads to the disappearance of this zone and thus to the absence of distal regions. The later the stage at

which it is removed, the more nearly complete is the limb. A prediction
from this mechanism is that if the tips of two limbs of different ages are
exchanged, then both pieces should continue to develop as they would
have, and some of the resulting composite limbs will be missing or
duplicated. We have found this to be the case. For instance, if an entire
early bud is grafted onto the tip at a later stage (Fig. 5), one obtains two
arms in tandem, whereas the reciprocal combination lacks the radius

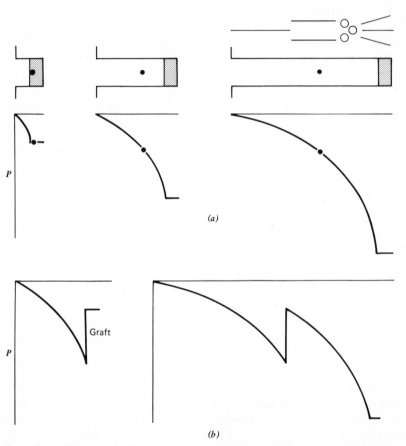

Figure 5. (*a*) Diagram to illustrate the concept of the progress zone in relation to the
development of the chick wing. All cells are assumed to be dividing, and the progress zone
is indicated by shading. The large dot marks the mean position of one typical cell lineage
at the three stages of growth. The value of *P* is shown on the assumption that it decreases
autonomously with time in the progress zone. (*b*) This illustrates the effect of grafting a
young progress zone onto an older stump. It assumes no interaction between the stump
and progress zone and predicts duplication of parts of the limb.

and ulna. We have carried out a variety of such experiments also using thinner tips, and on the whole a good correspondence between results and theory has been obtained.

If the three-dimensional pattern of muscle and cartilage is to be determined by positional information, then the other two axes, antero-posterior and dorsoventral, must be taken into account. There is substantial evidence from the work of Saunders and Gasseling[22] that a polarizing region close to the posterior border plays a crucial role. It appears that it is the source of a signal that specifies the anteroposterior positional value, possibly in a manner analogous to the head end of hydra and the micromeres in early sea-urchin development. If the polarizing region is grafted to an anterior site on the rim of the limb bud, a limb develops with reduplicated distal parts: the twinned elements are mirror images of one another.

The specification of positional information by autonomous change within a progress zone could account for a variety of regenerative and regulative systems involving epimorphosis. For example, amphibian limb regeneration involves blastema formation, which is analogous to the early chick-limb bud. One may thus regard blastema formation as the setting up of a progress zone from the cells at the cut surface, and this will ensure that the regenerate will automatically originate at the positional value corresponding to the level of the cut, and will go on to develop only more distal levels. In a sense, with this mechanism, polarity is determined by a temporal sequence and this immediately accounts for the phenomenon, seen in regenerates both from amphibian limbs and from insect imaginal disks, that the proximal face of a cut surface will regenerate distal structures to give mirror-image symmetry. There may also be a progress zone during retinal growth in amphibia, where, during most of the growth period, cell division is confined to a narrow ring near the outer edge.[23] Thus cells at the outer edge have divided more times than those nearer the center. This could provide a radial gradient in positional value that might be involved in the specification of the ordered neural connections between the retina and the tectum.

III. UNIVERSALITY

Certain features of pattern formation in the insect epidermis and in intercalary regeneration of the insect leg[24] seem to require all three parameters, P, S, and M. For example, polarity in the epidermis of certain insects is indicated locally by the direction in which hairs or bristles point, or by the ripple-like contours formed by the integument.

This polarity can be distorted by various operations, such as rotating a square region through 90 or 180°, or moving a piece to a different position along the anteroposterior axis. Most of these changes can be accounted for if it is assumed that polarity is determined by the sign of a gradient. In order to account for the observations, Lawrence et al.[15] postulate two gradients. One, which may correspond to S, is a diffusible substance, and the other is a more stable or remembered value, corresponding to P. It is the sign of the gradient in P that specifies polarity. P is set by S, but this occurs only at cell division (the similarity to the followup servomechanism proposed for hydra and the autonomous change in P in the progress zone should be emphasized). In addition, there is a homeostatic mechanism whereby the cells try to keep the local value of S equal to the P value.

In the case of intercalary regeneration, the relation between S, P, and M seems slightly different. Bohn has shown that it is possible to account for intercalary regeneration in the insect leg by assuming that when there is a discontinuity in the gradient, growth occurs until the original slope of the gradient is reestablished.[24] However, I would suggest a slightly different interpretation, which invokes the two gradients. At a discontinuity in the gradient, as S will diffuse so as to be below P on the one side and above P on the other, the precise distribution will depend on the strength of the homeostatic effect. If there is an additional rule that a progress zone is established if S is either greater or less than P, growth will continue until the gradient is restored.

It is quite encouraging that one can discuss pattern regulation in a variety of different systems from several different phyla in terms of positional information and using only the three parameters P, S, and M. To take an extreme view, I would suggest that P and S are the same in all systems. A central feature of positional information is that there is no unique relationship between the form of the gradient in positional value; the resulting pattern and the differences in pattern come mainly from interpretation. Thus all fields set up by the same mechanism would be indistinguishable, and the cells would behave according to position and genome. A number of experiments have been carried out along these lines; those involving genetic mosaics[8] are probably the most impressive. For example, there is a single-gene mutant in *Drosophila aristopaedia*, which results in the antenna of the fly forming a leg. There are mosaics in which parts of the antenna are wild-type and parts *aristopaedia*. In all cases the cells behave according to position and genome (see also Ref. 25). This conclusion holds for a wide variety of other cases.[7] The case for universality is also strengthened by the fact that all fields are about the same size, usually less than 1 mm or 50 cells in the maximum dimension, and that the time course of changes within fields is of the order of several hours. These space-time relations are consistent with a mechanism based on the diffusion of S.[16]

If the argument I have been presenting is valid, then a major part of pattern formation must lie in the interpretation of positional information. The development of pattern in an organism may be viewed as a hierarchy of positional fields, and the genetic network required for this becomes a problem of central importance, as Waddington[26] has emphasized. The concept of positional information is a very simple one and should help to reduce the apparent complexity of the mechanisms responsible for pattern formation. If simple universal mechanisms for interpretation are found, the development of pattern may begin to aspire to the elegance and universality of molecular genetics.

References

1. R. J. Britten and E. H. Davidson, *Science*, **165**, 349 (1969).
2. J. B. Gurdon and H. R. Woodland, *Curr. Topics Devel. Biol.*, **5**, (1970).
3. C. H. Waddington, *New patterns in genetics and development*, Columbia U. P., New York, 1962.
4. T. Gustafson and L. Wolpert, *Biol. Rev.*, **42**, 442 (1967).
5. T. E. Schroeder, *J. Embryol. Exp. Morphol.*, **23**, 427 (1970).
6. N. K. Wessels, B. S. Spooner, J. F. Ash, M. O. Bradley, M. A. Luduina, E. L. Taylor, J. T. Wrenn, and K. M. Yamada, *Science*, **171**, 135 (1971).
7. L. Wolpert, *Curr. Topics Devel. Biol.*, **6**, 183 (1971).
8. C. Stern, *Genetic Mosaics and Other Essays*, Harvard U. P. Cambridge, Mass., 1968.
9. A. M. Turing, *Phil. Trans. Roy. Soc. B*, **237**, 37 (1952).
10. J. Maynard Smith and K. C. Sondhi, *J. Embryol. Exp. Morphol.*, **9**, 661 (1961).
11. J. I. Gmitro and L. S. Scriven, *Symp. Int. Soc. Cell Biol.*, **5**, 221 (1966).
12. L. Wolpert, A. Hornbruch, and M. R. B. Clarke, *Amer. Zool*, **14**, 647 (1974).
13. S. Hörstadius, *Biol. Rev.*, **14**, 132 (1939).
14. H. Bode, S. Berking, C. N. David, A. Gierer, H. Schaller, and E. Trenker, *Roux. Arch. Entw.*, **171**, 269 (1973).
15. P. A. Lawrence, F. H. C. Crick, and M. Munro, *J. Cell Sci.*, **11**, 815 (1972).
16. F. H. C. Crick, *Nature*, **225**, 420 (1970).
17. L. Wolpert, M. R. B. Clarke, and A. Hornbruch, *Nature*, **239**, 101 (1972).
18. M. R. B. Clarke and L. Wolpert, to be published.
19. A. Gierer and H. Meinhardt, *Kybernetik*, **12**, 30 (1972).
20. O. K. Wilby and G. Webster, *J. Embryol. Exp. Morphol.*, **24**, 595 (1970).
21. D. Summerbell, J. H. Lewis, and L. Wolpert, *Nature*, **244**, 492 (1973).
22. J. W. Saunders and M. T. Gasseling, in *Epithelial-Mesenchymal Interactions*, R. Fleischmajer and R. E. Billingham, Eds., Williams and Wilkins, Baltimore, 1968, p. 78.
23. K. Straznicky and M. M. Gaze, *J. Embryol. Exp. Morphol.*, **26**, 67 (1971).
24. P. A. Lawrence in *Developmental Systems*, Vol. 2, B. J. Counce and C. H. Waddington, Eds., Academic, New York, 1973, p. 157.
25. H. A. Schneiderman and P. J. Bryant, *Nature*, **234**, 187 (1971).
26. C. H. Waddington, in *Developmental Systems: Insects*, Vol. 2, S. J. Counce and C. H. Waddington, Eds., Academic, New York, 1973, p. 499.

A MEMBRANE MODEL FOR POLAR ORDERING AND GRADIENT FORMATION

B. C. GOODWIN

*School of Biological Sciences, University of Sussex,
Brighton, Sussex, England*

CONTENTS

The occurrence of regulation and regeneration in unicellular and syncytial organisms[1,2] implies that whatever processes underlie these phenomena, they do not depend upon a partitioning of the system into cells. We are dealing here with the establishment of ordered spatial inhomogeneity over systems with little or no initial order, the stabilization of this order, and its subsequent morphogenetic interpretation, so our attention is drawn to the problem of constructing a model that is capable of demonstrating this behavior in a continuum. In this paper I consider an example, which has been named the wave-broom model. A more detailed treatment will be found in a forthcoming publication by Goodwin and McLaren.[3]

I. THE WAVE-BROOM MODEL

Since our model must apply to systems that are not partitioned into cells, and it must generate stable spatial order, it is natural to assume that membranes are somehow involved. Furthermore, since one of the basic features of the model will be a propagating activity wave of some kind, our attention is again directed to membranes, since these are excitable surfaces and can propagate activity waves. The best-known example is the nerve action potential. However, this ability is a virtually universal property of biological membranes, although most rates of propagation of activity waves are much smaller than that in myelinated nerve axons, which are highly specialized for rapid conduction.

The general process I have in mind is as follows. Suppose that we have enzymes E_1 and E_2 located on a membrane surface, and that these enzymes catalyze the metabolic reaction sequence shown in Fig. 1. This is the familiar metabolic control circuit that incorporates the essential features of the glycolytic oscillator. E_1 catalyzes the conversion of U into V, and E_2 that of V into W. V has the effect of releasing some molecular species from a bound (or otherwise constrained) state, and this activates E_1. The species involved is identified as potassium in the model, but this is only one of many possible candidates. W is a feedback inhibitor of E_1. So this metabolic system is intrinsically capable of oscillatory behavior.

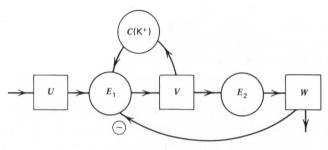

Figure 1.

Now let us consider the picture on the membrane. This is shown in Fig. 2. It is known that many enzymes are localized in or on membranes, and so I assume that E_1 and E_2 are so situated. Only E_1 is actually held in the membrane, E_2 being either bound or soluble. I assume that the potassium ion is normally adsorbed to sites on the membrane, and that the molecular species V can form a bound complex with a specific protein in the membrane, designated B.

An activity wave can be initiated by any event that causes a release of potassium ions, which will activate E_1 and start off the reaction sequence. Since V releases K^+, the wave of activation of E_1 spreads along the membrane in all directions from the point of initiation. Two consequences of this activity wave are desired: the first is the generation of a gradient in the spatial distribution of the substance V; and the second is periodic reinitiation of the wave from the point where it first occurred. These consequences may be realized by the following postulates. Assume that some time after the release of K^+ and the activation of E_1, the sites B for the binding of V become exposed (e.g., W could

Outside

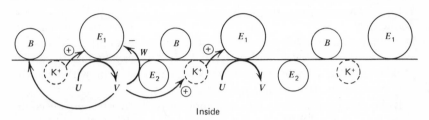

Inside

Figure 2.

act as an activator of B as well as an inhibitor of E_1, but this detail may be left unspecified). These sites revert to an unavailable condition by relaxation some time after being activated if they do not combine with V. The metabolite V will then be picked up by the membrane-bound protein B and held in the form of an association complex behind the wavefront, so that there is a spatial asymmetry with respect to the sites of V production by E_1 and its pickup by B. The fraction of V picked up will then have moved by a pure diffusional process through a certain distance in a direction opposite to that of the wave propagation. Thus bound V will accumulate at the point of initiation of the wave. This process is shown schematically in Fig. 3, where the propagating site of production of V is shown as a delta function that decays symmetrically along the one-dimensional spatial axis in time. The pickup function expressing the activation of the sites B and their decay is shown traveling along behind the wavefront.

Since we have assumed that V causes the release of K^+, it is a natural

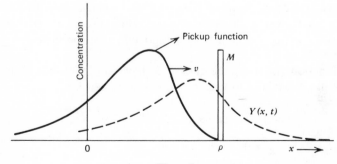

Figure 3.

consequence of the accumulation of V in some region of the membrane that an initiation of the wave will recur in this region, since there will be a finite decay rate of the B-V complex. As the wave recurs and a gradient of V is built up, the recurrence of the wave will increase in frequency until we have a rhythmically recurring event generating a steady gradient in the bound material V. In this way symmetry can be broken, and an initially uniform system, either with or without cells, may become spatially organized.

If we particularize the model to apply it to *Dictyostelium*, we could take U to be ATP, E_1 to be adenyl cyclase, $V =$ cyclic AMP, E_2 $=$ phosphodiesterase, and $W =$ AMP. Cyclic AMP initiates the process by causing a local release of K^+. Amoebae then break symmetry and become polarized. The accumulation of cyclic AMP could result in a periodic release of this substance from the amoeba when a threshold of depolarization is reached, or by some other release mechanism.

It is now possible to write down equations that describe such a process, and then to solve them under certain conditions to study the behavior of the model and compare it with biological processes. Let $X(\mathbf{x}, t)$ be the concentration of the diffusible form of the metabolite, and $Y(\mathbf{x}, t)$ be the concentration of the bound form, where $\mathbf{x}$ is the spatial dimension and t is the time. The equations are of the form

$$\frac{\partial X(\mathbf{x}, t)}{\partial t} = D \frac{\partial^2 X(\mathbf{x}, t)}{\partial \mathbf{x}^2} + \beta Y(\mathbf{x}, t) - f(\mathbf{x}, t) X(\mathbf{x}, t) + P(\mathbf{x}, t)$$

$$\frac{\partial Y(\mathbf{x}, t)}{\partial t} = -\beta Y + f(\mathbf{x}, t) X(\mathbf{x}, t) \tag{1.1}$$

X is subject to diffusion, with a diffusion constant D. We assume that the bound form Y is subject to decay at the rate βY. The rate of formation of Y is taken to be the product of the concentration of X and the concentration of binding sites, the latter being described by the function $f(\mathbf{x}, t)$, which also contains the rate constant for the process. $P(\mathbf{x}, t)$ is a forcing function of any kind that is to be included in the problem.

The function $f(\mathbf{x}, t)$ can be chosen to have any plausible form; it is convenient analytically to use, in one spatial dimension,

$$f(x, t) = k\alpha(vt - x)e^{-\alpha(vt - x)} \quad \text{for} \quad x \leqslant vt$$

$$= 0 \quad \text{for} \quad x > vt$$

Here v is the velocity of the wave, k is the product of a rate constant for the formation of the bound metabolite and a concentration parameter, and α is a parameter that determines the shape of the function.

I will now simplify the problem so that a simple analytical solution can be obtained and we can get some insight into the behavior of the variables. Let us suppose that there is a pulse of X that is produced at the point $x = \rho$ on the membrane, and that this decays without propagating. Let us assume further that the amount of X that is picked up and becomes Y is small compared with the amount of X produced. Then we can ignore $f(x,t)X(x,t)$ in the first equation. Let us also assume that Y is very stable, so we can ignore $\beta Y(x,t)$ in both equations. Then we can write down the solution for X immediately:

$$X(x,t) = \frac{M}{2\sqrt{\pi Dt}} e^{-(\rho-x)^2/4Dt}$$

where M is the amount of X in the pulse. We suppose that the pickup function $f(x,t)$ propagates in the normal way, so that $Y(x,t)$ is just the solution of the equation

$$\frac{\partial Y(x,t)}{\partial t} = \frac{k\alpha M(vt+\rho-x)}{2\sqrt{\pi Dt}} e^{-\alpha(vt+\rho-x)} e^{-(\rho-x)^2/4Dt}$$

that is,

$$Y(x,t) = \int_0^t \frac{k\alpha M(vt+\rho-x)}{2\sqrt{\pi Dt}} e^{-\alpha(vt+\rho-x)} e^{-(\rho-x)^2/4Dt} \, dt$$

where we have taken $Y(x,0)=0$. The picture corresponding to this process is shown in Fig. 3.

We may now ask questions such as how far the peak of Y is from the point ρ after the process is finished, how much Y there is in comparison with M, and how long the process takes to be essentially completed. Then we can look at the more general problem.

If we write $\beta = (\rho-x)/v$, $\gamma = (\rho-x)^2/4D$, the integral takes the form

$$Y(x,t) = \frac{k\alpha vM}{2\sqrt{\pi D}} \int_{\tau_0}^t (\tau+\beta) e^{-\alpha v(\tau+\beta)} e^{-\gamma/\tau} \tau^{-1/2} \, d\tau$$

where the lower limit is given by

$$\tau_0 = 0 \qquad \text{for} \quad \beta \geqslant 0$$

$$= -\beta \qquad \text{for} \quad \beta < 0$$

Now we are interested in Y when t is large, so we can use the standard form

$$\int_0^\infty x^{\nu-1} e^{-(\beta/x + \gamma x)}\, dx = 2\left(\frac{\beta}{\gamma}\right)^{\nu/2} K_\nu\left(2\sqrt{\beta\gamma}\right)$$

where K_ν is a Bessel function of imaginary argument. Thus evidently we get the result

$$Y(x,t) = \frac{k\alpha v M}{2\sqrt{\pi D}}\left\{2\left(\frac{\gamma}{\alpha v}\right)^{3/4} K_{3/2}(2\sqrt{\gamma\alpha v}) \right.$$

$$\left. + 2\beta\left(\frac{\gamma}{\alpha v}\right)^{1/4} K_{1/2}(2\sqrt{\gamma\alpha v})\right\} e^{-\alpha v\beta}$$

These Bessel functions have simple closed forms,

$$K_{1/2}(x) = \sqrt{\frac{\pi}{2x}}\, e^{-x}$$

$$K_{3/2}(x) = \sqrt{\frac{\pi}{2x}}\, e^{-x}\left(1 + \frac{1}{x}\right)$$

It is somewhat simpler to write the solution in terms of a variable whose origin is the point $x = \rho$, and to measure backwards from this point in relation to the diagram in Fig. 3. So we take $\xi = \rho - x$ as this new spatial variable, and then we find the solution

$$Y(\xi, \infty) = \frac{kM}{4}(\alpha v D)^{-1/2} e^{-(\alpha + \sqrt{\alpha v/D})\xi}\left[1 + \xi\left(2\alpha + \sqrt{\frac{\alpha v}{D}}\right)\right] \qquad (\xi \geqslant 0)$$

The appropriate units to use are microns (μ) and seconds. α has units μ^{-1}, v has units μ/sec, and D has units μ^2/sec. So $k/\sqrt{\alpha v D}$ has units μ^{-1}, and the units are all right. The shape of this distribution is essentially like the pickup function, but it does not have the value 0 at $\xi = 0$. It has the general shape shown in Fig. 3.

Now we may ask some questions about distances, assigning some plausible values to the parameters. We take $v = 4$ μ/sec, $D = 4 \times 10^2$ μ^2/sec, which corresponds to a molecule about the size of cyclic AMP;

and $\alpha = \frac{1}{16}$, which means that the peak of the pickup function occurs 16 μ behind the wavefront. With $v = 4$, this means that the time for the sites to become receptive is about 4 sec. This may be somewhat short, but it will do to give us some order-of-magnitude estimations.

The peak of $Y(\xi, \infty)$ occurs at a distance $\left(\alpha + \sqrt{\alpha v / D}\,\right)^{-1}$ from the point of origin of the pulse M. For the values above, this is about 4.75 μ. The amount of Y is found by integrating over the spatial variable. This turns out to be about 25% of the amount in the pulse, so our assumption about Y being small in relation to X is not a good one. For the value given to D, the decay time is a few seconds. Thus in 4 sec, the pickup function will have travelled past the origin of the pulse of M, and this pulse will have decayed to about $\frac{1}{150}$ of its original value. We can then say that the substance produced at the point ρ has been displaced through about 4.75 μ in 4 sec. This gives us some idea of the "transport" rate such a process is capable of.

Complicating the problem slightly, we can look for a solution of $X(x, t)$ in the case of a propagating delta function, still making the assumption that X is considerably larger than Y.

In this case the form is

$$X(x,t) = M \sum_{\pm} \int_0^{\pm vt} \frac{\pm\, d\xi}{2\sqrt{\pi D(t \mp \xi / v)}} \exp\left(\frac{-(x - \xi)^2}{4D(t \mp \xi / v)}\right)$$

This integral is taken over the spatial variable ξ from the origin up to the wavefront in both directions. If we write $\eta = t - \xi / v$, $\sigma_1 = (x + vt)^2 / 4D$, $\sigma_2 = (x - vt)^2 / 4D$, and $\rho = v^2 / 4D$, this integral becomes

$$\frac{Mv}{2\sqrt{\pi D}} \left(e^{2v(x+vt)/4D} \int_0^t \eta^{-1/2} e^{-(\sigma_1/\eta + \rho\eta)}\, d\eta \right.$$

$$\left. + e^{-2v(x-vt)/4D} \int_0^t \eta^{-1/2} e^{-(\sigma_2/\eta + \rho\eta)}\, d\eta \right)$$

This leads to the solution

$$X(x,t) = \frac{Mv}{2\sqrt{\pi D}} \left[e^{v(x+vt)/2D} \sum_{n=0}^{\infty} \frac{(-\rho)^n}{n!} \sigma_1^{n+1/2} \Gamma\left(-(n+\tfrac{1}{2}), \frac{\sigma_1}{t}\right) \right.$$

$$\left. + e^{-v(x-vt)/2D} \sum_{n=0}^{\infty} \frac{(-\rho)^n}{n!} \sigma_2^{n+1/2} \Gamma\left(-(n+\tfrac{1}{2}), \frac{\sigma_2}{t}\right) \right]$$

For large σ_i/t, this is approximately

$$X(x,t) \approx \frac{Mv}{2\sqrt{\pi D}}\left(\sigma_1^{-1}e^{-(x+vt)/4Dt}t^{3/2} + \sigma_2^{-1}e^{-(x-vt)^2/4Dt}t^{3/2}\right)$$

$$= 2Mv\sqrt{\frac{D}{\pi}}\ t^{3/2}\left(\frac{e^{-(x+vt)^2/4Dt}}{(x+vt)^2} + \frac{e^{-(x-vt)^2/4Dt}}{(x-vt)^2}\right)$$

This shows us how the solution decreases away from the wavefront. The graph looks qualitatively as shown in Fig. 4.

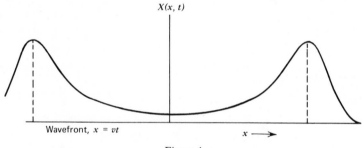

$X(x, t)$

Wavefront, $x = vt$

$x \longrightarrow$

Figure 4.

To find $Y(x,t)$ we must perform the integration

$$Y(x,t) = \alpha \int_{x/v}^{t}(v\tau - x)\, e^{-\alpha(v\tau - x)}X(x,\tau)\, d\tau$$

The pickup starts when the wavefront reaches the point x, which is at time x/v, and proceeds to time t. This integration is not an easy one, but it is considerably simplified by writing $x = pvt$, where $0 \leqslant p \leqslant 1$, and then expanding $\Gamma(-(n+\frac{1}{2}), \sigma_i/\tau)$ to the first order in the variable $\sigma_i/\tau = v^2[(1 \pm \rho)/4D]\tau$. For the parameter values used above, this is less than $2 \times 10^{-2}\tau$, so an expansion to first order will give a good approximation for $\tau \leqslant 20$ sec, say. In 20 sec the wave will have travelled 80 μ, which would be the whole length of a fucus egg, and a substantial part of an amphibian embryo. So this approximation will give useful information. The general shape of $Y(x,t)$ as obtained by this method is as shown in Fig. 5.

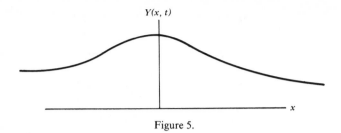

Figure 5.

This is the result of one wave propagation over a finite distance, with the simplifying assumptions made above. The system with periodic repetition of the wave from the origin, decay of the bound form at the rate β, and no restrictions on the relative magnitudes of X and Y can be solved only by computation, which is in progress. We can, however, say something about the expected solution.

Evidently the process described behaves as a memory wave that leaves a graded trace of its activity, the gradient being in the form of the bound metabolite Y. The origin will have the maximum value of the bound metabolite, which we may refer to as a morphogen, and under particular parametric conditions we may expect a steady state to be established, with a monotonic gradient of Y over the cell or the embryo. If we are considering an embryo that is undergoing cleavage into cells, then we need to suppose that every cell is in communication with every other cell, so that wave propagation can occur and substances can diffuse from one cell to the next. Embryonic cells have been shown to be in electrical communication with one another, and substances of the size of cyclic AMP can certainly diffuse from cell to cell. So there is no difficulty in applying the model to either cellular or acellular systems.

There are several very interesting questions that need to be answered regarding the behavior of the wave-broom model. There must be some optimal relationship between v, D, α, and the period of the repeated wave, which generates a gradient that peaks well at the origin and extends substantially from the origin. If the frequency of the periodic wave is very high, then effectively the concentration of X remains the same everywhere and no gradient in Y can result. However, for D values about $10^2\ \mu^2/\text{sec}$, which is the range for molecules of the type we are interested in, the frequency of the wave and its velocity would both have to be implausibly high to result in a stationary, flat distribution of X everywhere.

Another very important question is the dimensionality of the problem. Most membranes can be regarded as two-dimensional surfaces, so the problem should be presented in two spatial dimensions. This does not add significantly to the analysis, but again those questions will finally have to be resolved by computation.

II. THE ESTABLISHMENT OF A METRIC

So far we have considered only the problem of polarity, which breaks symmetry and establishes an origin for a spatial axis. It is now necessary to consider the next step in the developmental process, which is the establishment of some metric on this axis, so that cells or regions of the embryo "know" where they are in relation to the whole and can behave accordingly. Furthermore, this metric must be adjustable to the size of the whole, since large or small embryos of any particular species are perfectly scaled in relation to their size. Thus embryos are pattern-invariant, size-independent systems, a property referred to as regulation. Not only do embryos adjust to variations in size; they also adjust to variations in rate. Development takes different times at different temperatures, but the final organism is always the same. So both space and time are scaled.

At this point we have a choice of paths to follow regarding scaled metrics in the embryo (or the unicellular organism, since ciliate protozoa behave exactly as embryos do in regard to space and time regulation). The classical procedure is to regard the gradient of a substance, such as Y, as the carrier of information. In this case it is the concentration of Y at any point that determines how that part of the embryo will behave. Then it is necessary that this gradient have positive slope throughout the dimensions of the embryo, and that the maximum and minimum values be independent of size. The wave-broom model is one that tends to have these properties, since the amount of material transported with every wave is very largely independent of the length of the axis, once a certain minimum has been exceeded. However, it is by no means evident that the model as described produces a gradient with positive slope over a variety of axial lengths. It seems likely that modifications would have to be introduced to achieve this result.

One such modification, which is suggested by the stoichiometry of most allosteric binding processes, is that the protein in the membrane that binds X has multiple sites, say $n > 1$, with cooperative behavior. Then the concentration of Y as a function of X will have sigmoid characteristics of the type which arise in relation to cooperative behavior, as shown in Fig. 6, the relationship being

$$Y = \frac{aX^n}{1 + LX^n}$$

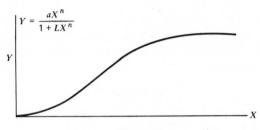

Figure 6.

Such a nonlinearity would have the effect of exaggerating whatever gradient is generated by the linear model, described by Equation 1.1. Then in place of the terms $f(x,t)X(x,t)$ we would have

$$\frac{f(x,t)X^n(x,t)}{1 + LX^n(x,t)}$$

making the equations nonlinear. The decay term for Y would give a term $n\beta Y(x,t)$, since for every molecule of protein that relaxes, n molecules of X would be produced. The variable Y now refers to the concentration of n-meric proteins, which could be referred to as "morphogens," that is, substances responsible for generating changes of epigenetic state in the embryo. Such equations could only be studied by computer. However, this would be of interest for two reasons. The first is to see if regulation of a gradient occurs "naturally" in such a model, without further complication. The other is to test the multimeric model against some well-established experimental results.

The experiments I have in mind are those performed primarily by Locke[4] on the insect cuticle. In this system one can see in a very elegant manner certain consequences of perturbing a gradient system in defined ways. A theoretical model has been proposed by Lawrence et al.[5] to account for these observations on the basis of stable homeostatic properties of cells; that is, the cell is the unit in their model, and it stabilizes the gradient by virtue of homeostatic performance. The behavior of their model under perturbation is consistent with the experimental observations. However, the basic assumption of their theory, that cells are capable of maintaining any one of a number of different concentrations of a single metabolite, the particular value being "set" by a concentration gradient, is not very plausible biochemically speaking. Thus it is of some value to investigate the stability properties of an alternative model in which the nonlinearity is of the type shown in Fig. 6. This model makes some predictions about the expected shape of the gradient itself, which can also be tested against experiment. These studies are in progress.

References

1. V. Tartar, Morphogenesis in Stentor, *Adv. Morphogenesis*, **2**, 1 (1962).
2. W. Herth and K. Sander, Mode and timing of body pattern formation in the early embryonic development of cyclorrhaphic dipterans, *Wilhelm Roux Archiv.*, **172**, 1 (1973).
3. B. C. Goodwin and D. L. McLaren, Morphogenic gradient formation on membranes. To be published.
4. M. Locke, The development of patterns in the integument of insects, *Adv. Morphogenesis*, **6**, 33 (1967).
5. P. A. Lawrence, F. H. C. Crick, and M. Munro, A gradient of positional information in an insect, Rhodnius, *J. Cell Sci.*, **11**, 815 (1972).

PERIODICAL SIGNALS IN THE SPATIAL DIFFERENTIATION OF PLANT CELLS

B. NOVAK*

*Faculté des Sciences, Université Libre de Bruxelles,
Belgium*

The differentiation of the cells toward a characteristic three-dimensional-pattern is usually considered as a time sequence of processes that are spatially dependent. The cellular differentiation can be realized only if the polarity of the cells has developed.[1,2]

The differentiating system undergoes a series of changes that generate differences along the array of units without changing the number or the position of the units. The origin of polarity can be studied at the simplest possible level of organization by considering just one unit, which naturally must be a cell.

The initial symmetrical state of a unit becomes unstable at a certain time, and consecutive inequalities are created within the unit, giving rise to an asymmetrical configuration. The purpose of this chapter is to look into which factors might influence the mechanisms of polarity formation. The main feature of this process is the appearance of a hypothetical line connecting two opposite regions or subspaces of the unit. In biological nomenclature this line is called the polarity axis connecting the apical and basal poles.[1] After a certain period of time, the polarity axis is stabilized, and the first morphological protuberances are observable at the surface of the cell. The cytoplasm appears also to be highly regionalized, and the first signs of the structural and biochemical polarity are detected in the system: In the brown alga *Fucus* the organelles and sulfated polysacharides are accumulated in the basal area 12 hr after fertilization.[3,4] Also, the existence of a gradient of mRNA concentration was reported in the cytoplasm of *Fucus*[5] and that of *Acetabularia mediterranea*.[6]

One part of the cell, called the basal pole, develops into a root, and the apical pole develops into a plant. A regulatory mechanism must be

*Also at Laboratoire de Photobiologie, Université de Liège.

postulated to prevent other points of the plasma membrane from becoming apexes if one has already been developed. The apical pole exerts control over the rest of the membrane area of the cell, a process that resembles apical dominance of high plants.[1]

The role of such a control mechanism has best been expressed by introducing the concept of a morphogenetic field.[7]

Now, the initiation of the polarity appearing in the cell includes the following general features:

1. The direction of the changes (polarity axis).
2. The localization of the reference point (apex or rhizoid).
3. The sense of the changes (apex→base).

These facts together mean that a vectorial morphogenetic field or an endogenous coordinate system have been developed inside of the cell.

The difference along the polarity axis can, for example, be realized as (cf. Ref. 8 and Wolpert's paper in the present volume, p. 253):

1. Space-dependent structural changes on the membrane, which are eventually detectable as a gradient of the membrane potential.
2. Gradients of substances inside the system.
3. Unidirectional active transport of a particular substance (in the case of charged particles, an electric current).
4. A spatial gradient of efficiency of the membrane active pump.[9, 10]
5. Sustained periodical signals unidirectionally conducted.[9–11]

We may now pose the question: Which factors influence the orientation of the polarity axis? At the molecular level, the information for synthesis of a particular molecule surely comes from DNA molecules according to the classical schema of molecular genetics.[12] This genetic information is then translated to the substances that are delivered to the initially symmetrical cytoplasm in the case of *Fucus* eggs. However, in the case of the green alga *Acetabularia* it was shown that the posterior stalk segments regenerate well without a nucleus. This finding indicates that long-living morphological substances are present in the cytoplasm long before the differentiation actually takes place. Thus, the gene activity and the space differentiation are two processes occuring in separate time intervals.[13] The substances responsible for the polar development are apparently delivered to the cytoplasm in an inactive form. They must be activated to the active form by an mechanism triggered by the local inequalities of the membrane function. It must therefore be concluded that the fluctuation due to external factors that occurs in the cytoplasm specifies the genetic information in the space. The creation of the polarity axis most probably involves a random perturbation of the initially symmetrical boundary conditions of the membrane, which is stabilized in the cell and amplified in course of the time.

Weiss[14] has made a suggestion that the pattern responsible for the early differentiation is contained not in the cellular volume but on the surface—on the cell membrane. Provided that the perturbation of the membrane function (generated spontaneously by the system or induced from outside the system during the sensitive phase of development) overcomes the threshold, the vectorial morphogenetic field inside the system will be consecutively stabilized and actively maintained by the cell as it progresses toward the differentiated state.

In addition to the influence of external factors, there is evidence that the formation of polarity may also be caused by random internal changes at the cytoplasm. The randomness of the polarity-axis orientation in space has been shown by a great number of experiments conducted with zygotes of brown algae of fucaceous species growing in a homogeneous medium (Fig. 1).[15–18] The *Fucus* egg is a very suitable object for these types of experiments, because it looks like a sphere of 70 μ in diameter with a homogeneous cytoplasm just after fertilization. The orientation of the polarity axis is totally randomly distributed in all

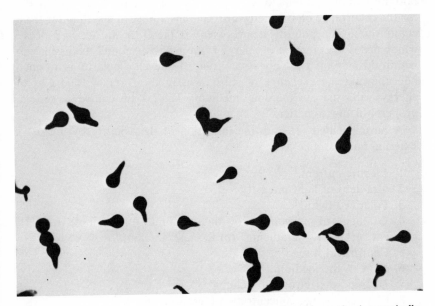

Figure 1. Population of *Fucus* eggs after unequal cleavage, observed microscopically. The polarity orientation is randomly distributed among all possible directions in the germinating plane.

possible directions in the germinating plane for a population of cells; no favored direction of the germination with respect to a fixed coordinate system has been observed. Apparently each cell actively develops itself toward the unsymmetrical state according to its individual randomly distributed perturbation of the boundary conditions. The probability of finding the outgrowth point of a zygote is the same at any orientation in the two-dimensional coordinate systems of the plane parallel to the slide bearing a population of growing cells.

The creation of polarity is one of the most general observations in biology. In terms of classical physics it indicates more "order" within the system, and in thermodynamical terms a decrease of entropy during the development. Turing[19] has developed an elegant theory of chemical morphogenesis for a homogeneous chemical system including autocatalytic chemical reactions and diffusion for two morphogenetic substances. When the system is randomly perturbed from its homogeneous steady state, the concentrations of both substances appear to develop spontaneously a space-dependent steady state. Prigogine[20-23] called such dynamical behavior "a symmetry breaking instability" resulting from a space-dependent perturbation when diffusion is taken into account. I quote the words of Prigogine because the development of the polarity in both his system and ours looks very much like a simple biological model for the occurrence of symmetry-breaking instability in a system far from equilibrium. The polarity development is therefore an active process, which needs the creation of energy from metabolism and its dissipation within the system. The only system that can develop polarity is an open one allowing the exchange of substances in the form of food and of energy with the surrounding medium. This is of course a general property of living matter.[24]

A most peculiar effect is the induction of the polarity by means of external factors, such as

1. Electric current.[15,25,26]
2. Gradient of light intensity.[27,18]
3. Polarized light.[16]
4. Gradient of substances of the medium, for example (a) H^+ gradient,[28-31] (b) K^+ gradient,[32] (c) RN-ase gradient.[33]
5. Centrifugal force.[34,35]
6. Flow in the medium.[32]
7. Dc and ac electrostatic field.[9,36,37]
8. Gradient of egg concentration ("group effect").[32]
9. Temperature gradient.[38]

All these physical or chemical factors force the cell to develop its polarity in a predestined direction (or sense), if the gradient of the external factor does not change the sign of its slope. Common to all the above-mentioned external factors used for polarity induction is their vectorial character. In the space where these physical or chemical quantities are defined, a vectorial field exists, which is determined by the direction of the maximal steepness of the particular quantities, that is, their gradient. The external vectorial fields have their effect because the membrane also has vectorial properties: Its structure and function have a polarity going from outside to inside the cell. The results of all induction experiments are the same: The polarity axis is oriented parallel to the gradient of the applied fields (Fig. 2). In the sense of dynamical theory an orientation of the polarity axis parallel to the external field can be looked upon as a forced response. The external fields just reset the boundary conditions causing the oriented perturbation of a particular cellular parameter. If this oriented perturbation is large enough and is established at the right moment, when according to the cell's inherent properties the substances needed for the polarity development are already present—during the so called induction phase —and is applied for a certain period of time, then the polarity of the cell is induced. The oriented perturbation required for the induction of the polarity must be of such a character that it results in the creation of an endogenous vectorial field, which is spontaneously amplified by the cell.

How the external field establishes the endogenous vectorial field can best be shown when the model cell is placed in a homogeneous electrical field of intensity E_0. Let us consider an ideal spherical cell—a model for a *Fucus* egg, with radius R, specific conductance of cytoplasm g_2 (Ω^{-1} cm^{-1}), and membrane conductance g_m (Ω^{-1} cm^{-2}), which is immersed in a medium of specific conductance g_0 (Ω^{-1}cm^{-1}). The origin of the spherical coordinate system r, ϑ, φ coincides with the center of the cell (Fig. 3). The modified Poisson equation for this case was solved, assuming no free space charge. The angular dependence of the membrane potential V at the outer surface of the membrane, with the origin of the coordinate system taken as a reference point, is given by[9,36]

$$V(R,\vartheta) = -\left(1 + \frac{1 - g_e/g_0}{2 + g_e/g_0}\right) E_0 R \cos\vartheta + U_m \qquad (1)$$

where U_m is the resting membrane potential, and g_e is equivalent conductance of cytoplasm and plasma membrane in series.

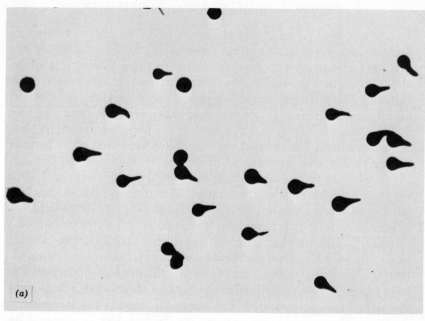

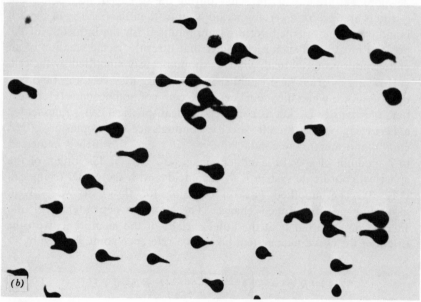

Figure 2. Population of *Fucus* eggs after first cleavage. (*a*) Unipolar distribution of the polarity axes parallel to the applied dc electric field. A homogeneous electric field is directed from left to right. (*b*) Bipolar distribution of the polarity axes parallel (but in both senses) to the external ac electric field. Voltage pulses of both polarities, 30 msec long were applied at intervals of 100 msec.

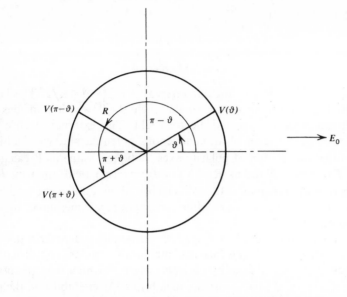

Figure 3. Cross section of the model cell. $V(R,\vartheta)$, $V(R,\pi+\vartheta)$ and $V(R,\pi-\vartheta)$ are membrane potentials calculated at different points on the cellular surface (see text). E_0 and R are the intensity of the external homogeneous field and radius of the cell, respectively.

If we take the experimental data of Weisenseel and Jaffe[39] for the conductance g_m of *Fucus* membrane, it follows that $g_e = 6 \times 10^{-7}$ (Ω^{-1} cm^{-1}). If the external medium is seawater ($g_0 = 5.6 \times 10^{-2}$ Ω^{-1} cm^{-1}), then the ratio $g_e/g_0 \ll 1$, and Equation (1) reduces to

$$V(R,\vartheta) \doteq -1.5 E_0 R \cos\vartheta + U_m \qquad (2)$$

The membrane potential is perturbed in the following way: The right hemisphere is depolarized and the left one is hyperpolarized. A transcellular electric field exists across the cytoplasm, which causes translocation of the charged particles in the direction of the potential gradient and with a sense depending on the sign of their charge. The particles are actively displaced at a rate proportional to the driving forces, and they move back because of diffusion at a rate proportional to the diffusion constant. The inverse concentration gradients of positive and negative ions maintained by the potential gradient increase till a steady state is reached. By comparing the potential at point (R,ϑ) and at point $(R,\pi+\vartheta)$ or $(R,\pi-\vartheta)$, the membrane potential difference ΔV is calculated from (2):

$$\Delta V(R,\vartheta) = V(R,\vartheta) - V(R,\pi+\vartheta)$$

$$= -3E_0 R \cos\vartheta \qquad\qquad (3)$$

which is satisfied in the angular interval $-\pi/2 \leqslant \vartheta \leqslant \pi/2$. The largest potential gradient is achieved for the angle $\vartheta = 0$, which coincides with the direction of the applied electric field. The external electric field appears to be a trigger of an endogenous potential gradient leading to the unified orientation of polarity axes. A typical value of V calculated for a *Fucus* egg of radius $R = 35$ μ with external field intensity $E_0 = 2$ V/cm is in the range of -18 mV, which is about 30% of the membrane potential occurring 3 hours after fertilization, as measured by Weisenseel and Jaffe.[39]

As a consequence of the spatially dependent depolarization and hyperpolarization of the plasma membrane, polar conformational changes of the membrane biopolymers occur,[40] which then modify the active transport of ions across the membrane. After stabilizing the polar differences of the active transport, the cell itself is able to sustain the potential gradient without external electric field.[41,9] In this way the polar transcellular transport of charged particles is established in the cellular region adjacent to the depolarized membrane section, and there occurs an excess of negative as well as positive charges in the region adjacent to hyperpolarized membrane area. The later cellular division partitions the polarized cell sections into two daughter cells having different compositions.[2] The previous polarization of the cytoplasm is a necessary assumption for the existence of position-dependent enzyme activities as measured in regenerating enucleate posterior stalk segments (EPSS) of *Acetabularia*.[42,43]

For direct measurement of the generation of the membrane potential gradient, a giant alga *Acetabularia mediterranea* was used instead of the small fucaceous egg (diameter 70 μ). The giant size, together with its unicellular character, has made this species of green marine alga the most suitable object for studying the regeneration of polarity. Several multicellular organisms (e.g. *Hydra*) were recently used for the study of axial polarity.[44] Compared to this organism, *Acetabularia* looks like an extremely simple system—just one cell, of cylindrical form, about 5 cm long. The adult *Acetabularia* is strongly polarized, having a nucleus at the basal pole and a cup at the apical pole, as well as functional regionalization of the cellular parts.[45,13,46] Haemmerling[47,13] has proposed an elegant method for preparing from an adult cell an EPSS that has lost its polarity. The absence of polarity after the treatment was proved for a population of the EPSSs, which regenerated at either end with the same probability. Placing the EPSSs in a cuvette divided into separate compartments A, B, C, D, the author has measured the development of a potential gradient between compartments A and D[9,10] (Fig. 4).

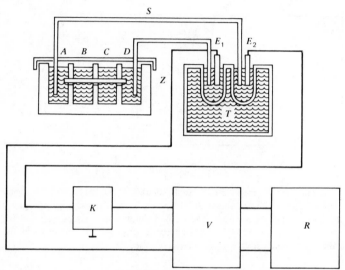

Figure 4. Electric circuit for the measurement and registration of the transcellular potential. A, B, C, D: compartments of the cuvette; S: agar-KCl bridge; E_1, E_2: calomel electrodes; Z: cell; T: thermostatization; K: voltage compensation; V: electrometer; R: registration.

The potential gradient between the ends of EPSSs was always zero in darkness, and in the further experiments the onset of the potential gradient was monitored after the EPSSs were illuminated with white light of constant intensity (Fig. 5). The most surprising effect was observed several hours after the onset of the illumination. With a delay of 5 to 10 hr after beginning of illumination, spontaneous activity was registered (Fig. 6). The spontaneous spikes were found to originate in that compartment of the cuvette that contained one end of the EPSS, and the pulses persisted for at least 2 days after onset of illumination (Fig. 7).

The sustained spontaneous activity has the following properties:

1. Spontaneous pulses occur about 15 hr earlier than the steady-state potential gradient.

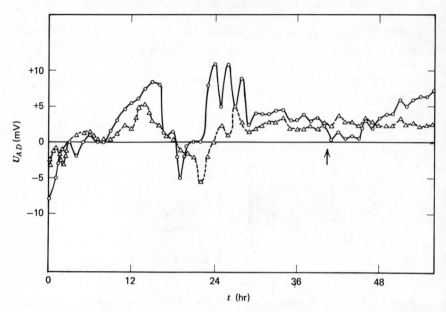

Figure 5. The time course of the external potential of two posterior stalk segments that regenerate apically in compartment A. The potential does not change its sign after 24 hr from the onset of the illumination. The arrow ↑ denotes the moment (about 41 hr), when the regeneration can be observed with an optical microscope. The cell was illuminated with white light of intensity $I = 1.2 \times 10^4$ erg cm^{-2} sec^{-1}.

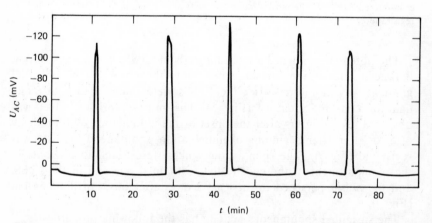

Figure 6. Unidirectional spontaneous pulses of the posterior stalk segment of *Acetabularia mediterranea* during the regeneration. The refractory time was in the range from 15 to 25 min. The intensity of the constant white light was 1.2×10^4 erg cm^{-2} sec^{-1}.

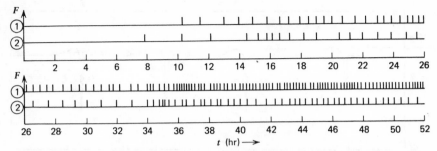

Figure 7. Sequence of the spontaneous impulses of two regenerating posterior segments of *Acetabularia mediterranea*. Steady light of intensity $I = 1.5 \times 10^4$ erg cm^{-2} sec^{-1} was used. The large time scale allows representation of the pulses by a series of the short lines (cf. Fig. 6).

2. The shape and the amplitude (about -100 mV) of the pulses remain constant. This indicates the existence of sustained relaxation oscillations localized at the regenerating end.

3. The repetition rate of the pulses increase from 3 to 6 pulses/hr over a period of several hours. When the maximal rate is reached it stays almost constant.

4. If the first depolarization pulse arises at a particular end of an EPSS, then all following pulses originate there. Apparently the refractory period is shortest in the pacemaker area of the EPSS.

5. There were no spontaneous pulses observed when the EPSSs were measured during the dark period or after adding—10^{-4} moles of DNP. It is concluded that the occurrence of the pulses is coupled to the energy influx into the differentiating system.

6. Spontaneous current pulses were observed under the voltage-clamp condition (Fig. 8), if the potential difference between compartments A and C containing both ends of the EPSS was held at a constant value in the range 10 to 50 mV (Fig. 9).

These sustained spontaneous current pulses correspond to the potential spikes described above, since the repetitive frequency and the shape of both appear to be identical. Just like the voltage pulses, they persist for at least 2 days, have the same repetition rate, and originate at the end that apparently regenerates in the future.

These spontaneous spikes are so far the first detectable signs of the polarity of the EPSS. The first spike, which occurs at a particular end of the EPSS, is the first experimental evidence of symmetry-breaking instability. The spontaneous activity presumably causes the polar transport of particular substances along the cytoplasm. This activity

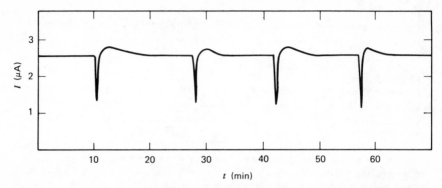

Figure 8. Unidirectional spontaneous current pulses during the voltage-clamp experiment. The regenerating posterior stalk segment of *Acetabularia mediterranea* was externally clamped at $U_e = 50$ mV and illuminated with steady light of intensity 1.5×10^4 erg cm^{-2} sec^{-1}. The refractory time was found to be in the interval from 15 to 25 min.

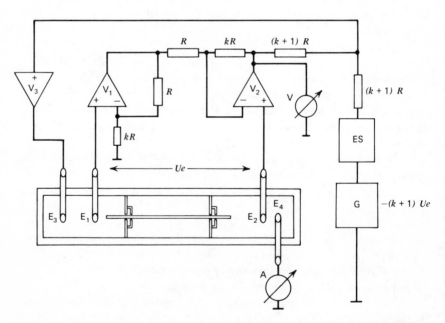

Figure 9. Circuit for the voltage clamp experiments. V_1, V_2, V_3: operational amplifiers; E_1, E_2: voltage electrodes; E_3, E_4: current electrodes; ES: electronic switch; G: generator; V: voltmeter; A: microammeter.

292

creates an endogneous potential gradient along the EPSS in accordance with an existing concentration gradient. From the experiments of Saddler[48,49] and Gradmann,[50] it is known that the relative high steady-state membrane potential, -170 mV, is driven by a large active influx of Cl^- ($\phi_a = 200$ to 850 pmole cm^{-2} sec^{-1}) counteracted by a passive diffusional efflux of the same size. The mechanism of creation of spontaneous pulses and of the steady-state potential involves the active electrogenic Cl^- pump.[51] It was concluded by the author that the spikes arise from the sudden decrease of the active influx of Cl^- caused by the rapid increase of the membrane conductance. From the voltage-clamp experiments the maximal increase of the membrane conductivity, Δg_m, at the tip of the spike was calculated to be 2.6×10^{-3} Ω^{-1} cm^{-2}. The following phenomenological equation describes quantitatively the relation between voltage and current in *Acetabularia* (cf. Refs. 52, 53):

$$U_{AM} = U_m + F\frac{\phi_a}{g_m} \tag{4}$$

where U_{AM} is the resting membrane potential measured at compartment A,

U_m the diffusion controlled membrane potential,

g_m the membrane chord conductance,

ϕ_a the active influx of Cl^-, and

F the Faraday constant.

From data published by Saddler[48,49,54] and Gradmann,[50] it can be concluded that the diffusion potential, $U_m = -90$ mV, is controlled only by diffusion of K^+. Using the value[49,50] $g_m = 10^{-3}$ Ω^{-1} cm^{-2}, we may calculate the active influx of Cl^- maintaining the resting potential from (4) to be 825 pmole cm^{-2} sec^{-1}. This value confirms the experimentally estimated influx in the range of 200 to 850 pmole cm^{-2} sec^{-1} given by Saddler.[48,49]

The potential U_{AP} during the spikes is approximated by

$$U_{AP} = U_m + F\frac{\phi_{ap}}{g_m + \Delta g_m} \tag{5}$$

where ϕ_{ap} is the active influx during the spikes, and Δg_m is the increase of chord conductance during the spikes. Substituting Δg_m into (5), we determine the active influx of the Cl^- at the tip of the spikes to be 50 pmole cm^{-2} sec^{-1}, which is about 6% of the steady-state active influx. If we apply Equation (4) to compartments A and C, the steady-state

potential difference between both ends of EPSS is given by the following expression:

$$U_{AC} = U_{AM} - U_{CM}$$

$$= F\frac{\phi_a - \phi_c g_{ma}/g_{mc}}{g_{ma}} \qquad (6)$$

where U_{AC} is the registered potential difference,

ϕ_a, ϕ_c are the active influxes of Cl^- into compartments A and B,
respectively, and

g_{ma}, g_{mc} are the membrane chord conductances at compartments A
and C, respectively.

It is easy to see that the active influx of ϕ_a must be greater than $\phi_c \times$
g_{ma}/g_{mc} if the registered potential difference U_{AC} is greater than zero.
Assuming that the difference in conductivities is relatively small, so that
$g_{ma} \doteq g_{mc}$, equation (6) can be simplified to

$$U_{AC} = F\frac{\Delta\phi_{ac}}{g_{ma}} \qquad (7)$$

where $\Delta\phi_{ac} = \phi_a - \phi_c$. For $U_{AC} = 5 \times 10^{-3}$ V, Equation (7) yields $\Delta\phi_{ac} = 50$
pmole cm^{-2} sec^{-1}. The difference in steady-state active transport, $\Delta\phi_{ac}$,
is of the order of the lowest transport occurring during the spike.
Subtracting (4) from (5), we conclude that the potential gradient along
the EPPS (between compartments A and C) during the spike at compartment A may be calculated from

$$U_{AP,C} = F\left(\frac{\phi_{ap}}{g_{ma} + \Delta g_{ma}} - \frac{\phi_c}{g_{mc}}\right) \qquad (8)$$

The potential difference during the spike calculated from (8), $U_{AP,C} =$
-85 mV, is in good agreement with measured values for spontaneous
pulses. Using the constant-field assumption,[55] we may calculate the
permeability coefficient P_{Cl} as 2.42×10^{-8} cm/sec for the steady-state
potential. The permeability coefficient decreases during the spike to the
value 1.48×10^{-9} cm/sec. The ratio of permeability coefficients, $1:16.4$,
resembles that of the conductivities for inward and outward currents,
$1:17$, measured under the voltage-clamp condition.[50]

The fact that spontaneous activity is observed only upon illumination
pointed to the causal relationship between the spontaneous activity and
the polarity formation. The triggering mechanism of the spontaneous
impulses is not yet known.

The relatively long refractory time of the oscillation phenomenon indicates some unknown cyclical biochemical reaction localized at the membrane (cf. Ref. 56), which periodically causes an abrupt depolarization via a decrease of the active Cl^- transport in the regenerating zone. The observed refractory clock period lies in the interval from 20 sec to half an hour, and is compatible with the theoretical treatment of the spatiotemporal organization of developing systems by Goodwin and Cohen.[11,57] These authors assumed the existence of two signals conducted from the common origin with different velocities in the multicellular system. The phase shift between them, which is a function increasing monotonically with displacement from the point of pulse origin, provides the positional information. The phase shift idea is not strictly applicable to *Acetabularia* because this sytem consists of one cell, where only one kind of spontaneous pulse was observed. It supplies sufficient information for the parts of this cylindrical cell. Their relative position is specified by the increasing phase of the pulse along the cellular stalk. The spread of the pulse defines a vectorial field for the part of the cell receiving the unidirectional information in the form of the pulse sequence coming from one origin. The sense of the propagation thus determines the polarity of the cell.

Up to this point we have been concerned with the emergence of polarity and positional information on the scale of a single cell. We close this report by presenting a few results concerning the problem of pattern formation on a supercellular scale. (see also the chapter by Hess et al. in this volume, p. 137.)

It is at present well established that sustained periodic signals control the aggregation of a priori identical cells of *Dictyostelium discoideum*.[58,59] The pacemaker centers exert an influence on the surrounding cells. If the cell is reached by a signal that overcomes the threshold, it moves towards the center. The characteristic radial streams of amoeboid cells were observed at this stage of differentiation.[58] A model system of aggreagating cells was simulated on a computer under the following assumptions[60]:

1. The cells are initially randomly distributed in a two-dimensional array.

2. A periodic chemical reaction including autocatalysis of the second order is supposed in every cell volume.

3. The reaction product, which is degraded at a constant rate in the outer medium, makes contact with the other cells.

4. The phases of the cellular chemical oscillations are randomly distributed in the cell population.

5. The phase of the cellular oscillation is perturbed by diffusion of a

substance from the aggregating center.

6. The phase perturbation triggers the chemotactic movement against the concentration gradient of the cells.

The spatiotemporal development of the model system is shown in Fig. 10. The system of amoebae develops from a random distribution into the characteristic pattern of radial streams of cells aggregating toward the centers.

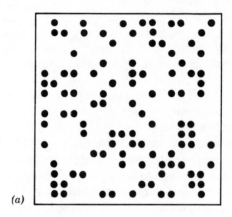

(a)

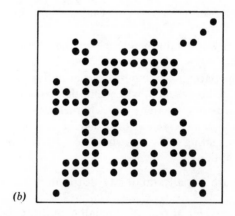

(b)

Figure 10. Two-dimensional array (20×20) for a computer simulation of the cellular aggregation. (*a*) Randomly distributed circles denote the postions of amoebae at the mesh points of the array at time $t_0 = 0$ (arbitrary units). (*b*) Distribution of amoebae at time $t_1 = 47.3$ (arbitrary units). The system evolves to an aggregated steady state after the computation of at least 47.3 time units. For details see the comprehensive publication by Novak and Seelig.[60]

Acknowledgments

I would like to express my gratitude to Professor P. Glansdorff, Professor I. Prigogine, and Professor G. Nicolis for their stimulating discussions and encouragement. My grateful thanks for the typewriting are due to Mrs. L. Février. This work was completed during the tenure of an EMBO fellowship.

References

1. C. W. Wardlaw, *Phylogeny and Morphogenesis*, Macmillan, London, 1952.
2. E. Bünning, *Entwicklungs- und Bewegungsphysiologie der Pflanze*, Springer, Berlin and New York, 1953.
3. R. S. Quatrano, Rhizoid formation in Fucus zygotes: Dependence on protein and ribonucleic acid synthesis, *Science*, **162**, 468 (1968).
4. R. S. Quatrano, Sulfation of fucoidan in Fucus embryos. I. Possible role in localisation, *Develop. Biol.*, **30**, 29 (1973).
5. S. Nakazawa, Regional concentration of cytoplasmatic RNA in Fucus eggs in relation to polarity, *Naturwissenschaften*, **53**, 138 (1966).
6. K. Zetsche, Steuerung der Zelldifferenzierung bei der Grünalge Acetabularia, *Biol. Rundschau*, **6**, 97 (1968).
7. C. M. Child, Patterns and problems of development, Chicago U. P., 1941.
8. L. Wolpert, Positional information and the spatial pattern of cellular differentiation, *J. Theoret. Biol.*, **25**, 1 (1969).
9. B. Novak, An electrophysiological study on the spatial differentiation of plant cells, Ph. D. Dissertation, Univ. of Tübingen, Germany, 1972.
10. B. Novak and F. W. Bentrup, An electrophysiological study of regeneration in Acetabularia mediterranea, *Planta*, **108**, 227 (1972).
11. B. C. Goodwin and M. H. Cohen, A phase-shift model for the spatial and temporal organization of developing systems, *J. Theoret. Biol.*, **25**, 49 (1969).
12. J. Brachet, The role of nucleic acids in morphogenesis, *Progr. Biophys. Molec. Biol.*, **15**, 97 (1965).
13. J. Haemmerling, Nucleo-cytoplasmatic interaction in Acetabularia and other cells, *Annual Rev. Plant Physiol.*, **14**, 65 (1963).
14. P. Weiss, *Principles of development*, Henry Holt, New York, 1939.
15. E. J. Lund, Electrical control of organic polarity in the egg of Fucus, *Botan. Gaz.*, **76**, 288 (1923).
16. L. Jaffe, Tropistic responses of zygotes of the Fucaceae to polarized light, *Exp. Cell Res.*, **15**, 282 (1958).
17. L. Jaffe, Localization in the developing Fucus egg and the general role of localizing currents, *Adv. Morph.*, **7**, 295 (1968).
18. W. Haupt, Die Induktion der Polarität bei der Spore von Equisetum, *Planta*, **49**, 61 (1957).
19. A. M. Turing, The chemical basis of morphogenesis, *Phil. Trans. Roy. Soc. Lond.*, **B237**, 37 (1952).
20. I. Prigogine, Dissipative structure in chemical systems, in *Proc. Fifth Nobel Symp., Fast reactions and primary processes in chemical kinetics*, Wiley-Interscience, New York, 1967, p. 371.
21. I. Prigogine and G. Nicolis, On symmetry breaking instabilities in dissipative systems, *J. Chem. Phys.*, **46**, 3542 (1967).
22. I. Prigogine and R. Lefever, Symmetry breaking instabilities in dissipative systems. II., *J. Chem. Phys.*, **48**, 1695 (1968).

23. I. Prigogine, R. Lefever, A. Goldbeter, and M. Hershkowitz-Kaufman, Symmetry breaking instabilities in biological systems, *Nature*, **223**, 913 (1969).

24. E. Schroedinger, *What is life?*, Cambridge U. P., Cambridge, England, 1944.

25. E. J. Lund, *Bioelectric field and growth*, University of Texas Press, Austin, Tex., 1947.

26. F. W. Went, Eine botanische Polaritätstheorie, *Jb. wiss. Bot.*, **76**, 528 (1932).

27. D. M. Whitaker, The effect of unilateral ultra-violet light on the development of the Fucus egg, *J. Gen. Physiol.*, **24**, 263 (1941).

28. D. M. Whitaker, The effect of hydrogen ion concentration upon the induction of polarity in Fucus eggs. I. Increased hydrogen ion concentration and the intensity of mutual inductions by neighbouring eggs of Fucus furcatus, *J. Gen. Physiol.*, **20**, 491 (1937).

29. D. M. Whitaker, The effect of hydrogen ion concentration upon induction of polarity in Fucus eggs. II. The effect of diffusion gradients brought about by eggs in capillary tubes, *J. Gen. Physiol.*, **21**, 57 (1937).

30. D. M. Whitaker, The effect of hydrogen ion concentration upon the induction of polarity in Fucus eggs. III. Gradients of hydrogen ion concentration, *J. Gen. Physiol.*, **21**, 833 (1937).

31. D. M. Whitaker, The effect of pH on the development of ultracentrifuged Fucus eggs, *Proc. Nat. Acad. Sci.*, **24**, 85 (1938).

32. F. W. Bentrup and L. Jaffe, Analysing the group effect: rheotropic responses of developing Fucus eggs, *Protoplasma*, **65**, 25 (1968).

33. S. Nakazawa, Polarity determination in Fucus eggs by localized exposure to RN-ase, *Bot. Mag. Tokyo*, **83**, 325 (1970).

34. D. M. Whitaker, Determination of polarity by centrifuging eggs of Fucus furcatus, *Biol. Bull.*, **73**, 249 (1937).

35. E. W. Lowrance and D. M. Whitaker, Determination of polarity in Pelvetia eggs by centrifuging, *Growth*, **4**, 73 (1940).

36. B. Novak and F. W. Bentrup, Induction of Fucus egg polarity by electric a.c.-fields, *Naturwissenschaften*, **57**, 549 (1970).

37. B. Novak and F. W. Bentrup, Orientation of Fucus egg polarity by electric a.c. and d.c. fields, *Biophysik*, **9**, 253 (1973).

38. E. W. Lowrance, Determination of Polarity in Fucus eggs by temperature gradients, *Proc. Soc. Exp. Biol. Med.*, **36**, 590 (1937).

39. M. H. Weisenseel and L. F. Jaffe, Membrane potential and impedance of developing fucoid eggs, *Develop. Biol.*, **27**, 555 (1972).

40. E. Neumann and A. Katchalsky, Hysteretic conformational changes in biopolymers induced by high electric fields, *Proc. First Europ. Biophys. Congr.* **6**, 91 (1971).

41. L. Jaffe, On the centripetal course of development of the Fucus egg, and self-electrophoresis, *Develop. Biol. Suppl.*, **3**, 83 (1969).

42. K. Zetsche, Regulation der UDPG–Pyrophosphorylaseaktivität in Acetabularia. I. Morphogenese und UDPG–Pyrophosphorylase-Synthese in kernhaltigen und kernlosen Zellen, *Z. Naturforsch.*, **23b**, 369 (1968).

43. K. Zetsche, Regulation der UDPG–Pyrophosphorylaseaktivität in Acetabuleria. II. Unterschiedliche Synthese des Enzyms in verschiedenen Zellregionen, *Planta*, **89**, 244 (1969).

44. J. Hicklin, A. Hornbruch, and L. Wolpert, Inhibition of hypostome formation and polarity reversal in Hydra, *Nature (Lond.)*, **221**, 1268 (1969).

45. J. Haemmerling, Über formbildende Substanzen bei Acetabularia mediterranea, ihre räumliche und zeitliche Verteilung und ihre Herkunft, *Arch. Entw. Mech.*, **131**, 1 (1934).

46. Ch. Haemmerling and J. Haemmerling, Über Bildung und Ausgleich des Polari-tätsgefälles bei Acetabularia, *Planta*, **53**, 522 (1959).

47. J. Haemmerling, Regenerationsversuche an kernhaltigen und kernlosen Zellteilen von Acetabularia Wettsteinii, *Biol. Zentralblatt*, **54**, 650 (1934).

48. H. D. W. Saddler, The ionic relations of Acetabularia mediterranea, *J. Exp. Bot.*, **21**, 345 (1970).

49. H. D. W. Saddler, Membrane potential of Acetabularia mediterranea., *J. Gen. Physiol.*, **55**, 802 (1970).

50. D. Gradmann, Einfluss von Licht, Temperatur und Aussenmedium auf das elektrische Verhalten von Acetabularia crenulata, *Planta* (*Berl.*), **93**, 323 (1970).

51. H. D. W. Saddler, Spontaneous and induced changes in the membrane potential and resistance of Acetabularia mediterranea, *J. Membr. Biol.*, **5**, 250 (1971).

52. A. L. Hodgkin and A. F. Huxley, A quantitative description of membrane current and its application to conduction and excitation in nerve, *J. Physiol.*, **117**, 500 (1952).

53. H. Kitasato, The influence of H^+ on the membrane potential and ion fluxes of Nitella, *J. Gen. Physiol.*, **52**, 60 (1968).

54. H. D. W. Saddler, Fluxes of sodium and potassium in Acetabularia, *J. Exp. Botany*, **21**, (68), 605 (1970).

55. D. E. Goldman, Potential, impedance and rectification in membranes, *J. Gen. Physiol. Lond.*, **27**, 37 (1943).

56. A. Katchalsky and R. Spangler, Dynamics of the membrane processes, *Quart. Rev. Biophys.*, **1**, 127 (1968).

57. M. H. Cohen, Models for the control of development, *Control mechanisms of growth and differentiation. Symposium of the Society for Exp. Biol.*, **25**, 455 (1971).

58. G. Gerisch, Cell aggregation and differentiation in Dictyostelium, in *Current Topics in Develop. Biol.*, **3**, 157 (1968).

59. G. Gerisch, Periodische Signale steuern die Musterbildung in Zellverbänden, *Natur-wissenschaften*, **58**, 430 (1971).

60. B. Novak and F. F. Seelig, A chemical model for aggregation of slime mold. A computer case study, *J. Theor. Biol.*, (1975).

STRUCTURE AND TRANSPORT IN
BIOMEMBRANES

LIANA BOLIS

Institute of General Physiology, University of Rome, Italy

CONTENTS

I. INTRODUCTION

The development of studies and research in cell permeability dates back to the last century, before a structural view of the membrane had been considered.[1]

It was soon seen that *lipophilic* compounds cross cell barriers faster than *hydrophilic* ones. This was based on experimental work of Overton[2,3] and later of Collander[4,5] and others.[6,7] Still later, attention was brought to other systems of transport into the cell, beginning with Osterhout,[8,9] who introduced the *carrier* concept to explain the accumulation of some substances in *Valonia* cells. Since then, together with the development of the concept of the membrane, much research has been concerned with studies of different possible ways in which the cell operates to exchange ions and metabolites between the inside and the outside of the cell.

A major advance was the suggestion by Rosenberg and Wilbrandt[10] that transport is connected with a specific *enzyme activity*.

An important function of the *membrane* is to provide mechanisms for the maintenance of large concentration differences between the interior of the cell and its environment. In animals whose body fluids are closed off from the environment, the cell membranes continuously control the concentrations of molecules and fluids in the intra- and extracellular milieu. They make possible the existence of phases that are closely adjacent, but at the same time *far from equilibrium* with one another.

Biological membranes can maintain such states only because, although they are extremely thin, they nevertheless offer a high resistance to diffusion of the water-soluble substances dissolved in the intra- and extracellular fluids. The latter fact minimizes the expenditure of energy, in the form of "active" transport, required to build up the system in the steady state away from equilibrium. It is, on the other hand, an important characteristic of membranes that although they offer a matrix in which dissipative *metabolic processes* controlling cellular environment are embedded, their own internal organization appears essentially as a stable equilibrium type of structure.

The *quantitative study of cell permeation kinetics* was developed particularly by Collander[5] and Jacobs.[11, 12] The first theoretical examination of the kinetics was made by Davson and Danielli,[13] Rosenberg,[14] Wilbrandt,[15] Bolis et al.,[16] and others,[17] and led to a classification into three types of membrane permeation:

1. *Simple activated* diffusion (most molecules).
2. *Facilitated* diffusion (special molecules).
3. *Active* transport (special molecules).

Simple activated diffusion through the cell membrane also implies an activating effect of the structure of the membrane on the movement of solutes, for instance by way of lipid solubility. It closely resembles diffusion in water and other fluids. It shows a dependence on molecular weight and volume, and on partition between hydrocarbon and water, but not on stereochemistry and other molecular details; an example is the entry of alcohols.

Facilitated diffusion, on the other hand, involves special mechanisms enabling permeation to be more rapid than it would be by simple diffusion. Such mechanisms are very closely related to structural details. No energy is put into facilitated diffusion other than thermal energy, so that the steady state reached by facilitated diffusion is the same as that reached by simple diffusion. This is the case of D and L *mono-saccharides*: in general the D form penetrates much faster than the L form, although both reach the same final concentration.

Facilitated diffusion represents a transport with characteristic kinetics and properties, which are described below, and which lead to the conclusion that it is mediated by a carrier.

Wilbrandt,[18] LeFevre,[19] Widdas,[20a, b] Stein,[21] and Miller[22a, b] largely expressed the kinetic parameters of this transport: This kind of transport is dependent on a component of a molecular group in the membrane. There is an interdependence of individual particles during the movement, which is the basis of most of the characteristic carrier features, such as *countertransport, competitive inhibition,* and *competitive accelaration*. Under conditions of high *saturation*, the interdependence is maximal, and this fact suggests a limited concentration of carriers in the membrane.

The chemical nature of the specific carrier supposed for the transport of so many different molecules through facilitated diffusion is not known. It has often been considered a protein: The high stereospecificity of some carrier systems suggest that the specific recognition requires properties of the carrier similar to those found in *enzymes*. The mechanisms of transport usually fit the Michaelis-Menten kinetics, but *allosteric models* also could be operative here. One possible mechanism is that the carrier masks the hydrophilic group of a molecule and renders its substrate soluble in the membrane.

Recently the carrier system has been considered to consist of a sequence of intermolecular reactions that allows the permeant to diffuse through the membrane.[23,24]

II. ACTIVE TRANSPORT

Active transport shows a dependence on molecular structure and kinetics similar to those found with facilitated diffusion. But in addition the cell provides energy that permits *electrochemical potential* gradients to be built up across the cell membrane.

For instance, Na is kept at a low level inside cells, despite a high Na concentration outside, which, together with the negative potential inside, provides a strong electrochemical gradient driving Na into the cell. When the cell energy supply is depleted or ATP production blocked by specific inhibitors, then the Na inside the cell increases. This, together with other evidence, shows that Na is actively transported out of the cell. Such a system is often described as a *"pump"* that pumps Na outwards.

We can divide active transport into *primary* and *secondary active transport*.

In the case of primary active transport of a substance, a difference in electrochemical potential may be generated without coupling to other fluxes, except to the advancement of the driving metabolic reaction.[25] Sodium pumping appears to be an excellent example of primary active transport.

On the other hand we have secondary active transport when a given substance moves against its electrochemical gradient without being directly entrained by the pump mechanism; in terms of irreversible thermodynamics, this transport is expressed solely by the cross coefficients of solute fluxes without directly involving metabolism. Such is assumed to be the case in many tissues, which transport organic solvents in a way that depends on an electrochemical gradient of an electrolytic ion. This is the case for the *transport of sugar* in the intestine.[26,27,28]

In addition. we must consider the possibility of coupling between solute *flow and bulk flow of the* solvent: The solvent flow may then be driven by a hydrostatic pressure gradient or by an osmotic one. In contrast with secondary transport, this kind of coupling is not stoichiometric, but depends on a direct interaction between solutes and water, whereas in the other cases, a carrier is supposed to be essential in the transport. This kind of solute-solvent coupling is very important in biological systems such as the intestine and kidney.

The coexistence of these different types of transport showed that the membranes should be similar in their structure. This was established by 1940,[13] and subsequently intensive work was done on both kinetics and structure to obtain more detail about the function of the membrane.

III. STRUCTURE OF BIOLOGICAL MEMBRANES

The free energy expended in the stabilization of biological membranes arises from lipid-lipid, lipid-*protein*, and protein-protein interactions, which provide the cohesive forces responsible for the structure. A series of structural pictures of the membranes have been presented in the past fifty years; a vast number of experimental data are available today.

The classical view of the membranes is that put forward by Davson and Danielli[29,30] and revised later.[13] In this model the primary structure is a continuous phase that is *a lipid bilayer*. The bilayer interacts with proteins on both sides through polar and nonpolar forces. The bilayer controls the permeation of molecules that penetrate by simple diffusion. Specific permeations are provided by *enzymelike* molecules, which extend through the thickness of the membrane and provided a polar pathway for molecular movement (Fig. 1).

Much recent work—for example, that of McConnel[31,32] and of Frye and Edidin[33]—has been directed toward determining the liquidity of biological membranes. This is reflected in the recent reviews by Gitler,[34] Singer,[35] and Oldfield.[36] This recent work indicates that not only membrane lipids, but also membrane proteins are relatively diffusible in the two dimensions of the bilayer.

It seems probable that *excitation* phenomena, as general properties of

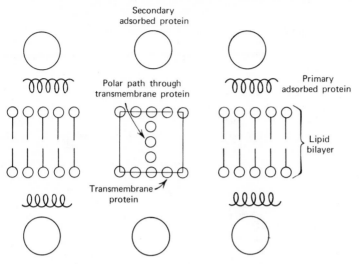

Figure 1. The continuous phase of the membrane is the bilayer, and the proteins are the disperse phase. Some proteins extend through the thickness of the membrane and provide mechanisms for facilitated diffusion and active transport, using a polar path in a transmembrane protein. There are two protein layers adsorbed on the bilayer. The primary adsorbed layer has both polar and nonpolar interactions with lipid. The secondary layer has mainly polar interactions with other membrane components. (After Danielli, in Ref. 13).

all cells, depend on an environment flexible enough (or fluid enough) to allow conformational changes of proteins at the active site to take place. It seems likely that the "*gating mechanism*" is based on a *cooperative process* in which quite a large number of membrane molecules may play a part. The precise form of the cooperation is still unclear. Recently Keynes[37,38] has discussed the applicability of this view to nerve membranes.

One may be tempted to compare the idea of dissipative structures recently developed by Prigogine et al.[39,40,41,44] with certain aspects of membrane function and structure. In order to establish such a connection with dissipative structures on a firmer basis, one will need specific models for each particular problem. Some attempts have been made in this direction. Thus, the influence of *nonequilibrium constraints* in nerve excitation is discussed in detail in the paper by Lefever and Deneubourg[43] in this volume (p. 349). Another characteristic example is the induction problem discussed in the papers by Nicolis (p. 29)[45] and by Ortoleva and Ross (p. 49).[46]

References

1. W. Pfeffer, *Osmotische Untersuchungen*, Engelmann, Leipzig, 1877.

2. E. Overton, Über die allgemeinen osmotischen Eigenschaften der Zelle, ihre vermutlichen Ursachen und ihre Bedeutung für die Physiologie, *Vjschr. Naturf. Ges. Zuriche*, **44**, 88 (1899).

3. E. Overton, Beiträge zur allgemeinen Muskel- und Nervenphysiologie, *Pflüg. Arch. Ges. Physiol.*, **92**, 115, 346 (1902).

4. R. Collander, Über die Permeabilität von Kollodiummembranen, *Soc. Sci. Fennica Comment. Biol.*, **2**, (6), 1 (1926).

5. R. Collander, The permeability of plant protoplasts to non-electrolytes, *Trans. Faraday Soc.*, **33**, 985 (1937).

6. R. Höber, Membrane permeability to solutes in its relation to cellular physiology, *Physiol. Rev.*, **16**, 52 (1936).

7. R. Collander and H. Barlund, Permeabilitätsstudien in *Chara ceratophylla*, *Acta Bot. Fennica*, **11**, 1 (1933).

8. W. J. V. Osterhout, Permeability in large plant cells and in models, *Ergebn. Physiol.*, **35**, 967 (1933).

9. W. J. V. Osterhout, How do electrolytes enter the cell? *Proc. Nat. Acad. Sci., U.S.A.*, **21**, 125 (1935).

10. Th. Rosenberg and W. Wilbrandt, Enzymatic processes in cell membrane penetration, *Int. Rev. Cytol.*, **1**, 65 (1952).

11. M. H. Jacobs and D. R. Stewart, A simple method for the quantitative measurement of cell permeability, *J. Cell. Comp. Physiol.*, **1**, 71 (1932).

12. M. H. Jacobs, The measurement of cell permeability with particular reference to the erythrocyte, in *Modern Trends in Physiology and Biochemistry*, E. S. G. Barron, Ed., Academic, New York, 1952.

13. H. Davson and J. F. Danielli, *The Permeability of Natural Membranes*, Cambridge U. P., Cambridge, England, 1943.

14. Th. Rosenberg, Membrane transport of sugars, a survey of kinetical and chemical approaches, *Biol. Path.*, **9**, 795 (1961).

15. W. Wilbrandt, Carrier Mechanisms, in *Biophysics and Physiology of Biological Transport*, L. Bolis, V. Capraro, K. R. Porter, and J. D. Robertson, Eds., Springer, Wien, New York, 1967, p. 299.

16a. L. Bolis, P. Luly, V. C. Becker, and W. Wilbrandt, An analysis of the apparent parameters of the glucose transport system in red blood cell membranes, *Biochim. Biophys. Acta*, **318**, 289 (1973).

16b. L. Bolis, P. Luly, B. A. Pethica, and W. Wilbrandt, The temperature dependence of the facilitated transport of $D(+)$-Glucose across human red cell membrane, *J. Membrane Biol.* **3**, 83 (1970).

17. W. Wilbrandt and Th. Rosenberg, The concept of carrier transport and its corollaries in pharmacology, *Pharmacol. Rev.*, **13**, 109 (1961).

18. W. Wilbrandt, Carrier Diffusion, in *Biomembranes*, Vol. 3, F. Kreuzer and J. F. G. Slegers, Eds., Plenum New York, London, 1972, p. 79.

19. P. G. LeFevre, Evidence of active transfer of certain nonelectrolytes across the human red cell membranes, *J. Gen. Physiol.*, **31**, 505 (1948).

20a. W. F. Widdas, Facilitated transfer of hexoses across the human erythrocyte membranes, *J. Physiol.*, **125**, 163 (1954).

20b. W. F. Widdas, Aspects of competitive inhibitors, in *Biomembranes*, Vol. 3, F. Kreuzer and J. F. Slegers, Eds., Plenum, New York, 1972, p. 101.

21. W. D. Stein, *The movement of Molecules across Cell Membranes, Vol. 6 Theoretical and Experimental Biology*, Academic, New York, 1967.

22a. D. M. Miller, The kinetics of selective biological transport. III. Erythrocyte-monosaccharide transport data, *Biophys. J.*, **8**, 1329 (1968).

22b. D. M. Miller, The kinetics of selective biological transport. IV. Assessment of three carrier systems using the erythrocyte-monosaccharide transport data, *Biophys. J.*, **8**, 1339 (1968).

23. W. R. Lieb and W. D. Stein, Quantitative predictions of a non-carrier model for glucose transport across the human red cell membrane, *Biophys. J.*, **10**, 585 (1970).

24. P. G. LeFevre, A model for erythrocyte sugar-transport based on substrate, conditioned "introversion of binding sites," *J. Mol. Biol.*, **12**, (1973).

25. T. Hoshiko and B. D. Lindley, Phenomenological description of active transport of salt and water, *J. Gen. Physiol.*, **50**, 729 (1967).

26. P. F. Curran, Solute solvent interactions and water transport, in *Role of Membranes in Secretory Processes*, L. Bolis, R. D. Keynes, and W. Wilbrandt, Eds., North Holland, Amsterdam, 1972.

27. S. G. Schultz and R. Zalusky, Ion transport in isolated rabbit ileum. II. The interaction between active sodium and sugar transport, *J. Gen. Physiol.*, **47**, 1043 (1964).

28. E. Heinz, Ed., Na-linked transport of organic solutes, Springer-Verlag, Berlin, 1972.

29. J. F.Danielli and H. Davson, Theory of permeability of thin films, *J. Cell. Comp. Physiol.*, **5**, 495 (1934).

30. J. F. Danielli, Protein films at the oil-water interfacies, *Cold Spring Harbor Symposia Quant. Biol.*, **6**, 190 (1938).

31. P. Devaux and H. McConnel, Lateral diffusion in spin-labelled phosphatidylcholine multilayers, *J. Am. Chem. Soc.*, **94**, 4475 (1972).

32. H. McConnel, Physics and chemistry of spin labels, *Quart. Rev. Biophys.*, **3**, 91 (1970).

33. L. D. Frye and M. Edidin, The rapid intermixing of cell surface antigens after formation of mouse-human heterokaryons, *J. Cell Sci.*, **7**, 319 (1970).

34. C. Gitler, *Ann. Rev. Biophys. Bioenergetics*, **1**, 51 (1972).

35. S. J. Singer and G. L. Nicolson, The fluid mosaic model of the structure of cell membranes, *Science*, **175**, 720 (1972).

36. E. Oldfield and D. Chapman, Dynamics of lipids in membranes: heterogeneity and the role of cholesterol, *FEBS Letters*, **23**, 285 (1972).

37. R. D. Keynes, Evidence for structural changes during nerve activity and their relation to conduction mechanisms, in *Neurosciences* Second Study Program, F. O. Schmidt, Ed., Rockefeller Univ. Press, New York, 1970.

38. R. D. Keynes, Organization of sodium channels in nerve membranes, *Comparative Physiology and Biochemistry of Transport*, Eds. K. Block, L. Bolis, S. E. Luria, F. Lynenen, N. H. P. C. 1974, p.283.

39. P. Glandsdorff and I. Prigogine, Thermodynamic theory of structure, stability and functions, Wiley-Interscience, New York, 1971.

40. I. Prigogine and G. Nicolis, Biological order, structure and instabilities, *Quart. Rev. Biophys.*, **4**, (2,3), 107 (1971).

41. I. Prigogine, R. Lefever, A. Goldbeter, and M. Herschkowitz-Kaufman, Symmetry-breaking instabilities in biological systems, *Nature (Lond.)*, **223**, 913 (1969).

42. R. Lefever and G. Nicolis, Chemical instabilities and sustained oscillations, *J. Theor. Biol.*, **30**, 267 (1971).

43. R. Lefever and J. L. Deneubourg, this volume p. 349.
     R. Blumenthal, J. P. Changeux, and R. Lefever, Membrane excitability and dissipa-
     tive instabilities, *J. Membrane Biol.*, **2**, 351 (1970).
45. G. Nicolis, this volume p. 29.
46. P. Ortoleva and J. Ross, this volume p. 49.

ION TRANSPORT THROUGH
ARTIFICIAL LIPID MEMBRANES

P. LÄUGER

Department of Biology, University of Konstanz, Germany

The basic invention of Nature in the assembly of cell membranes is the use of "amphiphilic" molecules, molecules that consist of both a hydrophilic (or polar) and a hydrophobic (or apolar) part. Molecules of this type are well adapted to form ordered aggregates in an aqueous environment such as the cellular milieu. In the case of a long-chain fatty-acid anion, an energetically favorable state is achieved with the formation of a spherical micelle in which the hydrocarbon chains form the interior of the aggregate, whereas the polar end groups are in contact with the aqueous medium. In nature, however, lipids with two hydrocarbon chains, like lecithin, are common. In this case another structure is favored, in which the lipid molecules are arranged in a two-dimensional order. This structure may be represented as a combination of two monomolecular layers and is usually referred to as a "bilayer."

The interior of a lipid bilayer membrane is made up by the hydrocarbon chains of the lipid molecules and therefore represents a medium of low dielectric constant. This means that the energy which is required to bring a small ion, such as sodium or potassium, from the aqueous phase into the membrane is many times the mean thermal energy. With other words, a lipid membrane represents an extremely high barrier for the passage of ions.

The shape of this energy barrier may be calculated, if the membrane is considered as a thin homogeneous film of dielectric constant ε_m interposed between two homogeneous media of dielectric constant ε. The energy of the ion in the membrane is then given by the electrical image forces that act upon a charged sphere near a boundary between media of different dielectric constant. The image force always tends to repel the ion from the phase with the lower ε value. This means that the potential energy curve of the ion in the membrane has the shape of a symmetrical barrier with a peak in the middle of the membrane.[1]

But biological membranes are more or less permeable to ions such as Na^+ or K^+, and therefore we have to assume that mechanisms exist by which the activation energy of the ion transport is drastically reduced. Two limiting cases for such a transport mechanism may be imagined, fixed pores and mobile carriers. A pore may be represented by a protein molecule that is built into the membrane structure and in which a special distribution of amino acids gives rise to a hydrophilic pathway through the membrane, or a pore may be formed by an aggregate of several interacting protein molecules. A carrier, on the other hand, is a molecule that binds the ion at one membrane-solution interface, then migrates to the opposite interface and releases the ion into the aqueous solution. The concept of a carrier that facilitates the transport of ions across a hydrophobic barrier is rather old, but the existence of ion carriers in biological membranes has been a mere hypothesis for a long time.

This situation has now changed, since compounds such as valinomycin, monactin, and enniatin B have been isolated and characterized.[2] Valinomycin and enniatin B are depsipeptides; they are built up by amino acids and α-hydroxy acids in alternating sequence. Monactin and the other macrotetrolides are cyclic compounds that contain four ether and four ester bonds.

All these substances share a common structural property: They are macrocyclic compounds, which contain both hydrophilic and hydrophobic groups. The polar carbonyl groups of valinomycin may interact with an alkali ion, so that a complex is formed in which the central cation is surrounded by six oxygen atoms, which form a cage. In other words, the interior of the complex offers to the ion an environment that is similar to the hydration shell of the ion in aqueous solution. The exterior of the complex is strongly hydrophobic. This picture is supported by x-ray experiments and by spectroscopic studies.[2]

The ability of valinomycin and other macrocyclic compounds to increase the potassium permeability of biological membranes was first demonstrated in experiments with mitochondria and erythrocytes.[3] For the interpretation of these results it has been proposed that these antibiotics act as carriers for K^+. But detailed information on the transport mechanism could not be obtained from these experiments, as a consequence of the great complexity of biological membranes. In the past years, artificial *model membranes* have therefore been used extensively in studies with macrocyclic ion carriers.[4–8]

The electrical resistance of a lipid bilayer membrane under normal conditions, say in a KCl solution, is very high, up to $10^8 \ \Omega \ cm^2$. If small amounts of valinomycin or monactin are added to the system in the presence of K^+, the resistance of the bilayer drops by several orders of magnitude. The increase in the membrane conductivity is strictly proportional to the antibiotic concentration in the aqueous phase over several decades. For low concentrations of valinomycin we may assume that a simple distribution equilibrium exists between the aqueous phase and the membrane—in other words, that the valinomycin concentration in the membrane is proportional to the concentration in the aqueous phase. The proportionality between conductivity and concentration then means that the smallest transport unit is a single valinomycin molecule.

Below $[K^+] = 10^{-1}$ M, the membrane conductivity is proportional to $[K^+]$. This finding suggests that the charge carrier in the membrane is a $1:1$ complex between valinomycin and K^+. At higher K^+ concentrations a deviation from linearity occurs in the sense that the conductance increases more slowly than $[K^+]$. This saturation behavior is characteristic of carrier systems at high concentrations of the transported particle where a substantial fraction of the carrier is in the complexed form.

If valinomycin and monactin are carriers in the classical sense, then we may describe the ion transport mediated by these compounds by the following reaction scheme. The transport occurs in four distinct steps: (1) the association of an alkali ion M^+ from the aqueous solution with a carrier molecule S in the membrane surface, (2) the translocation of the complex MS^+ from the left-hand to the right-hand interface, (3) the dissociation of the complex and the release of the ion into the aqueous phase, and (4) the back transport of S. We therefore have four different rate constants for the description of the carrier system: the rate constants for the association (k_R) and dissociation (k_D) of the complex, and the rate constants k_{MS} and k_S for the translocation of the complex and free carrier through the membrane.

From this picture several questions arise. We want to know the time scale of these events, and, more specifically, we may ask: Is there a rate-limiting reaction in the overall transport, or do all reactions take place at comparable speed? Or: Is the ion specifity of the carrier determined by thermodynamic factors alone (stability constant of the complex) or also by kinetic parameters (rate constants)? For an answer to these questions we have to make a detailed kinetic analysis of the carrier system. At first this seems to be a rather difficult task, because we have to determine not only the four rate constants, but also the concentration of carrier molecules in the membrane. However, such an analysis becomes possible if steady-state conductance data are combined with the results of electrical relaxation measurements.

Very useful information may be obtained simply by measuring the current-voltage characteristic of a bilayer membrane in the presence of the carrier. This seems at first surprising, because in most systems in which the conductance is ionic, the current-voltage curve is simply a straight line, and a straight line obviously does not contain much information. In the case of the lipid membrane, however, the shape of the current-voltage curve is determined by the relative rates of the individual transport steps.

For instance, if we assume that the rate of dissociation in the interface is very high, then the overall transport rate is limited by the translocation of the complex across the membrane. The translocation involves a jump over an activation energy barrier. As mentioned before, this barrier is determined by the electrical image forces that act on the charged complex. In the presence of an external voltage the barrier height is reduced by an electrostatic energy term. It is well known that such a situation leads to an exponential dependence of the translocation rate constants on voltage. In this limiting case, where the slowest step is the translocation, the electrical current will therefore increase exponentially with voltage.

On the other hand, if the dissociation in the interface is rate limiting, then the complex MS^+ will accumulate at one interface at large values of the voltage. The current is then limited by k_D and becomes independent of voltage.

The steady-state conductance of the membrane in the presence of the carrier may be described by the following relations. We introduce the specific membrane conductance λ, which is given by the ratio of current density I to the voltage V:

$$\lambda = \frac{I}{V}$$

In the limit of small V, λ reduces to the ohmic conductance λ_0:

$$\lambda_0 = \left(\frac{I}{V} \right)_{V \approx 0}$$

The formal analysis of the carrier model[9, 10] then yields a rather simple relation for the ratio λ/λ_0:

$$\frac{\lambda}{\lambda_0} = \frac{2}{u}(1+A)\frac{\sinh(u/2)}{1 + A\cosh(u/2)}$$

$$u \equiv \frac{V}{RT/F}$$

(R = gas constant, T = absolute temperature, F = Faraday constant). λ/λ_0 depends only on the reduced voltage u and on a parameter A, which is equal to a combination of the ion concentration c_M and the rate constants:

$$A = \frac{2k_{MS}}{k_D} + c_M \frac{k_R k_{MS}}{k_D k_S}$$

$$\lambda_0 = \frac{F^2}{RT} N_S k_{MS} \frac{c_M k_R / k_D}{1 + A}$$

The ohmic conductivity λ_0 depends in addition on N_S, the interfacial concentration of neutral carrier molecules at equilibrium. The parameter A may be directly obtained from the current-voltage characteristic of the membrane. If A is measured as a function of c_M, one can determine the single combinations k_{MS}/k_D and $k_R/k_{MS}k_D k_S$. In addition, a third quantity, $k_S/N_S/c_S$, may be obtained from steady-state experiments. Thus, stationary conductance measurements are not sufficient for a complete kinetic analysis of the system. However, the additional information that is required may be obtained from electrical relaxation experiments.

Relaxation techniques have been widely used in chemical kinetics for the evaluation of rate constants. This method, however, is not restricted to chemical reactions, but may also be used for the kinetic analysis of transport processes in membranes. The principle of the method is well known: The system is disturbed by a sudden displacement of an external parameter such as temperature or pressure, and the time required by the system to reach a new stationary state is measured. In our case the external variable that is suddenly changed is the electrical field in the membrane. Immediately after the voltage jump a capacitive current is observed, which decays with a time constant equal to the product of the cell resistance and the membrane capacitance. The capacitive spike limits the time resolution of the system; under favorable circumstances, the resolution is of the order of 1 μsec. After the disappearance of the capacitive transient, the membrane current approaches a stationary value I_∞. The initial current I_0 is obtained by extrapolation to time zero. Therefore, such an experiment yields two independent pieces of information, the relaxation time τ and the relaxation amplitude $\alpha = (J_0 - J_\infty)/J_\infty$. Both τ and α may be obtained from the theory of the carrier model.[7]

The physical reason for this relaxation process is easily understood: An instant after the voltage has been applied, the concentration of the charged complex at the two interfaces still has its equilibrium value and is the same on both sides. Under the influence of the electric field, charged complexes jump across the membrane, which gives a certain initial current. In the steady state, however, the concentrations of the complex at the two interfaces have become unequal, and correspondingly the stationary current is different from the initial current.

Together with the stationary conductance data, the additional information contained in the experimental values of τ and α is sufficient to calculate the individual rate constants of the carrier system. The result for valinomycin-K^+ is[7]

$$k_R \approx 5 \times 10^4 \, (M^{-1})(\text{sec}^{-1})$$

$$k_D \approx 5 \times 10^4 \, \text{sec}^{-1}$$

$$k_S \approx 2 \times 10^4 \, \text{sec}^{-1}$$

$$k_{MS} \approx 2 \times 10^4 \, \text{sec}^{-1}$$

From the numerical values of the rate constants a number of interesting conclusions may be drawn. First, we realize that k_D, k_S, and k_{MS} are of the same order of magnitude. This means that the translocation of free carrier and complex across the membrane and the dissociation of the complex in the interface accur at comparable rates, between 10^4 and 10^5 sec^{-1}. Or we may say that the jump time of the carrier molecule across the membrane is about 50 μsec. For comparison, the diffusion time of a molecule of the size of valinomycin in water over a distance of 70 Å (the thickness of the membrane) is about 0.2 μsec.

At present, we do not know to what extent Nature really uses carriers for ion transport. On the other hand, for certain biological transport systems we are rather sure that carriers are not involved. An example is the sodium channel of the nerve membrane. From electrical fluctuation measurements and other experiments we know that about 10^8 sodium ions per second pass through the open channel. This number is higher by four orders of magnitude than the maximal transport rate of a single carrier molecule of the valinomycin type. It is therefore almost impossible that the sodium channel is operated by a carrier mechanism. But, of course, the transport rate could well be explained by a pore mechanism.

As yet, the chemical identification of membrane proteins that may act as pore molecules has not been successful. But fortunately, there exist a number of simple molecules that may be used for the study of pore mechanisms.

Such a molecule is gramicidin A, a linear peptide, built up out of 15 mostly hydrophobic amino acids. If gramicidin A is introduced into a biological membrane, the membrane becomes cation permeable. In this respect gramicidin is similar to the macrocyclic ion carriers. But soon becomes clear that the mechanism of action of gramicidin is quite different from that of the macrocyclic carriers.

First, the question arises what conformation such a molecule could assume in a lipid membrane. An attractive possibility is a helix. In order to act as a pore, such a helix must have in its center a polar channel of the appropriate diameter, and besides this, the length must be sufficient to bridge the thickness of the membrane. As Urry has shown,[11] such a helical conformation does indeed exist. The length of the helix is only about 15 Å, much smaller than the hydrophobic thickness of the membrane ($\approx$40 to 50 Å). Urry therefore proposed that the pore consists in a dimer of gramicidin A. The dimer has a length of about 30 Å, and could be sufficient to bridge the hydrophobic core of the membrane if a local thinning of the membrane is assumed.

It is interesting to investigate whether the proposed structure agrees with the observable transport properties of gramicidin A. For this purpose it is again convenient to use artificial lipid membranes. An instructive experiment has been carried out by Hladky and Haydon[12]: If an extremely small amount of gramicidin is added to the membrane and if a voltage is applied, then the current shows discrete fluctuations. The detailed analysis of these fluctuations suggests that a single fluctuation corresponds to the formation and the disappearance of a single pore. If the dimer hypothesis is correct, then the switching on of the gramicidin channel corresponds to the association of two monomers. Accordingly, the switching off corresponds to the dissociation of the dimer.

This immediately leads us to the question of the kinetics of pore formation—in other words, we want to know how many association processes take place per second within one square centimeter of the membrane, and what the probability is that a given dimer will dissociate. Again, one may use the electrical relaxation technique in order to obtain information about the channel kinetics. In order to understand why a voltage jump may displace the equilibrium between monomers and dimers of gramicidin A in the membrane, we have to look once more at the geometrical situation in the membrane. The thickness of the hydrocarbon core is about 50 Å, but the length of the dimer is only 30 Å. For the formation of the dimer it would therefore be more favorable if the membrane were thinner. Indeed, if the chain length of the lipid is decreased, so that the thickness of the membrane is diminished, channel formation is enhanced.[12] If a voltage is applied to the membrane, charges of opposite sign accumulate at the membrane surfaces. This gives rise to a pressure (like the force that attracts the plates of a charged plate condenser), and the membrane is slightly compressed. It can be shown by independent electrical capacitance measurements that the thickness of the membrane decreases by a few percent at a voltage of 100 mV. This effect is not very large, but is sufficient to displace the monomer-dimer equilibrium considerably.

For the application of the relaxation method we further need an indicator for the concentration of the dimer in the membrane. This requirement is easy to fulfill: The unmodified lipid portion of the membrane is an almost perfect insulator, and also the monomers are nonconducting. The conductance is therefore proportional to the number of dimers in the membrane—in other words, we may use the membrane current as an indicator for the dimer concentration.

If k_R and k_D are the association and dissociation rate constants, respectively, of the monomer-dimer reaction, and N_m is the monomer concentration (moles/cm^2) in the membrane after the voltage jump, then the relaxation time τ of the reaction is given by[13]:

$$\frac{1}{\tau} = k_D + 4k_R N_m$$

If τ is measured as a function of N_m, then k_D and k_R may be obtained separately. But here a difficulty arises: The monomer concentration N_m in the membrane is not known. (Gramicidin is added to the aqueous phases and distributes between water and the membrane, but the exact concentration that is taken up by the membrane is unknown.)

This difficulty may be circumvented in the following way. The monomer concentration may be replaced by the dimer concentration N_d and by the equilibrium constant $K = k_R/k_D$; N_d in turn is related to the single-channel conductance Λ (which may be obtained from the fluctuation experiments) and to the stationary membrane conductance λ^∞, which is reached at long times:

$$\lambda^\infty = \Lambda N_d$$

This gives

$$\frac{1}{\tau} = k_D + \text{const}\sqrt{k_R \lambda^\infty}$$

Therefore, if relaxation measurements are carried out at different gramicidin concentrations and $1/\tau$ is plotted as a function of the square root of the stationary conductance λ^∞, a straight line should result. This is indeed found experimentally.[13] That fact that the experimentally observed dependence of τ on λ^∞ agrees with the theoretical prediction strongly supports the dimer model. The opening and closing of a pore therefore correspond to the formation and disappearance of a dimer. For a dioleoyllecithin membrane at 25°C the following values of k_R and k_D are obtained:

$$k_R = 2.4 \times 10^{14} \text{ cm}^2/(\text{mole})(\text{sec})$$

$$k_D = 1.6 \text{ sec}^{-1}$$

$1/k_D$ is the mean lifetime of a dimer in the membrane, about 0.5 sec. This value agrees with the lifetime that is obtained from the current fluctuations.[12] Once a pore is formed, it will last about half a second. The value of k_R may be illustrated in the following way: If the monomers are present in the membrane in such a concentration that the mean distance is 1000 Å, then a given monomer will associate with another monomer about four times per second.

We may compare this reaction rate with the theoretical limit of a diffusion-controlled reaction. The monomers may undergo translational diffusion in a two-dimensional space (the membrane). The upper limit of the association rate is reached if every encounter leads to the formation of a dimer. The diffusion rate of a monomer depends on the effective viscosity η of the membrane, which may be estimated from spin-label measurements and other experiments. If we use this estimated value of η, then the diffusion-controlled association rate in the above example becomes about 4000 sec^{-1}, or about 1000 times as high as the actual association rate. This factor of 10^3 is plausible, because we have to assume that the two monomers must come into a precise mutual position before the H-bonds that stabilize the dimer can be formed.

To conclude, we may say that the gramicidin system gives us an example how the ion permeability of a lipid membrane may be controlled by an electric field. Immediately after the field is switched on, the ion permeability is small, but it reaches a much higher final current with a characteristic time-constant. Under the influence of the electrical field, part of the inactive (monomeric) gramicidin is converted into the active form of a pore. One may speculate that at least in some cases the reversible association of peptide subunits in the cell membrane may be used for the control of transport processes.

Macrocyclic ion carriers and peptides of the gramicidin type are examples how relatively simple molecules may carry out rather specific functions in membranes. In the present, more advanced, stage of evolution, the ion channels in biological membranes are probably proteins. But one should bear in mind that properties such as ion specifity and even control of permeability by electric fields may be achieved at a much lower level of complexity.

References

1. B. Neumcke and P. Läuger, *Biophys. J.*, **9**, 1160 (1969).
2. M. M. Shemyakin, Yu. A. Ovchinnikov, V. T. Ivanov, V. K. Antonov, E. I. Vinogradova, A. M. Shkrob, G. G. Malenkov, A. V. Evstratov, I. A. Melnik, and I. D. Ryabova, *J. Membrane Biol.*, **1**, 402 (1969).
3. B. C. Pressman, E. I. Harris, W. S. Jagger, and I. H. Johnson, *Proc. Natl. Acad. Sci. U.S.A.*, **58**, 1949 (1967).
4. P. Mueller and D. O. Rudin, *Biochem. Biophys. Res. Commun.*, **26**, 398 (1967).
5. T. E. Andreoli, M. Tiefenberg, and D. C. Tosteson, *J. Gen. Physiol.*, **50**, 2527 (1967).
6. G. Eisenman, S. Ciani, and G. Szabo, *Fed. Proc.*, **27**, 1289 (1968).
7. G. Stark, R. Benz, B. Ketterer, and P. Läuger, *Biophys. J.*, **11**, 981 (1971).
8. P. Läuger, *Science*, **178**, 24 (1972).
9. P. Läuger and G. Stark, *Biochim. Biophys. Acta*, **211**, 458 (1970).
10. G. Stark and R. Benz, *J. Membrane Biol.*, **5**, 133 (1971).
11. D. W. Urry, *Proc. Natl. Acad. Sci. U.S.A.*, **68**, 672 (1971).
12. S. Hladky and D. A. Haydon, *Biochim. Biophys. Acta*, **274**, 294 (1972).
13. E. Bamberg and P. Läuger, *J. Membrane Biol.*, **11**, 177 (1973).

PHYSICOCHEMICAL PROBLEMS IN EXCITABLE MEMBRANES

Y. KOBATAKE

*Faculty of Pharmaceutical Sciences, Hokkaido University,
Sapporo, Japan*

CONTENTS

Recent physicochemical studies on squid giant axons have revealed that the process of excitation of nervous tissues is accompanied by a conformational change of macromolecules constituting the membrane, triggered by cooperative cation exchange at fixed negative sites in the membrane.[1,2] Changes in thermal[3] and optical[4,5] properties of nervous tissues during excitation support this view. There is serious ambiguity, however, in interpreting the results of physical measurements carried out on nerves, since there are many undefinable quantities in a system consisting of an excitable membrane and its natural environment, and since the molecular structures and their conformation in the living membrane are not entirely known at present. Therefore, it is necessary, as a counterpart of investigations on living tissues, to study the transport processes in a model system in which a definite transformation of the molecular conformation in the membrane takes place with variation of the external conditions or with external stimuli. It is worthwhile to check whether or not the transition of membrane conformation is the real key to the occurrence of excitation in the membrane system.

In this article, we describe excitability, stability, and phase transitions in three kinds of excitable membranes—a living membrane and two model systems—and discuss the relation between the excitation process and the dissipative structure of membrane systems.

I. STABILITY AND EXCITABILITY OF SQUID
GIANT AXONS

The thickness of the plasma membrane of a living cell is a mere 100 Å, which is covered with overwhelmingly thick connective tissues and the Schwann cells; also, the chemical and physical structure of nerve membrane is not known to us. For these reasons, it is difficult to detect directly the transformation of membrane macromolecules during the excitation of nervous cells. However, if we assume that a transition of state of the membrane system occurs during the process of excitation, some prototype changes in physical properties of the system associated with the transition may be expected. For example, a coalescence or giant fluctuation of a physical property must be observed in a thermo-dynamical system immediately before the transition point.[6] Similarly, in a membrane system such as we are considering, the instability and/or giant fluctuations of the membrane potential may be observed when various forms of perturbations (e.g., a salt-condition, temperature, or electrical distrubance) are applied across the membrane.

The following results were obtained by using giant axons of squid (*Doryteuthis bleekeri*), caught in the northern part of the Sea of Japan near Sapporo. The giant axon was perfused internally by the two-cannula method as explored by Tasaki et al.[7] The potential-pipette Ag-AgCl electrode, filled with 0.6 M KCl solution was introduced into the center of the 15-mm-long perfused zone of the axon through the outlet cannula. The external fluid medium was grounded with a large coil of Ag-AgCl wire imbedded in agar gel. A high input-impedance preamplifier, a dual-beam oscilloscope, and an oscilloscope camera were used for recording. When internal application of electric current pulses was required, a platinum wire 50 μm in diameter was inserted into the perfused zone of the axon through the outlet cannula. The external fluid medium was circulated rapidly by the use of a device consisting of large reservoirs of solutions, polyethylene tubing for leading the solution to the perfused chamber, a baffle for achieving uniform application of the solution, and a suction pipe for maintaining a constant level of the fluid in the chamber. The fluid in the suction pipe was electrically isolated from the suction pump by means of a trap.

Figure 1 shows the stability of the resting and excited states of a squid giant axon. The internal perfusion fluid contained 15 meq/l of NaF (pH 7.3), and the external fluid medium was 100 mM CaCl$_2$; no salt of alkali-metal ions was added to the medium. The upper oscillograph trace in record B shows that an all-or-none action potential was pro-duced in response to a pulse of electric current of about 0.5 μA passing

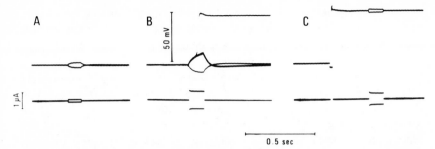

Figure 1. Demonstration of excitability and instability in the membrane of a squid giant axon internally perfused with a 15 mM NaF solution and immersed in an external medium of 100 mM CaCl$_2$ solution containing no 1:1-type electrolyte. The upper oscillograph trace represents changes in the membrane potential, and the lower trace shows the applied membrane current. Records A and C show the stability of the membrane in the resting (A) and active (C) states. Temperature, 19°C.

outwards through the perfused part of the axon membrane. Record A shows that the axon membrane is stable against small perturbations caused by the application of pulses of weak electric current. Unless the external electric perturbation does not exceed the threshold value, the electric response returns to the original value along a decay curve when the perturbation is removed. The state of the membrane during the plateau of an action potential is also shown to be stable against external electric perturbation, as seen in record C.

With a view toward testing the stability of the axon membrane against perturbations of a different kind, axons were exposed to external media with varying ionic compositions. In both equilibrium and non-equilibrium thermodynamics, it is well known that the fluctuation in various measurable quantities of the system becomes unstable at or near its transition point.[6, 8] The following observations show that a macro-scopic fluctuation can actually be observed near transition points in the squid axon membrane. Axons used in these observations were internally perfused with a dilute solution of CsF and were immersed in a slowly circulating external fluid medium containing CaCl$_2$ and NaCl. Initially, the concentration of Na$^+$ in the medium was far from the critical level for the transition, where the fluctuation of the membrane potential was very small. With increase of Na$^+$ concentration in the external medium, the fluctuations in potential increased gradually (see records B and C). As can be seen in record C, the amplitude of the fluctuation in the membrane potential became quite large, and finally the membrane underwent an abrupt transition to its excited state. There was a certain degree of periodicity in the potential fluctuation, that is, the fluctuation synchronized to some extent. In axons internally perfused with a dilute solution of CsF, the period was of the order of 1 to 2 sec, depending on the salt species in the external and internal perfused fluids. It is interesting to note that these giant fluctuations vanished completely after a large potential jump. However, the similar large fluctuation of the membrane potential and its synchronization was observed also in the excited state, when the system approached the critical point for repolarization. Sometimes, the membrane potential oscillated between excited and resting states, as seen in Fig. 3. These giant fluctuations and/or oscillations of the emf observed with perfused squid axons are attributable to the nonuniform distribution of active and resting patches along the membrane surface,[2] and also indicate that a close parallelism exists between the process of nerve excitation and the coalescence of the thermodynamical transition.

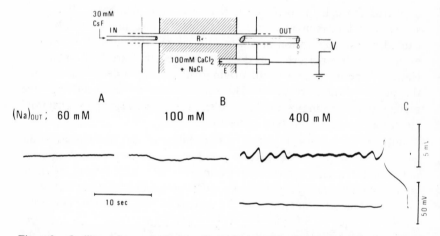

Figure 2. Oscillograph records showing fluctuations in the membrane potential produced by the addition of NaCl to the external medium. The electrolyte compositions of the internal and external solutions are given. In record C, a second oscillograph trace at a low voltage sensitivity (recorded simultaneously) is shown. Temperature, 19°C.

Keynes et al. showed that the excitation process of a nervous tissue accompanies an exothermic reaction.[3] This implies that the excitation must be enhanced by a lowering of the surrounding temperature. On the contrary, an abrupt transition from the excited to the resting state of the membrane could be induced by raising the temperature of the external fluid medium. The diagram at the top of Fig. 4 shows the experimental

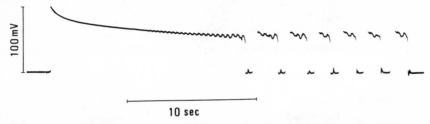

Figure 3.   Record showing the fluctuation and oscillatory changes of the membrane potential in the excited state of a squid giant axon. The axon was perfused internally with a 15 mM NaF solution, and externally with a solution containing 100 mM CaCl₂ and 100 mM NaCl. Temperature, 18.5°C.

arrangement used. The axons were internally perfused with a dilute solution of either CsF or NaF. The medium contained both CaCl₂ and NaCl. The temperature of the medium was changed slowly and uniformly by flowing precooled or prewarmed solutions. The temperature in the medium was measured with a thermocouple. Two examples of records presented in Fig. 4 show the typical results obtained. Record A was taken from an axon internally perfused with a dilute CsF solution, and record B from an axon perfused with NaF. In these records, the horizontal deflections of the oscillograph beam represent the output of the thermocouple, and the vertical deflection displays the membrane potential of the axon under study. When the axon was cooled gradually, the membrane potential changed only slightly at the onset of cooling. However, at a certain temperature (about 5°C under the present experimental conditions), the potential in the axon interior was found to jump upwards by more than 70 mV (see the arrows directed upwards). Further lowering of the temperature did not produce any abrupt change in the membrane potential. When the temperature of the surrounding fluid was raised gradually from this stage, the membrane potential changed slowly at the outset. Again, at a certain temperature, the potential was found to change abruptly, as illustrated by the arrows directed downward in the figure. Note that the critical temperature for this downward transition was much higher than that for the upward transition. In other words, there was a pronounced hysteresis in the state of the axon membrane brought about by a cyclic change in the temperature of the external medium. This hysteresis loop can be produced repeatedly. However, if a small electrical perturbation—for example, a subthreshold alternating current of relatively low frequency—is superposed on the potential, the hysteresis loop disappears. A typical example

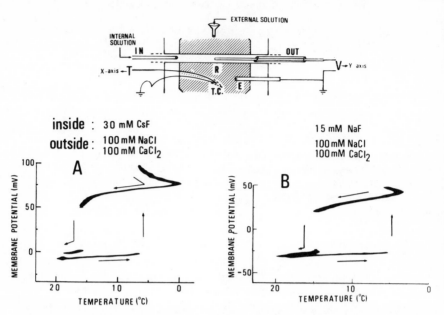

Figure 4. Oscillograph records showing hysteresis in the membrane potential associated with a cyclic change in the temperature. The electrolyte compositions of the internal and external solutions are given. The arrangement of the inlet (IN) and outlet (OUT) cannulae, the recording (R) and ground (E) electrodes, and the tip of a thermocouple (TC) are illustrated on the top. A period of approximately 3 min was required to complete one cycle of temperature change.

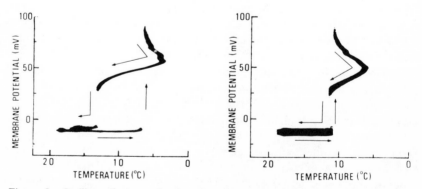

Figure 5. Oscillograph trace showing that the pronounced hysteresis loop in the membrane potential as a function of temperature disappears when a subthreshold ac disturbance is applied across the membrane. The experimental condition was the same that employed in Fig. 4A.

is presented in Fig. 5. A pronounced hysteresis loop similar to that observed in Fig. 4 has been demonstrated in transitions induced by changing the external electrolyte composition.[9]

All of these experimental results presented above support the view that a kind of the phase transition in a thermodynamical system is involved in the process of excitation of the nerve membrane.

II. EXCITABILITY AND STABILITY OF A MODEL MEMBRANE ACCOMPANYING A CONFORMATIONAL CHANGE

As shown above, the process of excitation of living tissues is plausibly explained in terms of a phase transition of the membrane system. However, we have had no definite evidence of the transition of the molecular conformation in the membrane. Therefore, it is worthwhile to show that various electrophysiological functions of living excitable membranes can be reproduced by using a model system in which a transformation of molecular conformation in the membrane takes place with the variation of environmental conditions or with external stimuli.

In our previous studies,[10,11] it was shown that a synthetic lipid analogue, dioleylphosphate (DOPH), impregnated in a Millipore filter paper, changed its conformation at a certain critical concentration of the external solution of 1:1-type electrolyte. Above and below the critical concentration c_t, the DOPH-Millipore membrane displayed an appreciable difference in its characteristics. Thus the steady-state processes (e.g., the membrane potential, ion permeability, electrical capacitance, and the dc resistance of the membrane) were found to change rather discontinuously at c_t when the external salt concentration was varied continuously. It was shown that the DOPH-Millipore membrane has two distinct states under a given set of the external salt concentrations, depending upon the history of the membrane: one is hydrophobic in character, and the other is hydrophilic. These two states are interchangable with a slight perturbation of the external condition. Therefore, it is possible to check whether or not various transient processes observed in living membranes can be reproduced by the transition between two states of the present model membrane when certain external stimuli are applied across the membrane.

A Millipore filter paper composed of cellulose ester with 2-μm nominal pore size was soaked in a benzene solution of DOPH with an appropriate concentration, and the adsorbed quantity of DOPH in the membrane, Q, was adjusted between 2 and 3 mg/cm^2. Under this condition, the critical salt concentration c_t was about 40 mM for KCl.

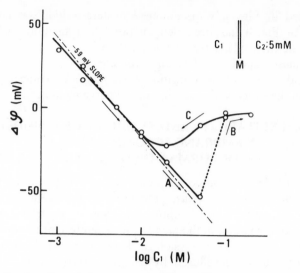

Figure 6.   Steady values of the membrane potential $\Delta\varphi$ as a function of $\log C_1$ for a DOPH-Millipore membrane of $Q = 2.18$ mg/cm^2. C_2 was fixed at 5 mM KCl, and C_1 was varied progressively as shown by arrows A, B, and C. Temperature, 27°C.

Figure 6 shows a typical example of the concentration dependence of the membrane potential observed with a DOPH-Millipore membrane in various concentrations of KCl solution. Here the observed membrane potential $\Delta\varphi$ was plotted against $\log C_1$ under the condition that the concentration of KCl on one side of the membrane, C_2, was fixed at 5 mM, and that in the other compartment, C_1, changed progressively. When C_1 was lower than c_t, the slope of the plot of $\Delta\varphi$ against $\log C_1$ gave approximately the ideal value, that is -59 mV per tenfold variation of C_1. However, when C_1 arrived at c_t, $\Delta\varphi$ increased discontinuously, and further increase of C_1 led to a continuous increase toward zero of the observed $\Delta\varphi$, as illustrated by arrow B in the figure. If C_1 was decreased progressively from this stage, $\Delta\varphi$ decreased smoothly, following the arrow marked C, and manifested a hysteresis loop, as seen in the figure. The simultaneous measurement of the membrane potential and the impedance of the membrane revealed that the electrical resistance in state A was independent of the external salt concentration, and of order 10^6 Ω cm^2. Furthermore, no ion permeation was detected in this state. From these and other characteristics of the membrane,[12] the transport process in state A is deduced to be nonionic, and hence the membrane is regarded as an oil membrane or a hydrophobic membrane. On the other hand, states B and C are considered to correspond to a charged membrane, since the concentration dependence of the membrane potential is represented quantitatively by the fixed-charge theory of the membrane potential,[13] and since both the ion permeability and the dc conductance of the membrane are roughly proportional to the salt concentration in the external solution in these states.[12] The electrical resistance of state C decreases about a factor of 10^{-2} in comparison with that of state A. These results imply that the oil membrane in state A becomes a leaky one with charged pores, and the DOPH is transformed into a number of micelles or bilayer leaflets in the filter paper, when the salt concentration in the membrane arrives at the critical value c_t. This inference is consistent with the discrete change in the electric capacitance of the membrane at c_t with continuous variation of the KCl concentration,[11] and also accords with the capability of bilayer formation of DOPH in a relatively concentrated solution. The transformation of DOPH in the filter paper has been confirmed by a shift of the emitted light spectrum of a fluorescent dye (ANS) adsorbed on the membrane. In dilute solutions where both C_1 and C_2 are lower than c_t, the stable state of the DOPH-Millipore membrane is A. The membrane potential in state C increases slowly toward A when the membrane is allowed to stand in this salt condition for several hours. However, state C is stable with respect to small external perturbation. In this respect, state C is considered to be a metastable state in the thermodynamic sense. On the other hand, when the concentrations in the external solutions, C_1 and C_2, are higher than c_t, state B or C is more stable than A, but state A is still stable in comparison with the surrounding nonsteady state. At an appropriate intermediate concentration where c_t is between C_1 and C_2 ($C_1 < c_t < C_2$), it is possible to make the external conditions such that either A or C is more stable than the other. In this case, if the state of the membrane is shifted from A to C (or vice versa) by some external means, spontaneous recovery to the original state may be expected.

Figure 7 shows a spontaneous firing of spike-wise variations of the emf across the membrane. First, a sample membrane is immersed in a 1 mM KCl solution, and then the membrane is placed between two solutions of 10 mM and 60 mM KCl. The membrane potential oscillates spontaneously as seen in the figure. The size of a spike is about 20 mV, and the duration of each spike is about 1 min. The variation of the salt concentration in the external solutions leads to a change in duration and in size without varying the form of the spikes. It is evident that the oscillation in the membrane potential can be attributed to the variation

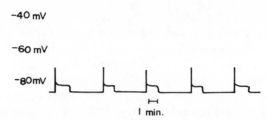

Figure 7. Spontaneous firing of the emf across a DOPH-Millipore membrane of $Q = 2.69$ mg/cm^2, when the membrane is subjected to a sudden change in the external KCl concentration. The experimental conditions are specified in the text. Temperature, 27°C.

of conformation of DOPH in the filter paper caused by the salt concentration, since the oscillation occurs only for the case in which c_t is between C_1 and C_2.

Figure 8 shows an example of the oscillation of the electric current observed under a fixed external voltage, where the membrane is immersed in a solution of 20 mM KCl. When the external voltage rises to a certain value, 2.0 V for this specific case, spontaneous firing of the current spikes is observed.[14] The form of each current spike is shown in the lower trace of Fig. 8. As seen in the figure, the time course of a current spike is the same as that found in Fig. 7. With an increase of the external voltage, the frequency of the spikes is increased, with no variation in the form and duration of the spike. Figure 9 shows an

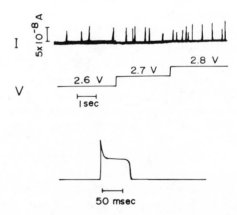

Figure 8. Traces of oscillation of the electric current observed under fixed applied voltages. The membrane is the same as in Fig. 7, and is immersed in a 20 mM KCl solution. The lower trace shows the time course of each current spike. Temperature, 27°C.

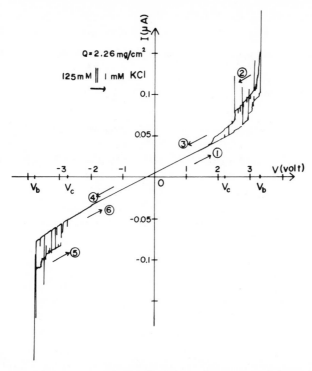

Figure 9. *I-V* relation for a DOPH-Millipore membrane of $Q=2.26$ mg/cm^2 placed between 125 mM and 1 mM KCl solutions. The external applied voltage V was changed successively following arrows 1 to 6. The meanings of V_c and V_b are explained in the text.

example of the current-voltage relation for a membrane placed between 1 mM and 125 mM KCl solutions. The other combinations of salt concentrations gave essentially the similar *I-V* relation. In this figure, the positive direction of the applied voltage is defined in such a way that the electric current is passing through the membrane from the concentrated to the dilute side. As seen in the figure, when V is smaller than the critical value V_c, the current I increases proportionally with V, and the electrical resistance is evaluated to be the order of 10^6 Ω cm^2, which is equal to that of state A in Fig. 6. When V is increased to V_c, (2.1 V in this specific membrane), the current undergoes spontaneous firing of spikes. If V rises to a certain value V_b (3.3 V for the present system), the current increases indefinitely, just as if the membrane had burst. If the applied voltage is decreased from this stage, the current decreases with V with a low electrical resistance, and shows an appreciable hysteresis loop. As seen in the figure, the current displays negative spikes in this state, that is, the current decreases transiently in spike fashion. However, when the external voltage V is decreased approximately to V_c, the system returns to the high-resistance state, where the oscillation of I ceases. A similar hysteresis loop in the I-V relation is observed when the direction of the external field is reversed. The critical voltage V_c and frequency of the current spikes in the negative field are different from those observed in the positive field. However, the shape of a current spike, depicted in Fig. 8, is not altered.

It is reasonable to consider that the spontaneous transient variation of current, that is, the generation of current spikes during a constant applied voltage, is related to the variation of DOPH conformation in the Millipore filter. As is well known, the application of an external electric field accelerates the thinning process of the formation of phospholipid bilayer leaflets. This thinning process under external electric field depends strongly on the salt concentration in the surrounding medium. Therefore, it is not unreasonable to consider that the current spikes observed in Figs. 8 and 9 may be attributed to the conformational change of DOPH in the void space of the filter paper, induced by the applied voltage.

The hysteresis loop in the I-V relation observed in Fig. 9 indicates that the membrane is transformed to a low-resistance state when it is subjected to a high external voltage (larger than V_b), and stays in that state until the applied voltage is decreased below V_c. This fact implies that the membrane returns from the metastable to the stable original state with time lapse when the external voltage is removed suddenly. Figure 10 shows a typical example of the potential variation in response to an external rectangular electrical stimulus. This figure refers to the case where the external voltage V is applied across the membrane from the concentrated to dilute side, and in which the membrane is placed between 1 mM and 125 mM KCl solutions. Different combinations of a membrane and other salt concentrations show essentially the same potential response so long as the applied voltage exceeds V_b. As seen in the figure, the peak of the generated potential reaches about 100 mV. This emf variation corresponds to the difference in the membrane potential of two steady states A and C observed in Fig. 6 for a given set of concentrations, indicating that the membrane is transformed from state A to state C transiently by the external electrical stimulus.

The stimulus inducing the variation of the emf need not be an electrical stimulus, so long as the salt concentration inside the membrane is changed by the stimulus. For example, when a rectangular pressure difference is applied across the membrane, so that the con-

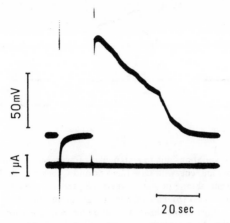

Figure 10.   A photographic trace of an action potential that is produced in response to an external electrical stimulus for a DOPH-Millipore membrane of $Q = 2.87$ mg/cm^2. The membrane is place between two KCl solutions of concentration $C_1 = 125$ mM and $C_2 = 1$ mM. Temperature, 27°C.

centrated solution flows into the membrane, the membrane potential increases transiently with time. Figure 11 shows the action potential produced by a rectangular pressure stimulus applied across the membrane from the concentrated to the dilute side. Here the system is the same as that used in Fig. 10. As seen in Fig. 11, even in the case of pressure difference a threshold value of stimulus strength is observed, that is, below the threshold value the potential varies exponentially with time toward the original state after the pressure stimulus is removed, while once the emf attains the threshold value the potential either flips up or flips down. No time variation of the potential is observed when the pressure is applied from the dilute to the concentrated side, nor when the concentrations in the two external solutions are lower than c_t. Therefore, the change in the emf in response to the external pressure stimulus is apparently due to the transformation of the conformation of DOPH molecules in the filter paper, as is consistent with the variation of the membrane potential due to the salt concentration observed in Fig. 6.

Figure 12 illustrates the change of impedance during excitation, monitored by ac of 10-kHz frequency, and compared with the variation of the membrane potentials. The impedance bridge was balanced at the resting state (state A). The impedance of the membrane does not change very much when the membrane is in the subthreshold state. If

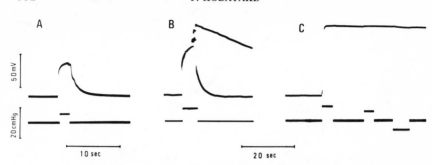

Figure 11.   Potential variations due to rectangular pressure stimuli for a DOPH-Millipore membrane of $Q = 2.89$ mg/cm^2 with $C_1 = 125$ mM and $C_2 = 1$ mM KCl aqueous solutions. (*a*) Response in the membrane potential when the pressure stimulus is smaller than the threshold value (8 cm Hg). (*b*) Potential variations at the threshold stimulus pressure, where the emf either flips up or flips down after the pressure stimulus is removed. (*c*) Variation of potential due to a pressure stimulus whose strength is larger than the suprathreshold value. Note that the plateau of the action potential is stable against external mechanical stimulus.

the membrane is excited, however, the impedance decreases by a factor of about 100 in comparison with that in the resting state. One may notice that these observations quite resemble those observed in an excitable living tissue.[2]

Figure 13 shows photographic traces of the emf fluctuations when a difference in pressure is applied across the membrane. With increasing pressure difference (directed from the more concentrated to the dilute side), the fluctuation in potential increases correspondingly. Further increase in pressure difference leads to an increase in fluctuation level, and causes stepwise increments of the transmembrane potential, including some transient firing to the excited state (see record C in Fig. 13). When the pressure difference reaches the threshold value, an abrupt increase in potential is induced. Similar discrete variations (stepwise increments) are also observed in the electrical resistance of the membrane when a constant electrical potential difference is applied across it.[14]

As in the case of the excitation process in living tissues, both resting and excited states are found to be stable against small external perturbation. Figure 14 illustrates the stability of the resting and excited states, where the external electrical perturbation does not have any effect on the steady state, and hence the potential returns to its respective steady-state value with removal of the external perturbation if the electrical perturbation does not exceed the threshold value. When the

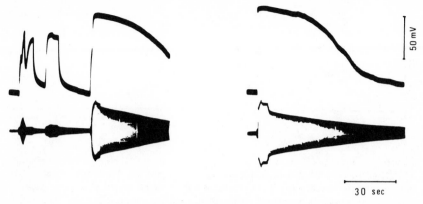

Figure 12. Variations of the membrane impedance in various states of the membrane caused by pressure stimuli for a DOPH-Millipore membrane of $Q=2.51$ mg/cm^2 with $C_1 = 125$ mM and $C_2 = 1$ mM KCl solutions. Upper traces show the variation of the emf, and lower traces show the unbalance of the impedance bridge, which was prebalanced at the resting state with 10 kHz ac. Note that the impedance of the membrane in the subthreshold state differs appreciably from that in the excited state.

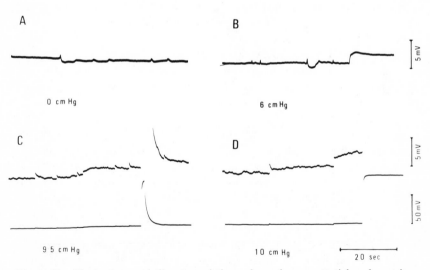

Figure 13. Fluctuations and discrete variations of membrane potential under various differences in pressure applied across the membrane from the concentrated to the dilute side. The membrane and the salt conditions are the same as those in Fig. 10.

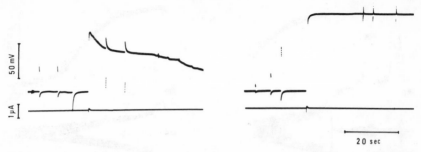

Figure 14. Stability of the resting and excited states of a DOPH-Millipore membrane. The membrane and salt conditions are the same as those in Fig. 11. Neither the electrical nor the pressure perturbations lead to the transition of the membrane state unless the emf across the membrane is brought to the threshold state by external stimuli (cf. Figs. 1 and 11).

membrane is in either the resting or the excited state, a mechanical perturbation applied across the membrane leads to no appreciable change in potential unless the state of the membrane is shifted to the threshold value by the pressure (see also Fig. 11).

If the conformational change in the DOPH-Millipore membrane accompanies either an exothermic or an endothermic reaction, the variation of the surrounding temperature should lead to the transition of DOPH conformation as found experimentally in the squid giant axons. Figure 15 shows the relation between $\Delta\varphi$ and temperature for a DOPH-Millipore membrane, indicating that the transition of the membrane conformation is enhanced by the lowering of the temperature. The hysteresis loop observed in Fig. 15 is in good accord with that of squid giant axons, shown in Figs. 4 and 5.

The results described here for the DOPH-Millipore membrane indicate that the change in conformation of molecules constituting the membrane due to external stimuli is a sufficient condition for the occurrence of the characteristic behavior of the stimulus-response relation observed in an excitable living membrane. In other words, if a phase transition from hydrophobic to hydrophilic states takes place in the membrane, all characteristic functions observed in living membrane are reproducible by the model membrane. However, as will be shown below, the transformation of the membrane macromolecules is not a necessary condition, at least for the occurrence of excitation in a membrane system. In fact, excitability is observed even in a system where no conformational change occurs in the membrane.

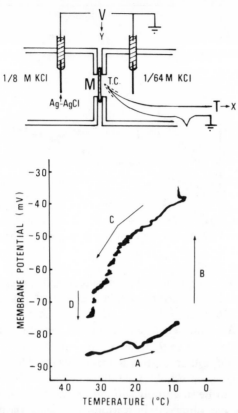

Figure 15. Hysteresis loop observed in the plot of $\Delta\varphi$ against temperature for a DOPH-Millipore membrane of $Q = 2.00$ mg/cm^2 with $\frac{1}{8}$ M and $\frac{1}{64}$ M KCl solutions. The temperature of the system was changed by the use of flowing warmed and cooled test solutions, measured by a copper-constantan thermocouple, and observed as the horizontal deflection of a synchroscope.

III. EXCITATION PROCESS IN TERMS OF DISSIPATIVE STRUCTURE IN A MEMBRANE SYSTEM

Consider a system in which two electrolyte solutions of different concentrations of C_1 and C_2 are separated by a sintered glass filter, and a difference of hydrostatic pressure is applied across the membrane from the concentrated to the dilute side. When the electric current is passing in the reverse direction, that is, from the dilute to the con-

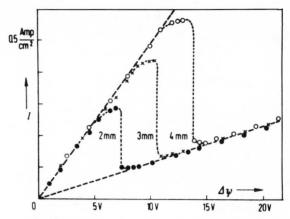

Figure 16. Experimental plot of I against $\Delta\varphi$ for the system of a glass filter and NaCl solutions of 0.01 and 0.1 M with various fixed pressure differences.[16]

centrated side, the variation of the transmembrane potential $\Delta\varphi$ with the intensity of the electric current gives a characteristic N-shaped curve, as shown in Fig. 16.[15, 16, 17] Initially, it is seen that I increases almost linearly with $\Delta\varphi$, but when $\Delta\varphi$ attains a certain value, there is a discontinuous decrease in I, followed by another linear increase, but with a smaller slope than that in the initial stage. Of special interest is the existance of a threshold potential, below and above which the system appears to be in entirely different steady states. Broadly similar transitions can be observed in various problems in non-equilibrium systems—for example, the instability of phonons in a semiconductor, the emission of coherent laser light, oscillation in gaseous plasma, and thermal instability in fluid systems (the so-called Bénard problem). A similar thing is also observed in living cells, especially in sensory cells and nervous tissues.

We can treat the present nonlinear I-V relation in terms of the dissipative structure of nonequilibrium systems, explored by Prigogine et al.[8, 18] The conservation laws of mass, momentum, and energy can be combined to lead to the following inequality without any reference to the relation between fluxes and forces:

$$d\Phi = \sum_i J_i dX_i \leqslant 0$$

Here J_i is the generalized flux, and dX_i is the variation of the force. In the derivation of the above equation, the well-known conditions for the stability of equilibrium—that is, the thermal stability, the mechanical stability, and the diffusion stability—have been introduced, and the boundary conditions for the system are assumed to be time independent. $d\Phi$ remains negative or zero upon perturbation of the force, with no restriction on the relation between forces and fluxes or on the phenomenological coefficients. The equality sign in the above equation refers to the steady state.

The basic problem associated with this equation is the study of integrability. Although an exact differential does not exist for $d\Phi$ in general, in many specific cases $d\Phi$ is an exact differential in the neighborhood of the steady state. Thus Φ retains its potential character in the vicinity of the steady state. The potential thus obtained is refered to as the kinetic potential or as the generalized entropy production (g.e.p.). A potential would induce ergodic behavior, and for any initial state the system would approach the state for which Φ reaches its extreme value. This implies that the transition between two stationary nonequilibrium states occurs when two states have the same value of the g.e.p. under a given set of boundary conditions.

Now we apply this theory to our present problem. Subject to the boundary conditions described above, together with appropriate assumptions and approximations for the system to be considered, the g.e.p. in the membrane can be calculated.[19] The results are shown in Fig. 17, where the g.e.p. is plotted against the electrical potential across the membrane with various fixed pressure differences. In the inset in the figure, the free energy of a Van der Waals gas in the condensation region is shown as a function of pressure. As seen in the figure, there is close parallelism between the transition of the two steady states and that of the equilibrium system. Figure 18 shows the theoretical relation between current and voltage calculated from the g.e.p. It should be noted that the theoretical I-V relation is in line with the experimental data shown in Fig. 16.

The reader is referred to the chapter by Turner (p. 63) for another application of these ideas.

The physical situation behind the flip-flop phenomenon is explained as follows: Under the external conditions studied here, the Poiseuille pressure flow transports fluid from the more concentrated to the less concentrated solution, while the electro-osmotic flow caused by the potential gradient tends to carry fluid in the opposite direction. With increasing potential, the electro-osmotic flow becomes appreciable, outweighs the pressure flow, and eventually changes the direction of mass flow from negative to positive. Calculations show that this change in

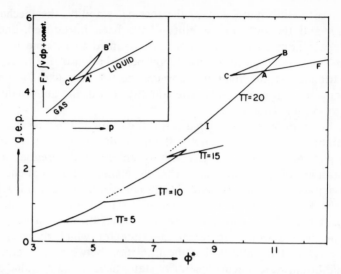

Figure 17. Generalized entropy production (g.e.p.) in a glass filter as a function of the reduced electric potential ϕ^* at various pressure differences Π with a fixed ratio of concentrations, $C_1/C_2 = 0.1$. The inset shows a schematic diagram of the Gibbs free energy against pressure for a Van der Waals gas when $T < T_c$.

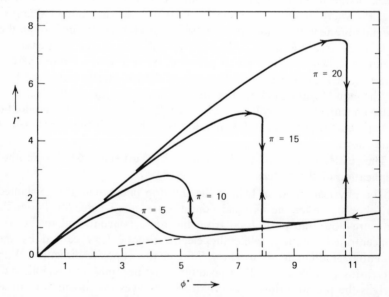

Figure 18. Theoretical curves for electric current I^* against electric potential Φ^*, at various pressure differences Π with $C_1/C_2 = 0.1$.

338

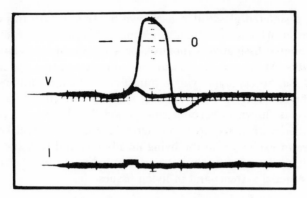

Figure 19.   An action potential observed in a glass filter-electrolyte system in response to an external electrical current stimulus.[20]

direction of mass flow occurs discontinuously when $\Delta\varphi$ reaches a certain value. Correspondingly, the average salt concentration in the membrane is lowered, that is, the membrane is occupied by the less concentrated solution on its left side. The effect of this change in concentration in the membrane is reflected in the dependence of I on $\Delta\varphi$.

If the pressure difference between the two sides of the membrane is allowed to vary at fixed current density, then the transmembrane potential, the mass flow, the electric resistance, and the pressure difference all oscillate with time.[15,20] Furthermore, if we select appropriate external and boundary conditions, the present glass membrane system can be shown to display a kind of action potential in response to an external electrical stimulus. A typical example for the action potential is shown in Fig. 19.

IV. CONCLUDING REMARKS

As seen in the experimental results on squid giant axons, the process of excitation of living tissues is plausibly explained by assuming a conformational change of membrane macromolecules. In fact, it has been shown that various functions of biological membranes can be reproduced by a model membrane of DOPH-Millipore, where a phase transition of lipid analog in the filter paper occurs in response to a change in the environmental conditions or external stimuli. However, similar functions to those of living excitable membranes are also reproduced by a sintered glass filter, as shown by Teorell.[20] In this case, the characteristic behavior of stimulus-response relationships is interpreted in terms of the dissipative structures, or a transition between multiple steady states. At present, it is hazardous to conclude that the excitation of living tissues is accompanying a transition of the membrane structure itself in the classical thermodynamical sense. A transition of states in a wider sense is shown to account equally well for the process of excitation. Further work is required to determine what is the real key to the occurrence of excitation in the living membrane, and seek the underlying principle that governs these transient nonlinear processes observed in various model systems and in living tissues.

References

1. I. Tasaki, *Nerve Excitatation*, Charles C. Thomas, Springfield, Ill., 1968.
2. Y. Kobatake, I. Tasaki, and A. Watanabe, *Adv. Biophys.*, **2**, 1 (1971).
3. J. V. Howarth, R. D. Keynes, and J. M. Ritchie, *J. Physiol.*, **194**, 745 (1968).
4. L. B. Cohen, R. D. Keynes, and B. Hille, *Nature*, **218**, 438 (1968).
5. I. Tasaki, A. Watanabe, R. Sandlin, and L. Carnay, *Proc. Natl. Acad. Sci. U. S. A.*, **61**, 883 (1968).
6. I. Prigogine, and R. Defay, *Chemical Thermodynamics*, Longmans Green, London, 1954.
7. T. Oikawa, C. S. Spyroporulos, I. Tasaki, and T. Teorell, *Acta Physiol. Scand.*, **52**, 195 (1961).
8. R. J. Donnelly, R. Herman, and I. Prigogine, Eds., *Nonequilibrium Thermodynamics*, Univ. Chicago Press, Chicago,, London, 1966.
9. I. Tasaki, T. Takenaka, and S. Yamagishi, *Am. J. Physiol.*, **215**, 152 (1968).
10. Y. Kobatake, A. Irimajiri, and N. Matsumoto, *Biophys. J.*, **10**, 728 (1970).
11. M. Yoshida, Y. Kobatake, M. Hashimoto, and S. Morita, *J. Membrane Biol.*, **5**, 185 (1971).
12. M. Yoshida, N. Kamo, and Y. Kobatake, *J. Membrane Biol.*, **8**, 389 (1972).
13. Y. Kobatake, N. Takeguchi, Y. Toyoshima, and H. Fujita, *J. Phys. Chem.*, **69**, 3981 (1965); Y. Toyoshima, Y. Kobatake, and H. Fujita, *Trans. Faraday Soc.*, **63**, 2814 (1967).
14. N. Kamo, T. Yoshioka, M. Yoshida, and T. Sugita, *J. Membrane Biol.*, **12**, 193 (1973).
15. Y. Kobatake and H. Fujita, *J. Chem. Phys.*, **40**, 2212, 2219 (1964).
16. U. F. Franch, *Elektrochem.*, **67**, 657 (1963).
17. P. Meares and K. R. Page, *Phil. Trans. Roy. Soc. Lond.*, **272**, 1 (1972).
18. P. Glansdorff and I. Prigogine, *Physica*, **30**, 351 (1964); P. Glansdorff and I. Prigogine, *Thermodynamic Theory of Structure, Stability and Fluctuations*, Wiley-Interscience, London, New York, 1971.
19. Y. Kobatake, *Physica*, **48**, 301 (1970).
20. T. Teorell, *J. Gen. Physiol.*, **42**, 831, 847 (1959); *Biophys. J.*, **2**, 27 (1962).

EXCITABILITY AND IONIC SELECTIVITY, COMMON PROPERTIES OF MANY LIPIDIC DERIVATIVES

A. M. MONNIER

Université Paris VI—9, Quai St Bernard, Paris, France

Dr. Kobatake, in the present Volume (p. 319), has demonstrated that what he calls a lipid "analog," (i.e., dioleylphosphate impregnating a Millipore filter) displays many features of excitability and ionic selectivity, without the aid of the additives required by bilayer membranes, such as cyclopeptides or specific protein material. This is precisely what our group has shown in recent years, on membranes made with many lipid derivatives.[1-7] For instance, a drop of an unsaturated fat (a drying oil such as linseed oil), deposited upon an oxidizing solution of MnO_4K, spreads as a thin membrane, showing interference colors. This membrane then undergoes oxidative polymerization. It becomes sufficiently strong and flexible to be removed and inserted between two aqueous compartments containing chloride solutions and electrodes. Under moderate fields (50 to 500 V/cm), the membrane shows considerable transient increases in conductance, thus imitating the essential characteristic of the "response" of a cell membrane. Many features of electrical excitability are observed on this artificial membrane, such as the frequency of responses as a function of the field, synchronization of the responses of neighboring sites, and impedance parameters. The membrane is markedly cation selective. The biionic-potential method of Sollner shows that is possesses very pronounced intercationic selectivity, K^+ being 8 to 20 times more permeant that Na^+ —a feature that is found on the resting cell membrane. In most cases, the more hydrated a cation, the less it traverses the membrane. But large organic cations, such as cationic drugs (ephedrine or amphetamine) are more permeant, by one or two orders of magnitude, than K^+. This is a remarkable pharmacological analogy. It results from the membrane being a mixed body consisting of adjacent hydrophilic and lipophilic zones. The invasion of organic cations is also enhanced by the fact that the membrane possesses fixed acidic groups and behaves as a cation exchanger, which has a marked preference for organic cations. These carboxylic groups result from the previous oxidation of the membrane.

Recently, Wallon has demonstrated that the membrane is remarkably photoexcitable when in contact with solutions containing various dyes.

Most of the above phenomena can be observed on membranes made from alkyds or glycerolmonooleate gels. All these membranes function without any additives, except in the latter case, which requires the addition of a small amount of fatty acid so as to form a few fixed acidic groups. The essential conditions for the excitability and ionic selectivity of a lipidic derivative appear to be: (*1*) semi-solid structure, (*2*) a slight hydration, (*3*) a few fixed acidic groups.

Thus, membranes formed with widely different lipidic derivatives readily behave as "analog" of the "passive" features of the cell membrane. The bilayer structure does not appear to be a necessary condition for the display of the above features. But the intimate processes in these various membranes are not yet completely clear. An attractive approach to their disclosure appears to be in the "conformational changes" found by Kobatake et al. in their dioleylphosphate membranes.[8]

References

1. A. M. Monnier and A. Monnier, Formation de membranes minces et stables par étalement de lipides non saturés sur des solutions oxydantes, *J. Physiol. Paris*, **56**, 410 (1964).

2. A. M. Monnier, A. Monnier, H. Goudeau, and A. M. Reynier-Rebuffel, Electrically Excitable Artificial Membranes, *J. Cell. Comp. Physiol.*, **66**, 147 (1965).

3. A. M. Monnier, Experimental and Theoretical Data on Excitable Artificial Lipidic Membranes, *J. Gen. Physiol.*, **41**, 26 s(1968).

4. A. M. Monnier, Données récentes sur les membranes lipidiques artificielles électriquement excitables, *Membranes à perméabilité sélective*, Editions du C.N.R.S., Paris (1969).

5. A. M. Monnier, Sélectivité Ionique et Réponses Electriques des Membranes Lipidiques Artificielles Excitables, *EEG Clin. Neurophysiol.*, Suppl. 31, 139 (1972).

6. A. M. Monnier, *Les modèles de la membrane cellulaire*, Institut de Biophysique de l'Université de Rio, Rio in press.

7. N. Kamo, T. Yoshioka, M. Yoshida, and T. Sugita, Transport phenomena in a model membrane accompanying a conformal change: Transient process in response to external stimuli. *J. Memb. Biol.* **12**, 193 (1973).

8. M. Yoshida, N. Kamo, and Y. Kobatke, Transport phenomena in a model membrane accompaning a conformational change. Membrane potential and ion permeability. *J. Memb. Biol.* **8**, 389 (1972).

THERMODYNAMIC CONSIDERATIONS ON THE BEHAVIOR OF EXCITABLE MEMBRANES

M. DELMOTTE, J. JULIEN, J. CHARLEMAGNE,
AND J. CHANU

*Laboratoire de Thermodynamique des Milieux Ioniques et
Biologiques,
Université Paris VII, 2 place Jussieu, Paris, France**

CONTENTS

The extreme diversity of the observable properties in biological membranes must not mask the fact that, in many cases, these properties remain understandable in the thermodynamic realm of dissipative structures. Considerations of this sort brought one of us (M.D.) to propose in 1969 a mode of reception of isolated photons by the retina and to suggest an amplification mechanism for their effects and a generation mechanism for the first receptor potential. Since then, our investigations in this direction have allowed us to widen this primitive reactive scheme. We here consider applying this amplification mechanism to other nervous transduction and conduction phenomena.

I. BIOLOGICAL DATA

Let us recall that in the initial phase, the photoreception starts at the level of the outer segments of cone and rod visual cells, which line the pigmentary epithelium of the eye. The rod outer segment, responsible

* Equipe de recherche associée au CNRS, ERA No. 370.

for acute vision, is cylindrical and made of about one thousand "flattened sac" piling. It is not impossible that these sacs only represent a multiply folded membrane. The cylindrical segment axis of each rod cell gives the direction of the piling and is parallel to that of the incident light. These sac membranes are principally made of rhodopsin.

It is now agreed that the initial act of photoreception consists in the absorption by the sac rhodopsin of only a fraction of the incident photons. On the basis of work by Wald et al.,[1] Delmotte,[2] has shown that each sac can absorb at best 3 quanta per second. This is a very small reference number. Consequently, in the following development, the flattened sac will be chosen as elementary photoreceptor site.

It is well known that, when it is absorbed, the photon transforms rhodopsin into opsin and retinal. But rhodopsin or a closely linked protein would have ATP-ase properties.[3] This ATP-ase activity would depend on the participation of certain ions: Mg^{++}, Na^+, K^+.[4, 5] Etingoff[6] associates this ATP-ase activity with the same kind of active transport in every nervous cell.

The existence of this kind of active transport is compatible with the extensive development of rhodopsin membranes in the rod outer segments, whether individualized or in multiply folded sacs. We have at this level two ionic solutions of different composition separated by a membrane rich in enzymes.

II. SUGGESTED MODEL

The former considerations allow us to propose a very schematic model of the flattened sac: an interior compartment i separated from the exterior solution e by the sac membrane. For the sake of simplicity, we suppose the transport of *only one* ionic component. The concentration I_e in the outer compartment remains constant. The membrane contains the enzyme E, ATP-ase, and also a chemical transport mediator. The concentration of the inner solution, I_i, is variable, but the ATP concentration is considered as constant in the interior compartment. The imposed values of I_e and the ATP concentration play the role of thermodynamic constraints, which keep the system out of equilibrium.

One of the transport phenomena is a simple diffusion of the I ion, explicited by a Fick-type law

$$J_d = D(I_i - I_e) \tag{2.1}$$

where the flow J_d contains a coefficient D with the dimensions of permeability. The other phenomenon is a chemical flow J_c due to a number of intermembranous reactions of autocatalytic type:

$$E + \alpha I_i \xrightarrow{k_1} E'$$

$$E' + ATP \xrightarrow{k_2} E'' + ADP$$

$$E'' + I_e \xrightarrow{k_3} E + (\alpha + 1)I_i + P \tag{2.2}$$

in which the enzyme E, first activated on the internal membrane surface, hydrolyzes ATP and loses its excess energy to ensure the transport of I.

The scheme conveniently describes the active transport of I in the case where $I_i > I_e$ ($\alpha > 1$). In the case where $I_i < I_e$, we can write the transport step [third step of the system (2.2)] as follows:

$$E'' + I_i \xrightarrow{k_3} E + I_e + \alpha I_i + P \tag{2.3}$$

which implies a competitive inhibition such as

$$E + \beta I_i \underset{k_4'}{\overset{k_4}{\rightleftharpoons}} E^{IV} \tag{2.4}$$

with $\alpha < \beta$. Other biological situations, which do not lead to the classical nervous-conduction overshoot, imply the participation of an effector H such that

$$E + H \underset{k_5'}{\overset{k_5}{\rightleftharpoons}} E^V. \tag{2.5}$$

III. MODEL THERMODYNAMIC STUDY AND BIOLOGICAL INTERPRETATION

In the range of a simple kinetic study, we can prove the possible existence of multiple steady states. On the other hand, the thermodynamic analysis of the system stability conditions by the normal-mode method or stability-criterion method shows the unstable character of some of these states.[7, 8] Depending on the conditions imposed on the system and the values of the kinetic parameters, we have either

1. a stable stationary state, or
2. *three statinary states*, which must consist of an unstable state between two stable states.*

In the resting state, the biological site is in a steady stable state, which can only be out of equilibrium. If the amount of energy associated with the exciting disturbance is sufficient to bring the elementary site beyond the unstable state, then the site will fall into the second stationary stable state. Such behavior is especially obvious in photoreception, where the exciting disturbance is due to absorbed incident photons. The return to the initial state, if it does happen, can be explained in many ways, for example, the exhaustion of energy reserves (ATP), the effect of an inhibited reaction (see above), or the competitive effect of another ion transport.

It is remarkable that such a physicochemical system shows *amplification* features. The starting of the reaction following the exciting disturbance consumes the available energy provided by the nonequilibrium conditions in the resting system.

IV. CONCLUSION

The scheme followed in this study has successively brought us to define the notion of an elementary biological site, to index thermodynamically conjugated flows and forces, and to analyze the numerous system states in terms of nonequilibrium considerations. We must consider such biological systems to be out of equilibrium, and almost always very far from equilibrium. Whether it is an elementary site or a complete organism, the system always undergoes *exchanges* and *dissipation*. We must specially emphasize that the properties characterizing the mode of action of excitable membranes (all-or-none response, amplification, action potential, *etc.*) can only appear far from equilibrium, in the range where nonlinear laws make possible the existence unstable steady states.

References

1. G. Wald, P. K. Brown, and I. R. Gibbons, The problem of visual excitation, *J. Opt. Soc. Am.*, **53**, 20 (1963).
2. M. Delmotte, Description thermodynamique de la photoréception rétinienne. Eventualité d'un mécanisme d'amplification et de codage, Thèse 3ème Cycle en Biophysique, Université de Paris, 1969. See also Processus de photoréception rétinienne et thermo-dynamique des phénomènes irréversibles, *Vision Res.*, **10**, 671 (1970).

*An interesting example of an excitable system has been studied by Blumenthal, Changeux, and Lefever. (Ref. 9).

3. M. A. Ostrovsky, Investigation of certain links of the photoenzymochemical chains of processes in photoreceptors, *Biophysics*, **10**, 519 (1965).

4. D. G. Scarpelli and E. L. Craig, The fine localisation of nucleoside triphosphate activity in the retina of the frog, *J. Biophys. Biochem. Cytol.*, **17**, 279 (1963).

5. Y. Sekoguti, On the ATPase activities in the retina and the rod outer segments, *J. Cell. Com. Physiol.*, **56**, 139 (1960).

6. R. N. Etingoff, Some aspects of biochemistry of photoreception (in Russian), in *The Primary Processes in Receptor Elements of Sense Organs*, Nauka, Moscow, 1966.

7. P. Glansdorff and I. Prigogine, Structure, Stabilité et Fluctuations, Masson, Paris, 1973.

8. G. Nicolis, Stability and dissipatives structures in open systems far from equilibrium, *Adv. Chem. Phys.*, **XIX**, 209 (1971).

9. R. Blummenthal, J.-P. Changeux, and R. Lefever, Membrane excitability and dissipative instabilities, *J. Membrane Biol.*, **2**, 351 (1970).

10. M. Delmotte and J. Chanu, Description d'un système ouvert illustrant un transport ionique dans les membranes biologiques, *Electrochem. Acta*, **16**, 623 (1971).

11. M. Delmotte, J. Julien, and J. Charlemagne, Rôle éventuel de l'inhibition dans les transports ioniques membranaires, *J. Chim. Phys., Paris*, **70**, 1663 (1973).

12. J. Julien, M. Delmotte, and J. Charlemagne, Transport ionique au niveau des membranes excitables. Point de vue thermodynamique, *J. Chim. Phys., Paris*, **70**, 1660 (1973).

ON THE CHANGES IN CONDUCTANCE AND STABILITY PROPERTIES OF ELECTRICALLY EXCITABLE MEMBRANES DURING VOLTAGE-CLAMP EXPERIMENTS

R. LEFEVER AND J. L. DENEUBOURG

*Faculté des Sciences, Université Libre de Bruxelles,
Bruxelles, Belgium*

CONTENTS

I. INTRODUCTION

It is a most remarkable natural phenomenon that small and local depolarizations of the electric field established across electrically excitable membranes can amplify in the form of action potentials propagating in a sustained fashion, at a rate of more than 20 m/sec and without

349

distortion. The all-or-none character of this behavior has always been a fascinating property, suggesting the existence of some kind of "phase transition" of the membrane taking place when it switches from the resting polarized state to the excited depolarized state.

The molecular basis of this transition has not yet been elucidated, but since the work of Hodgkin and Huxley (HH) and their followers (for references see the books by Hodgkin[1] and Katz[2]), we have had a quite precise picture of the drastic permeability changes and the ionic events that are associated with it. These authors also proposed a quantitative, phenomenological model based on a detailed analysis of the ionic currents recorded during voltage-clamp experiments as well as during the propagation of action potentials. The properties of this model have been widely confirmed, so that any more recent attempt to describe electrical excitation on a physicochemical basis has in general been evaluated with regard to its agreement with the HH results. Let us therefore first briefly summarize the essential characteristics of electrical excitation as they have been demonstrated by Hodgkin and Huxley. The basic equation describing the potential changes across the membrane is

$$C\frac{\partial V}{\partial t} = I(V) + \frac{a}{2R}\frac{\partial^2 V}{\partial x^2} \tag{1.1}$$

.The current density flowing through the membrane has been decomposed into a capacitive and a resistive component, respectively $C\partial V/\partial t$ and $I(V)$; $(a/2R)\partial^2 V/\partial x^2$ is then the current flowing along the axon in the direction of spatial coordinate x. The dependence of $I(V)$ on V is quite complicated. There are three principal separate systems carrying the current through the membrane: the so-called sodium and potassium channels, and some leakage channels associated with chloride and other ions present in the internal and external mediums. This can be expressed in the following manner:

$$I(V) = g_{Na}(V)(V - V_{Na}) + g_K(V)(V - V_K) + g_L(V - V_L) \tag{1.2}$$

g_{Na} and g_K are the sodium and potassium conductances, which depend sensitively on the voltage across the membrane; g_L is a constant; V_{Na}, V_K, V_L are the equilibrium voltages of Na, K, and the leakage current, depending on the ratios of the concentrations of these ions on both sides of the membrane.

The major characteristic properties of the sodium and potassium conductances can then be summarized as follows:

1. In response to an abrupt change of membrane potential, the *instantaneous* sodium and potassium currents are in first approximation linear functions of the applied voltage.

2. Sustained depolarizations are followed by an increase in sodium and potassium conductance (activation of Na and K channels). The time course of this increase in conductance obeys a high-order kinetics.

3. For depolarizations close to the threshold, the maximum conductance for sodium varies steeply as a function of potential: an *e*-fold increase of sodium conductance can be brought about by a change of membrane potential of the order of 5 mV.

4. The increase in g_{Na} following a depolarization is not sustained, but "inactivates" spontaneously. This inactivation process follows a simple exponential decay in time.

5. After inactivation, the sodium transport system cannot be rapidly activated even with large depolarizing stimuli: there exists a refractory period. The time lag necessary for the recovery to be complete is much larger (30 msec) than the duration of the action potential (2 msec).

The activation of potassium and sodium as well as the inactivation of sodium constituting independent processes, HH introduced as set of three parameters (m, n, h), referring respectively to the sodium and potassium activation and to the sodium inactivation, whose value is simply related to the conductances g_{Na} and g_K, and whose change in time is given by three independent first-order linear differential equations; the parameters of these equations are complicated functions of the external potential. Although this model was first derived in the case of the giant axon of *Loligo*, it has also proved to be a very useful tool for investigation and description, applicable to a large number of other electrically excitable membranes.

Many studies have investigated the properties of the HH equations in simulation experiments on analog or digital computers[3–8] and have demonstrated the rich set of mathematical phenomena that this model may exhibit. Its complexity, however, and the various nonlinearities present in the equations make its theoretical study rather tedious and discourage analytic approaches. Several ingenuous simplifications of the HH equations have been proposed, which retain the principal feature of nerve excitation: the existence of a threshold below which perturbations are rapidly damped, and beyond which they are amplified in a characteristic traveling wave. A system of this type that has been particularly well studied is the Nagumo equation,[9] for which several distinct regimes of propagation are known.[10, 11] These results have led to a detailed understanding, but in phenomenological terms, of the type of stability behavior that might be associated with nerve conduction and excitation.

In this chapter we attempt to provide a link between this behavior and a more precise physicochemical basis suggested by recent developments in various domains of membrane biology. Particularly, the artificial ionophores now offer a picture of the type of molecular structures that might selectively recognize ions and transport them through a membrane[12]; the allosteric transitions that account for the regulation of biochemical reactions illustrate how such transport might be controlled by a change of conformation.[13] On the other hand, the recent developments of nonequilibrium thermodynamics furnish a general conceptual framework[14] for the interpretation in physical terms of the stability problems raised by excitable membranes.

II. GENERAL HYPOTHESES OF THE MODEL

A. Structure of the Ionophores

The transporting systems of electrically excitable membranes are constituted of elementary units or *ionophores* selectively involved in the translocation of the ions. The ionophores both recognize the permeant ions and transport them selectively along their electrochemical gradient. The ions play the role of both a *ligand* and a *permeant*. The capacity for selective transport of the ionophores involved in the excitation process changes upon excitation. They are "versatile"[15] in the sense that they undergo a structural transition between at least two states, one for the membrane at rest (S), and the other for the membrane during excitation (R). These two states differ in their affinity for the ions and their rate of transport. By convention, the active R state is taken as the most permeable one.

B. Distribution of the Ionophores in the Membrane

Embedded in a lipid bilayer, the ionophores are organized in small-number aggregates, or in much larger ones that then may be assimilated to a continuous lattice structure. Interactions might be established between individual ionophores through a cooperative conformational coupling.

C. Effect of the Electric Field on the Conformational Equilibrium

The ionophore is a macromolecule or a macromolecular assembly that bears a number of charges or polar groups, which may be considerable and depend on the conformational state. For the time being, we shall suppose that the essential consequence of this situation is to make the conformational equilibrium between R and S field dependent because of a difference in permanent dipole moment: The orientation of the ionophore as a whole being rigid in the membrane, the reorientation of the polar groups when the field is changed induces the transition from one state to the other . The effect is such that a depolarization favors the permeable (R) state.

D. Effect of the Electric Field on the Ionic Transport

The number of charges that exert an electrostatic repulsion against the passage of the ions depends on the conformational state of the ionophores and their neighborhood. This number varies when the orientation of the polar groups attached to the ionophores is modified by a field variation.

E. Environment of the Ionophores

As a whole, the membrane is a coherent phase, which creates a permeability barrier between two volumic phases of different electrochemical potential; on both sides of the membrane, however, the ionophores are not in direct contact with the bulk solutions. There exists, in between a physical medium, or "equilibration layer," where the activity and diffusion coefficients of the ligand may be different from both that of the bulk solutions and that of the ionophores.

III. FORMULATION OF THE MODEL

A. Kinetics of the Transition Between the R and S Conformations

We assume that the ionophores present in electrically excitable membranes have a fixed axis of orientation, which spans the lipid layer, and along which their spatial structure is a repetition of similar segments. Each segment may exist in two conformational states characterized by different ionic permeabilities and different net dipolar moments. Models for such ionophores are the polycyclic hexapeptides described recently by Urry[12].

The transition of an ionophore from the state in which all its segments are in the S conformation to the state in which they all are in the R one appears as a complex process involving a considerable number of intermediate states. One may, however, expect that there are, at a minimum, two types of elementary processes between which it is necessary to distinguish: (1) the nucleation of a R segment within a fully S ionophore, (2) the growth of the number of segments in one conformation in comparison with the number of segments in the other conformation. The transition between the two extreme states of the ionophore can thus be written as

$$x_{0,n} \underset{l_{SR}}{\overset{l_{RS}}{\rightleftarrows}} x_{1,n-1} \underset{k_2}{\overset{k_1}{\rightleftarrows}} x_{2,n-2} \underset{k_2}{\overset{k_1}{\rightleftarrows}} \cdots \underset{k_2}{\overset{k_1}{\rightleftarrows}} x_{n,0} \qquad (3.1)$$

where n is the number of segments per ionophore; the $x_{i,j}$ are the molar fractions of ionophores per unit area with i segments in the S conformation and j segments in the R conformation; l_{RS}, l_{SR} are the rate constants associated with the nucleation step; and k_1, k_2 are the rate constants for the "growth" process. These constants are field dependent because of the existence of the nonvanishing dipole moments μ_R and μ_S characteristic of the R and S conformations

$$k_1 = k_1' e^{\mu_R E/RT}, \qquad k_2 = k_2' e^{\mu_S E/RT}, \qquad \frac{k_1}{k_2} = l_R e^{(\mu_R - \mu_S)E/RT} = l_R k \quad (3.2)$$

where E is the electric field across the membrane. We define the fraction $\langle r \rangle$ of ionophores that have at least one segment in the R conformation:

$$\langle r \rangle = \sum_{i=0}^{n-1} x_{i,n-i} \qquad (3.3)$$

The time change of the $x_{i,j}$ fractions can then be described by the following set of kinetic equations:

$$\frac{d\langle r \rangle}{dt} = l_{RS} x_{0,n} - l_{SR} x_{1,n-1}$$

$$\frac{dx_{i,n-i}}{dt} = k_1 x_{i-1,n+1-i} - (k_1 + k_2) x_{i,n-i} + k_2 x_{i+1,n-1-i} \qquad (i \geqslant 3) \quad (3.4)$$

together with the conservation relation

$$\sum_{i=0}^{n} x_{n-i,i} = 1 \qquad (3.5)$$

We consider that the nucleation reaction is rate limiting compared to the "growth" process, and thus have the inequality

$$k_1, k_2 \gg l_{RS}, l_{SR} \qquad (3.6)$$

which permits us to reduce (3.4) to a single kinetic equation:

$$\frac{d\langle r \rangle}{dt} = l_{RS}\left(1 - \langle r \rangle - \frac{l_{SR}}{l_{RS}}\frac{(1 - l_R k)}{(1 - l_R^n k^n)}\langle r \rangle\right) \tag{3.7}$$

and $n-2$ algebraic relations for the intermediate $x_{i,j}$ states:

$$x_{i,n-i} = \frac{(l_R k)^{i-1}(1 - l_R k)\langle r \rangle}{1 - l_R^n k^n} \tag{3.8}$$

It has often been suggested that cooperative structural interactions might exist between the ionophores and facilitate the transition from the nonpermeable to the permeable state.[13, 16] To take the possibility of such effects into account, we shall suppose here that the free energy of the nucleation step depends on the average conformational state of the membrane, and can be approximated in the following manner:

$$\Delta F = \epsilon - \eta U(x_{i,j}) \tag{3.9}$$

where η is a constant and $U(x_{i,j})$ is given by

$$U(x_{i,j}) = \sum_i i x_{i,n-i} = \frac{\langle r \rangle}{n(1 - l_R^n k^n)}\left(\frac{1 - l_R^n k^n}{1 - l_R k} - n l_R^n k^n\right) \tag{3.10}$$

represents the average fraction of segments in the R state per ionophore. On the other hand, the analogy with artificial ionophores suggests that the difference in dipole moment between the two conformations could be related to a change in orientation of only a small number of peptide bonds; in that case the orders of magnitude to expect for $\mu = \mu_R - \mu_S$ are probably only of a few tens of debyes. It is easy to see from Equation (3.7) that the conformational equilibrium will nevertheless exhibit a strong field dependence in such a system as long as the segments are sufficiently numerous. For example, when $l_R > 1$ and $n \gg 1$, Equation (3.7) can be rewritten as

$$\frac{d\langle r \rangle}{dt} = l_{RS}\left(1 - \langle r \rangle - \frac{l_{SR}}{l_{RS}}\frac{k^{1-n}}{l_R^{n-1}}\langle r \rangle\right) \tag{3.11}$$

and has the same form as if it corresponded to ionophores bearing a single large dipole of $(n-1)(\mu_R - \mu_S)$ D. An ionophore consisting of 20 segments in two conformations differing by a dipolar moment of 20 D may thus have a conformational equilibrium that depends on the field in the same way as if there were a single dipole of 400 D.

B. Derivation of the Current-Voltage Relation

Within the ionophore, the electrolyte is assumed to be completely dissociated; Na^+ and K^+ are transported as monovalent cations. The effect of the electrical field on the electrolytic dissociation (Wien effect) is neglected. Its influence on the ionic transport becomes significant when the rate-limiting step is the passage of the ions across the interface.[21] It has been shown that then, at least in the case of lipid bilayer membranes, considerable nonlinearity in the I-V curves may arise from the Wien effect. Here, however, we expect that the rate-limiting step in the ionic transport is the conformational change of the ionophore, and that it is this step, rather than an electrolytic dissociation reaction that controls the ionic flows. Nonlinearity in the I-V curves appears also as a result of the image forces acting on the ion during its transport across the membrane.[28] This additional complication is not taken into account here. For simplicity we also suppose that at the interfaces between ionophores, equilibration layers, and bulk solutions, the ionic concentrations vary in a discontinuous manner; the dielectric constant of the equilibration layers is taken as equal to that of the ionophore.

The transport of the ions is viewed according Eyring's rate theory as a series of discrete jumps through consecutive energy barriers. The electric field is constant through the membrane and influences the transport by altering the height of the energy barriers inside the membrane.

The permeability of a given ionophore will of course depend on the height of the energy barriers that it opposes to the translocation of the ions. These barriers in turn are related to the conformational state of the various segments forming the molecule; a detailed description of the translocation process would therefore require the consideration of all the intermediate states $x_{i,j}$ together with their particular energetic configuration with respect to the passage of the ions. This would yield very difficult and bulky algebraic expressions, so instead of considering the transport across each particular type of ionophore separately, we choose a single scheme for the energy barriers met by the ion in the membrane phase, and define the kinetic constants across these barriers as functions of the average conformational state of the membrane. In Fig. 1 we have represented the arrangement of barriers chosen. For simplicity, only one barrier is considered in the equilibration layer. The ionophore itself has

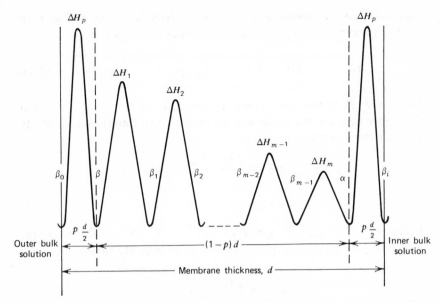

Figure 1. Location of the energy barriers in the membrane. The position of the equilibrium layers depends on the value of the parameter p: When $p = 0$, both sites are at the interface with the bulk solutions; when $p = 1$, both sites are at $d/2$. The β_j are the ionic densities at the various transition sites inside the membrane.

m barriers whose hieghts vary linearly in such a way that

$$\frac{\Delta H_{i-1}}{\Delta H_i} = h \tag{3.12}$$

where h is a constant. This organization has been suggested by Woodbury[17] and is preferred because it leads straightforwardly to linear instantaneous I-V curves when the condition $n \to \infty$ with $nh = zFV_{Na}/RT$ is satisfied. It is, however, only a mathematical convenience. In practice the I-V curves are already linear within a few percent over an interval of potential $-150 \text{ mV} < V < 150 \text{ mV}$, when $n \geqslant 3$.

Let us now consider the transition rate across a particular energy barrier. The electrical field, which is constant through the membrane, simply adds a constant term to the activation energy of the transition probability $k_{i \to i+1}$ from site i to site $i+1$. On the other hand, this activation energy is descreased when the average number of segments per ionophore in the R conformation increases (Section II.D). This property is expressed by the addition of a term $qU(x_{i,j})$, where q is a dimensionless parameter and $U(x_{i,j})$ is given by (3.10).* The total activiation energy at a given barrier can thus be written as

$$\Delta H_{i\to i+1} = \Delta H_1 + (1-i)h - \frac{zF}{RT}(p-1)\frac{\Delta V}{n} - qU(x_{i,j}) \quad (3.13)$$

where z, F, R, T have their usual meaning. $\Delta V = V_i - V_0$ is the potential difference between the inside and the outside. The transition probability $k_{i\to i+1}$ is now given by

$$k_{i\to i+1} = \epsilon_i \exp\left(\frac{(p-1)\chi}{n} + qU(x_{i,j})\right)$$

$$= \epsilon_i \Omega^U l \quad (3.14)$$

with $\epsilon_i = e^{-\Delta H_1 - ih}$, $\Omega = e^q$, $\chi = zF\Delta V/RT$, $l = e^{(p-1)\chi/n}$. In the equilibration layers we define the inward transition probability simply as

$$c_p = \epsilon_p \exp\left(-\frac{p\chi}{4}\right) = \epsilon_p k_p$$

so that we can finally write down the equations for the time change of the ionic concentrations α, β in the inner and outer equilibration layers and β_j inside the membrane:

$$\frac{d\beta}{dt} = k_a'\left[\gamma\left(k_p\beta_0 - k_p^{-1}\beta\right) - \langle r\rangle\Omega^U\left(l_\beta - l^{-1}\beta_1\right)\right]$$

$$\frac{d\alpha}{dt} = k_a''\left[\gamma\left(k_p^{-1}\beta_i - k_p\alpha\right) - \langle r\rangle\Omega^U\left(l^{-1}\alpha - l\beta_{m-1}\right)\right]$$

$$\frac{d\beta_j}{dt} = \epsilon_1\Omega^{\langle r\rangle}\left[\frac{\epsilon_j}{\epsilon_1}l\beta_{j-1} - \frac{(l^{-1}\epsilon_j + l\epsilon_{j+1})}{\epsilon_1}\beta_j + \frac{l^{-1}\epsilon_{j+1}}{\epsilon_1}\beta_{j+1}\right] \quad (3.15)$$

* As recently reported by Haydon et al.,[18] alametycin molecules incorporated in black lipid membranes reveal transport and structural properties of the type assumed here: (1) alametycin exists in a conducting and a nonconducting conformation; the transition from one state to the other corresponds to a discrete jump in conductance whose magnitude is constant when the alametycin molecule is isolated within the membrane; (2) under some conditions the molecules form small aggregates of 6 or 7 units; the permeability of an alametycin molecule within the aggregate then depends on the conformational state of its neighbors; it is bigger when the latter already are in the conducting state. Furthermore, Haydon et al. also demonstrated that the transition probability from the nonconducting to the conducting state increases when the neighbors already are in the conducting conformation.

where k_d', k_d'' are constants proportional to the number N of ionophores per unit membrane area, and $\gamma = \epsilon_p / \epsilon_1 N$. In order to be in agreement with the fact that the equilibrium potential corresponding to zero flux through the membrane is the sodium potential given in first approximation by Nernst's law, we have to require that the internal concentrations β_j equilibrate very quickly compared to α, β. Accord-ingly the inequality

$$k_d', k_d'' \ll \epsilon_1 \qquad (3.16)$$

needs to be satisfied, and the current through the membrane becomes

$$I_{Na} = \frac{\epsilon_n}{\epsilon_1} \frac{\langle r \rangle \Omega^U (l^n \beta - l^{-n} \alpha)}{l^{1-n} \left(1 + \epsilon \sum_{j=1}^{n-1} l^{2(n-j)} \right)} \qquad (3.17a)$$

or equivalently

$$I_{Na} = \frac{\epsilon_n}{\epsilon_1} \langle r \rangle \Omega^U \frac{l^{-1} (l^{2n} \beta - \alpha)(1 - e^2 e^h)}{1 - e^{2n} e^{nh}} \qquad (3.17b)$$

The equilibrium potential is then given by

$$V_{Na} = -\frac{RT}{F(1-p)} \ln \frac{\beta}{\alpha} \qquad (3.18)$$

and by analogy with HH we define the sodium conductance as

$$g_{Na} = \frac{I_{Na}}{V - V_{Na}} \qquad (3.19)$$

Taking (3.16) and (3.17) into account, we can reduce the set of equations (3.15) to a system of two equations:

$$\frac{d\beta}{dt} = k_d' \left[\gamma \left(k_p \beta_0 - k^{-1} \beta \right) - I_{Na} \right]$$

$$\frac{d\alpha}{dt} = k_d'' \left[\gamma \left(k_p^{-1} \beta_i - k_p \alpha \right) + I_{Na} \right] \qquad (3.20)$$

which together with Equation (3.7) describe the ionic transports through the membrane.

IV. PROPERTIES OF THE MODEL AND
EXPERIMENTAL PREDICTIONS UNDER
VOLTAGE CLAMP

In this section, we shall compare the results obtained on the basis of Equations (3.7) to (3.20) for the time change of the sodium conductance with the experimental results obtained by HH in voltage-clamp experiments.

It is clear from Equations (3.17a, b) that in the variation of the Na current that follows a depolarization of the membrane, two distinct processes happen sequentially: first, corresponding to the opening of the channels, we observe an increase of the function $\langle r \rangle \Omega^U$ in time, and thereafter occurs a variation of the ionic concentrations α, β in the equilibration layers, resulting from the discharge of this intermediate electrochemical gradient across the open channels. These two processes can be seen in Fig. 2, where we have represented theoretical curves calculated from Equations (3.20) and (3.7) for the time course of g_{Na} at various clamping potentials. The similarity between these curves and the experimental properties summarized in the introduction for the time course of the Na^+ conductance in the giant axon is striking. We now

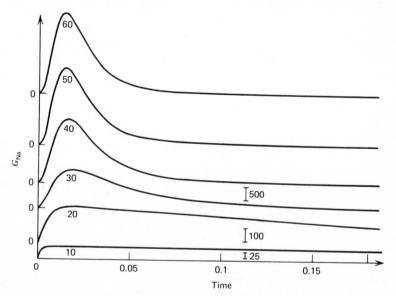

Figure 2.   Time course of the sodium conductance following depolarizations of 10, 20, 30, 40, 50, and 60 mV. The resting state corresponds to -70 mV, and the following numerical values have been given to the parameters: $k_d' = k_d'' = 1$, $q = 3.711$, $K = 0.089$, $\mu_S = 0$, $\mu_R = -600$, $p = 0.5$, $\gamma = 0.418$, $m = 1$, $l' = 200$, $\eta = 2.196$, $\beta_0 = 10$, $\beta_i = 1$.

consider the activation and inactivation processes of Na^+ in more detail.

A. Kinetics of the Sodium Activation

We call the function

$$g = \langle r \rangle \Omega^U \qquad (4.1)$$

the permeability function of the membrane. It has a role comparable to $g'm^3$ in the HH model. However, after a depolarization when n, the number of segments, is greater than one, g exhibits here a behavior that is twofold: An instantaneous variation of $U(x_{i,j})$ results from the jump of the value of $l_R k^*$; a slower response is associated with the time change of $\langle r \rangle$ and constitutes the activation process itself. In Fig. 3, as might be expected from the presence of the exponential factor in (4.1), it is clearly seen that the rise in conductance during the activation follows a high-order kinetics. It should be noticed, however, that the range of potentials over which the curves display a sigmoid shape depends simply on the value of q. By increasing q it becomes possible to obtain sigmoid curves even for small depolarizations and with conditions under which the resting state remains stable infinitesimally. If, on the contrary, q were chosen equal to zero, the only way to recover a sigmoid shape within the framework of our hypotheses would be to relate the rise of $\langle r \rangle$ to a phase transition or instability of the membrane such that $d\langle r \rangle / dt$, $d^2\langle r \rangle / dt^2 > 0$. In that case, however, there would exist a threshold in the course of the sodium activation, which has not yet been detected up to now.

Concerning the maximal conductances, we observe, in agreement with experiment, that for increasing depolarizations the position in time of the maxima first moves to the right and thereafter presents a reversed displacement to the left. On the other hand HH and also Dodge and Frankenhauser[20] have shown that the height of the peak as a function of V might vary very steeply, an e-fold increase of g_{Na}^{max} resulting from a change in membrane potential of 5 mV. Several authors have suggested[13, 16, 23] that this steep variation found its origin in the existence of structural cooperative interactions between channels. The model proposed here permits us to obtain some indications of this possibility. Without inactivation, the maximum conductance is simply proportional

---

* The instantaneous response of U may be neglected as long as $\langle r \rangle \ll 1$, that is, with the resting state as initial condition; on the contrary when $\langle r \rangle \approx 1$, a repolarization of the membrane leads to an instantaneous decrease of U, which produces an instantaneous significant fall of the permeability function.

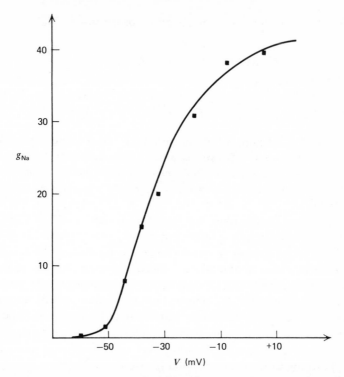

Figure 3. Fitting of the experimental values of Hodgkin and Huxley for g'_{Na} (squares) on the basis of Equations (3.5) to (3.15). The numerical values are $K = 0.39$, $\eta = +2.945$, $q = 3.737$, $\mu_R - \mu_S = -407$ D. In this fitting, we have tried to minimize the absolute value of $\mu_R - \mu_S$; it may therefore be considered as an estimation of the minimal dipole moment necessary in the case of a conformational equilibrium depending on the orientation of a single dipole. Much lower values become sufficient when the number of dipoles per ionophore increases.

to the equilibrium values of (4.1) as a function of V. This equilibrium is an intricate function of the parameters entering Equations (3.7) to (4.1). However, as a crude test we may consider the simplified case corresponding to Equation (3.11). Then $U(x_{i,j}) \to \langle r \rangle$, and we are left with the problem of finding values of the effective dipole moment $(n-1)$ $(\mu_R - \mu_S)$ and of the cooperativity parameter η that permit us to fit the experimental points satisfactorily. The result is reported in Fig. 4; it yields a value of about 400 D for the dipole moment, and -2.95 for η. This is only a rough approximation, and we shall not try to make it more precise. Let us simply say that in order to diminish the importance

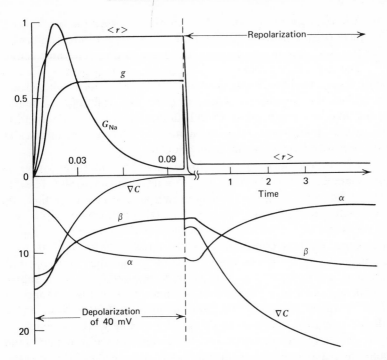

Figure 4. Time course of $\langle r \rangle$, g, g_{Na}, and the environmental functions α, β, and ∇c during a depolarization of 40 mV, followed by a repolarization to the resting state. All numerical values of the parameters are as in Fig. 2; for convenience, g_{Na} has been normalized by dividing through its maximum value at the peak, and for $t > 0.11$, the time scale has been contracted by a factor 40. It is interesting to notice that the surge that follows the repolarization decreases very rapidly in a time interval much shorter than the time interval needed by the membrane to go successively through the activation and inactivation processes.

of cooperativity one necessarily would have to look for bigger dipole moments. A more detailed discussion of these orders of magnitude would require a rigorous knowledge of the kinetic parameters involved, and in particular of the ratio k'_d / l_{RS}, which, as we shall see in the discussion of inactivation, largely determines the maximum height measured for the peaks.

B. The Spontaneous Inactivation of g_{Na}

The spontaneous inactivation of g_{Na} is, in general, considered as a process distinct from the reversal of the activation process. Among the mechanisms invoked one generally mentions "the movement of a blocking particle to a certain region of the membrane" (HH), the blocking of Na^+ transport by calcium ions,[24] and the presence of a third conformation of the Na^+ ionophore in addition to the resting state and active state[25]. The explanation adopted here, related to the presence of equilibration layers, was first indicated by Tasaki[19] in his book, who wrote, "The fall of the membrane potential from the peak of the action potential may be considered as a result of accumulation of the interdiffusing cations in and near the axon membrane." Let us now look at the implications of this mechanism in more detail.

Figure 2 clearly shows that the time course of Na^+ inactivation at various clamping potentials can be reproduced satisfactorily: The inactivation follows a simple exponential law as a function of time, in contrast with the high-order kinetics for activation, and vanishes for small variations of the potential, as in the experimental case. In Fig. 4 are compared the time dependences of the three fundamental functions of the theory: g_{Na}, the permeability function g, and the environmental functions α, β, $\nabla c = \beta l^{2n} - \alpha$. The plots correspond to a small suprathreshold depolarization (40 mV). It becomes evident that changes of the Na^+ gradient within the equilibration layers suffice to bring about an almost complete inactivation of g_{Na}. Interestingly, and in contrast with the generally accepted concepts, the permeability of the membrane remains high and constant while the NA^+ current vanishes; in other words, the sodium "channel" remains open during the inactivation.

A striking advantage of this interpretation of inactivation is that it furnishes a simple explanation of HH's two-pulse experiments (effect of a small conditioning pulse on the changes of Na^+ conductance following a suprathreshold consecutive pulse). They showed, for instance, that a conditioning pulse of small amplitude (8 mV) but lasting more than 20 msec might cause up to 40% reduction of the maximal change of sodium The permeability g and $\langle r \rangle$ return rapidly to their resting values, while g_{Na} remains very small, but α and β take a long time to recover their initial values; in the equilibration layers, Na^+ ions take a long time to reequilibrate with the bulk solutions. According to the model, the time necessary for the reequilibration would correspond to the refractory period. Most of the experimental findings relative to the process of Na^+ inactivation can thus be understood on the basis of limited diffusion conductance caused by a subsequent 40-mV pulse. For shorter conditioning pulses the decrease of g_{Na}^{max} follows a smooth exponential curve. This is indeed the behavior obtained here. In the upper part of Fig. 5 we have plotted $\langle r \rangle$, the permeability function g, as a function of V. In the lower part we have plotted the maximum conductance that would be

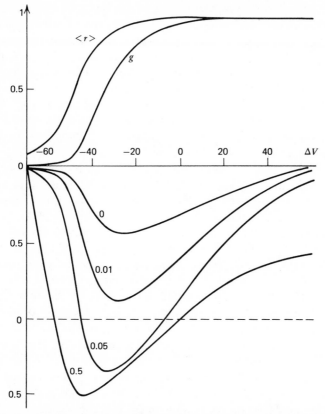

Figure 5. In the upper part of the figure, we have drawn the steady-state solutions for $\langle r \rangle$ and g as a function of V, as well as the corresponding normalized values of g'_{Na} found by Hodgkin and Huxley (squares). In the lower part of the figure we have reported, for values of p equal to 0, 0.01, 0.05, 0.5, the effect of a conditioning pulse, of amplitude given by the abscissa, on the maximum conductance of the membrane in a depolarization of 70 mV. The curves are normalized by dividing by the maximum conductance that would be attained in the absence of conditioning pulse. All parameters are as in Figure 2.

recorded in a depolarization of 70 mV preceded by a conditioning pulse of amplitude given by the abscissa. The curves are drawn for different values of the parameter p, which fixes the location of the equilibration layer. It is seen that the effect increases rapidly as p varies from 0 to 1. In particular, for $p = 0.5$, a depolarization of 8 mV produces a decrease in the peak conductance of more than 50%, although the permeability function g remains practically unchanged. Furthermore it is also predicted that if the conditioning pulse is of larger amplitude (15 mV $< V_c$

< 70 mV), the current recorded in the test pulse could be reversed and be *outward*.

Finally, if the equilibration layers cause inactivation by creating a diffusion barrier to the permeant ion, the recovery from inactivation should be much slower than the inactivation itself. This is indeed what is seen with exicitable membranes, where the refractory period lasts much longer than the action potential. The right-hand parts of the curves in Fig. 4 illustrate the variation in time of the fundamental functions of the theory following an abrupt termination of the clamp. between ionophores and bulk solutions. In this respect it is interesting here to underline the recent results of Dubois and Bergman,[26] who have demonstrated the existence on both sides of the membrane of an adsorption site on which the fixation appears as a preliminary rate-limiting step to the transport. On the other hand, it is known that calcium ions enhance inactivation. One simple interpretation of this effect would be that calcium alters the properties of the equilibration layers, or of the fixation sites studied by Dubois and Bergman, in such a manner that the diffusion of Na^+ becomes more difficult at their level. In other words, Ca^+ binding would decrease. A similar interpretation could also account for the prolonged action potentials observed by Rojas[27] after the treatment of the membrane with pronase.

In the coupling between activation and inactivation, the parameter whose value is critical is the ratio k'_d / l_{RS}. In Fig. 6, we have plotted as a function of potential, for a set of parameters, the theoretical maximum conductance as it is given by the equilibrium solutions of (4.7), and the actually observed maximum conductances for several values of k'_d / l_{RS}. It is manifest that when this ratio increases, the deviations between the theoretical curve and the values that would be observed increases drastically. Eventually, if $k'_d / l_{RS} \to \infty$, there would be no rise in conductance at all.

A test for the validity of our interpretation of sodium inactivation would be offered by the direct demonstration that during the inactivation phase the permeability of the membrane remains constant while the ionic concentrations in the equilibration layers vary. Such an experiment appears feasible as long as one possesses a physical parameter characteristic of the state and environment functions. Much more information certainly could be obtained concerning the latter quantities by the detailed study of the behavior of V_{Na} under various sets of conditions. On the other hand, adequate spectroscopic probes might be found that selectively bind to the ionophore or to the equilibration layer and represent the change of ionophore conformation with Na^+ concentration.

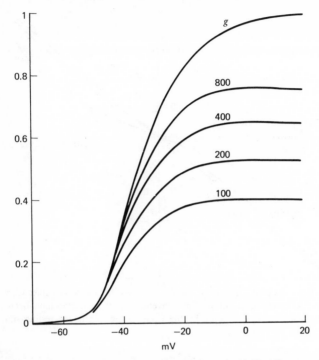

Figure 6. Maximum conductance as a function of the ratio k'_d/l'. All parameters are as in Figure 2. $k'_d = 1$; $l' = 100, 200, 400, 800$.

C. The Instantaneous Linearity of the I-V Curves.

When the number of energy barriers crossed by the ions inside the membrane increases, it has been shown[17] that the instantaneous I-V curves rapidly become linear. It can be seen in Fig. 7 that, in varying p, a similar result is easily obtained even if only one barrier in the membrane is considered. For $p = 0.5$ and with the resting state as initial condition, the I-V curves are linear within a few percent over a considerable range. This property is moreover preserved in the course of the transition towards the active state, as long as this process is sufficiently rapid with respect to the relaxation of the gradient of concentration. It must, however, be recognized that this is not sufficient to fully account for the experimental observations: Instantaneous linearity is required not only for the resting state, but for any state of the membrane as initial condition; it is verified whether the channels are activated or inactivated, and can also be observed in the case of the potassium conductance changes. One is therefore tempted to suggest

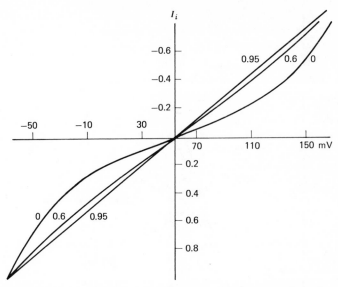

Figure 7. Instantaneous linearity of the $I-V$ curves, for different values of p and with the resting state as initial condition. For convenience, all the curves have been scaled by dividing the instantaneous response by the value of the flux in the resting state at -70 mV.

that the instantaneous linearity of the I-V curves is related to structural properties like those described by $U(x_{i,j})$ and that are capable of causing an instantaneous and important variation. In any case, a more detailed analysis of the field effect on the conformation equilibrium and of the fast charge motion characteristic of the capacitive currents is necessary.

V. STABILITY OF SPACE-CLAMPED ACTION POTENTIALS IN THE ABSENCE OF SODIUM INACTIVATION

In this last section we would like to consider some stability properties associated with the action potential itself. We investigate a simplified situation corresponding to a space-clamped nerve in the absence of sodium inactivation. It is indeed well known that sodium inactivation can be attenuated in various ways, and in particular by perfusing the nerves with pronase[27]; such a situation is therefore not unrealistic.

We suppose that the opening of the potassium channels may appropriately be described by an equation analogous to (3.11).

Furthermore, for the sake of simplicity, both for sodium and for potassium we neglect the effect of structural cooperativity. Under those conditions, the time change of the active fractions $\langle r \rangle$ and $\langle t \rangle$ of the sodium and potassium channels is described by the simple set of two equations:

$$\frac{d\langle r \rangle}{dt} = l_{RS}\left[1 - \langle r \rangle - \alpha\langle r \rangle e^{\mu_{Na}E/RT}\right] = F_r \qquad (5.1)$$

$$\frac{d\langle t \rangle}{dt} = l_{tS}\left[1 - \langle t \rangle - \alpha'\langle t \rangle e^{\mu_K E/RT}\right] = F_t \qquad (5.2)$$

where μ_{Na} and μ_K are the differences in dipole moment between the Na and K active and nonactive conformations; l_{RS}, l_{tS}, α, and α' are constants. On the other hand, since the equilibration layers play no role here, it is sufficient in order to describe the time change of V to consider the equation:

$$-C\frac{dV}{dt} = \langle r \rangle e^{q\langle r \rangle}(V - V_{Na}) + \langle t \rangle e^{q\langle t \rangle}(V - V_K) = F_V \qquad (5.3)$$

The steady states of the system can be obtained by solving the set of simultaneous equations

$$\langle r \rangle_0 = \frac{1}{1 + \alpha \exp(\mu_{Na}E_0/RT)} \; ; \langle t \rangle_0 = \frac{1}{1 + \alpha' \exp(\mu_K E_0/RT)}$$

$$V_0 = \frac{\langle r \rangle_0 e^{q\langle r_d \rangle_0}V_{Na} + \langle t \rangle_0 e^{q'\langle t \rangle_0}V_K}{\langle r \rangle_0 e^{q\langle r \rangle_0} + \langle t \rangle_0 e^{q'\langle t \rangle_0}} \qquad (5.4)$$

Their stability properties then depend on the nature of the normal modes of the following secular determinant:

$$\begin{vmatrix} \omega - \dfrac{\partial F_r}{\partial r} & 0 & -\dfrac{\partial F_r}{\partial V} \\[2mm] 0 & \omega - \dfrac{\partial F_t}{\partial t} & -\dfrac{\partial F_t}{\partial t} \\[2mm] -\dfrac{\partial F_V}{\partial r} & -\dfrac{\partial F_V}{\partial t} & -\omega - \dfrac{\partial F_V}{\partial V} \end{vmatrix} = 0. \qquad (5.5)$$

On the basis of (5.4) and (5.5) a stability diagram of the type shown in

Fig. 8 can straightforwardly be established. All other parameters being fixed, it is seen that as a function of μ_{Na} and α, the system may present at least four distinct domains. In domain I, (5.4) admits only one steady-state solution, corresponding to the resting state of the membrane, and which is a stable node. A similar stable node is also found in domains II, III, and IV, but furthermore we then have two other solutions, which in II correspond to saddle points, in III to a saddle point and an unstable focus, and in IV to a saddle point and a stable focus. Essentially, we thus see that in going from domain I to domain IV, we pass from membranes that present a unique steady and stable regime, corresponding to the resting state, to membranes that may exist either in a polarized stable state or in a depolarized stable state. The latter state is assimilable to a prolonged action potential.

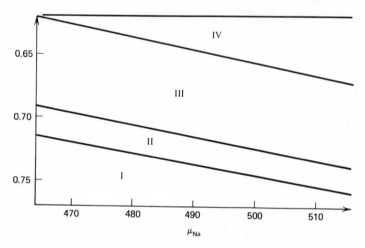

Figure 8. Stability diagram as a function of the difference in the dipole moment of sodium between the R and S states and of the conformational equilibrium constant α. All other parameters are fixed: $\alpha' = 0.661$, $\mu_K = 250$, $l_{tS} = 1$, $l_{RS} = 30$.

On the other hand, it is important to realize that the existence of multiple-steady-state regimes in a membrane is not necessarily directly related to the presence of a threshold or all-or-none effect during the action potential. In Fig. 9 we have plotted, for various depolarizations, the time course of V, obtained by solving equations (5.1) to (5.3) numerically in the case of a system that lies in domain II and presents

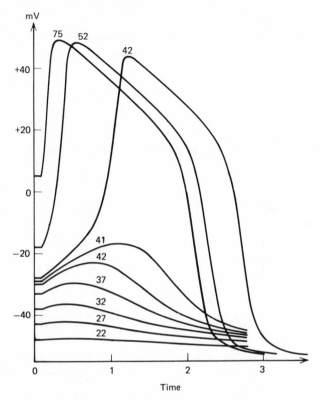

Figure 9. Space-clamped action potentials obtained in domain II. $\mu_{Na}=469$; $\alpha=0.708$. All other parameters are as in Fig. 8.

typical action-potential behavior. In Fig. 10, the maximum values of V are given as a function of the amplitude of the initial depolarizing stimulus. It is clear from this curve that the threshold for depolarization is reached and lies between 41 and 42 mV. On the contrary, the steady-state values of V correspond to 0.0017 mV. The parameter that critically controls the position of the threshold here is the ratio l_{RS}/l_{tS}, which measures the delay in potassium activation with respect to sodium activation: When l_{RS}/l_{tS} diminishes, the value of the threshold tends to increase. In Fig. 11 we have graphed the behavior in time of the fractions $\langle r \rangle$ and $\langle t \rangle$ following a depolarization of 75 mV; it can be seen that at the peak of sodium conductance, when 95% of the sodium channels are open, one has only 40% of the potassium channels open.

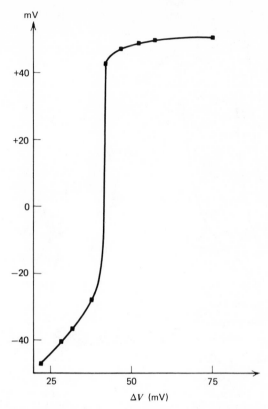

Figure 10. Maxima of the potential as a function of the depolarization. The squares correspond to the curves drawn in Fig. 9. All parameters have the same value as in Fig. 9.

VI. CONCLUSIONS

We have presented an interpretation of the changes of conductance observed during electrical excitation, based on the properties of ionophores, which may exist in two conformational states characterized by different permeabilities and dipole moments. Primarily, we have considered the case of sodium transport and shown that rather low dipole-moment differences between the conformational states (let us say, less than 100 D) permit us to account for the steep variation of the maximum conductances as a function of the applied depolarization; however, several dipoles per ionophore would then necessarily be implicated in the conformational transition. The existence of cooperative interactions among the ionophores might, on the other hand, both

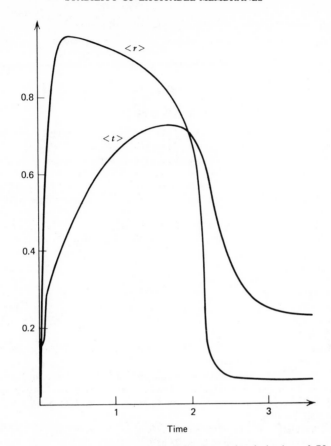

Figure 11. Time variation of $\langle r \rangle$ and $\langle t \rangle$ following a depolarization of 75 mV. All parameters are as in Fig. 9.

facilitate this conformational transition and increase the transport capacity of the ionophores; in this way, cooperativity permits us to account for the high-order kinetics characteristic of the activation process.

We also investigated in detail the behavior of our transporting systems when they are separated from the bulk solutions by the presence of equilibrium layers, which create a diffusion barrier hindering the ionic motion. On the basis of this simple hypothesis, the essential characteristics of sodium inactivation can easily be reproduced; several predictions have been made that should be experimentally testable.

Finally, we considered the type of stability problems that may arise in space-clamped nerves when the sodium inactivation process has been abolished. Surprisingly, it was shown that the absence of inactivation does not necessarily imply the occurrence of prolonged action potentials. Depending on the values of the parameters, we have, even for this simplified situation, quite complex stability diagrams, in which certain domains may contain multiple-steady-state regimes.

References

1. A. L. Hodgkin, *The Conduction of the Nervous Impulse*, Liverpoot U. P., 1967.
2. B. Katz, *Nerve, Muscle and Synapse*, McGraw-Hill, New York, 1966.
3. G. A. Bekey, and B. Paxon, Analog simulation of nerve excitation. Communication at the 2nd National simulation Conference, Houston, Texas, 1957.
4. R. Fitzhugh, *J. Gen. Physiol.*, **43**, 867 (1960).
5. K. S. Cole, A. H. Antosiewiecz, and P. Rabionowitz, *J. Soc. Ind. Appl. Math.*, **3**, 153 (1955).
6. K. S. Cole, *J. Soc. Ind. Appl. Math.*, **6**, 196 (1958).
7. R. Fitzhugh and H. A. Antosiewiecz, *J. Soc. Ind. Appl. Math.*, **7**, 447 (1959).
8. R. Fitzhugh, *Biophys. J.*, **2**, 11 (1962).
9. J. Nagumo, *Proc. IRE*, **50**, 2061 (1962).
10. H. Cohen, IBM technical report, (1971).
11. H. P. McKean, Jr., *Adv. Math.*, **4**, 209 (1970).
12. D. M. Urry, *Proc. Nat. Acad. Sci. U.S.A.*, **69**, 1610 (1972).
13. J. P. Changeux, J. Thiéry, Y. Tung, and C. Kittel, *Proc. Nat. Acad. Sci. U.S.A.*, **57**, 335 (1967).
14. P. Glansdorff and I. Prigogine, *Thermodynamic Theory of Structure, Stability and Fluctuations*, Wiley-Interscience, New York, 1971.
15. J. P. Changeux, R. Blumenthal, M. Kasai, and T. Podleski, Conformational transitions in the course of membrane excitation, in *CIBA Foundation Symposium on Molecular Properties of Drug Receptors*, R. Porter and M. O'Connor, Eds., J. & A. Churchill, London, 1970.
16. R. Blumenthal, J. P. Changeux, and R. Lefever, *J. Membrane Biol.*, **2**, 351 (1970).
17. J. W. Woodbury, *Adv. Chem. Phys.*, **21**, 601 (1971).
18. L. N. G. Gordon, and D. A. Haydon, *Biochem. Biophys. Acta.* **215**, 1014 (1973).
19. I. Tasaki, *Nerve excitation*, Ch. Thomas, Springfield, Ill., 1968.
20. F. A. Dodge and B. Frankenhaeuser, *J. Physiol. (Lond.)*, **148**, 188 (1959).
21. B. Neumcke, D. Walz, and P. Läuger, *Biophys. J.*, **10**, 172 (1970).
22. R. Lefever and J. P. Changeux, submitted to *T. Biophys. Chem.*
23. T. L. Hill and Yi-der Chen, *Biophys. J.*, **11**, 4565 (1971).
24. S. N. Fishman and M. V. Volkenstein, *Biochim. Biophys. Acta*, **241**, 697 (1971).
25. T. L. Hill, Comments on the theory of ion transport across the nerve membrane, in *Perspectives in Membrane Biophysics*, D. Agin, Ed., Gordon and Breach, New York, 1972.
26. R. Dubois and C. Bergman, private communication.
27. E. Rojas and C. Armstrong, *Nature, New Biol.*, **229**, 177 (1971).
28. B. Neumcke and P. Läuger, *Biophys. J* **9** 1160 (1969).

AUTHOR INDEX

Numbers in parentheses are reference numbers and show that an author's work is referred to although his name is not mentioned in the text. Numbers in *italics* indicate the pages on which the full references appear.

SUBJECT INDEX